"Most introductory statistics texts teach students how to apply specific tests in specific circumstances, with little room for generalizing knowledge to new settings. *Data Analysis* instead teaches students how to think like scientists, always framing hypotheses as formal comparisons between competing explanations. The first three editions were ahead of their time in their philosophical approach to data analysis, and this new edition retains and expands their unifying framework."

Kristopher J. Preacher, *Vanderbilt University, USA*

"I am delighted that both logistic regression and multilevel modeling are now included. Both topics are introduced using the authors' clear, useful, and integrative approach. Not only does the new material help me to teach this to my students better, it also helps me to understand the topics better!"

J. Michael Bailey, *Northwestern University, USA*

"I've relied on previous editions of *Data Analysis: A Model Comparison Approach to Regression, ANOVA, and Beyond* for years in my graduate-level data analysis courses. The book's clear, integrated approach to complex statistical models—coupled with its focus on practical application and ethical considerations—has made it an indispensable resource for both students and instructors. This latest edition continues to be a top choice for mastering advanced data analysis techniques."

Markus Brauer, *University of Wisconsin-Madison, USA*

Data Analysis

This essential textbook provides an integrated treatment of data analysis for the social and behavioral sciences. It covers all the key statistical models in an integrated manner that relies on the comparison of models of data estimated under the rubric of the general linear model.

The text describes the foundational logic of the unified model comparison framework. It then shows how this framework can be applied to increasingly complex models including multiple continuous and categorical predictors, as well as product predictors (i.e., interactions and nonlinear effects). The text also describes analyses of data that violate assumptions of independence, homogeneity, and normality. The analysis of nonindependent data is treated in some detail, covering standard repeated measures analysis of variance and providing an integrated introduction to multilevel or hierarchical linear models and logistic regression.

Highlights of the fourth edition include

- Expanded coverage of generalized linear models and logistic regression in particular
- A discussion of power and ethical statistical practice as it relates to the replication crisis
- An expanded collection of online resources such as PowerPoint slides and test bank for instructors, additional exercises and problem sets with answers, new data sets, practice questions, and R code.

Clear and accessible, this text is intended for advanced undergraduate and graduate level courses in data analysis.

Joshua Correll is a professor of psychology and neuroscience in the College of Arts and Sciences at the University of Colorado at Boulder. His research examines face processing, stereotypes, and data analysis.

Abigail (Abby) M. Folberg is an assistant professor of psychology in the College of Arts and Sciences at the University of Nebraska at Omaha. Her research examines the impacts of stereotypes and prejudice on marginalized group members as well as how individuals and organizations can reduce prejudice and discrimination.

Charles "Chick" M. Judd is Professor Emeritus of Distinction in the College of Arts and Sciences at the University of Colorado at Boulder. His research focuses on social cognition and attitudes, intergroup relations and stereotypes, judgment and decision-making, and behavioral science research methods and data analysis.

Gary H. McClelland is Professor Emeritus of Psychology at the University of Colorado at Boulder. A faculty fellow at the Institute of Cognitive Science, his research interests include judgment and decision-making, psychological models of economic behavior, statistics and data analysis, and measurement and scaling.

Carey S. Ryan is Professor Emeritus in the Department of Psychology at the University of Nebraska at Omaha. Her research interests include stereotyping and prejudice, group processes, and program evaluation.

Data Analysis

A Model Comparison Approach to Regression, ANOVA, and Beyond

Fourth Edition

Joshua Correll, Abigail M. Folberg, Charles M. Judd, Gary H. McClelland and Carey S. Ryan

Routledge
Taylor & Francis Group
NEW YORK AND LONDON

Fourth edition published 2026
by Routledge
605 Third Avenue, New York, NY 10158

and by Routledge
4 Park Square, Milton Park, Abingdon, Oxon, OX14 4RN

Routledge is an imprint of the Taylor & Francis Group, an informa business

First edition published by Harcourt Brace Jovanovich 1989
Third edition published by Routledge 2017

Library of Congress Cataloging-in-Publication Data
Names: Judd, Charles M. author | Correll, Joshua author | Folberg, Abigail M. author | McClelland, Gary H., 1947– author | Ryan, Carey S. author
Title: Data analysis : a model comparison approach to regression, ANOVA, and beyond / Joshua Correll, Abigail M. Folberg, Charles M. Judd, Gary H. McClelland, Carey S. Ryan.
Description: Fourth edition. | New York, NY ; Abingdon, Oxon : Routledge, 2025. | Previous edition by Charles M. Judd, University of Colorado, Gary H. McClelland, University of Colorado, Carey S. Ryan, University of Nebraska at Omaha. | Includes bibliographical references and indexes.
Identifiers: LCCN 2025012602 (print) | LCCN 2025012603 (ebook) | ISBN 9781032572093 hbk | ISBN 9781032572086 pbk | ISBN 9781003438335 ebk
Subjects: LCSH: Mathematical statistics
Classification: LCC QA276 .J83 2025 (print) | LCC QA276 (ebook) | DDC 519.5/36—dc23/eng/20250415
LC record available at https://lccn.loc.gov/2025012602
LC ebook record available at https://lccn.loc.gov/2025012603

ISBN: 978-1-032-57209-3 (hbk)
ISBN: 978-1-032-57208-6 (pbk)
ISBN: 978-1-003-43833-5 (ebk)

DOI: 10.4324/9781003438335

Typeset in Times New Roman
by Apex CoVantage, LLC

Access the Support Material: routledge.com/9781032572086

Contents

Preface

For many students raised in the tradition of model comparison, this book holds special significance. It played a distinct role in the evolution of our understanding. We had exposure to statistics before model comparison, opaque and dull; then we encountered this book and everything changed. It is thus with a sense of both humility and responsibility that three of those students (Joshua Correll, Abigail M. Folberg, and Carey S. Ryan) join in collaboration with the original authors to present the fourth edition of this text.

The first edition resulted from Charles "Chick" M. Judd's and Gary H. McClelland's teaching a graduate level course in data analysis at the University of Colorado, affectionately called, "The Chick and Gary Show." In subsequent editions (in 2008 and 2017), Carey S. Ryan joined as a co-author, and the content expanded to include linear mixed models and logistic regression. In this fourth edition, the five authors now add a separate chapter on issues related to power, and we introduce a new organization, clustering the material into three distinct sections, described later. We believe the resulting book provides a sound basis for understanding data analyses ranging from the relatively simple to the extremely complex.

There is much that is new and different in this fourth edition. At the same time, our basic model comparison approach and our way of thinking about data have not changed at all and remain the core of the book. Next, we clarify our assumptions and our approach.

Goals and Assumptions

Statistics courses, textbooks, and software are usually organized in the same way that a cookbook is organized. Each chapter contains recipes for different research designs or data structures. Many books even include a chart at the beginning, pointing the reader to various chapters and ostensibly different statistical procedures, depending on what their data look like and the kinds of questions that they wish to answer. As a result, social and behavioral scientists, as consumers of statistics, typically organize their statistical knowledge in much the same way: "With this sort of data, I must do this test. With that sort of data, I must do that test."

This book has been written under the assumption that this sort of organization for statistical knowledge has costs. To extend the cookbook analogy, cooks who rely on cookbooks do not know how to proceed when they wish to prepare a dish for which no recipe has been included. When students are confronted with data that do not fit nicely into one of the categories for which they have learned a statistical procedure, frustration and error frequently occur. The student may proceed to use the wrong test, making inappropriate assumptions about the structure of the data in order to use a well learned statistical procedure. We have encountered students who have bludgeoned their data to fit an orthogonal analysis of variance with equal sample sizes because it is the only test they know. We have also heard some colleagues moan that a given data analysis problem would be simple if only they had a particular specialized computer program that was available at their previous university.

A totally different sort of organization forms the basis of this book. Our focus is on how statistical procedures can be seen as tools for building and testing models of data. A consistent framework is used throughout the book to develop a few powerful techniques for model building and statistical inference. This framework subsumes the different statistical recipes normally found in the cookbooks, but it does so in a totally integrated manner. As we build and test models, we develop a consistent vocabulary to refer to our models and inference procedures. At each step, we also show the reader how our vocabulary and models can be translated back into the recipes of the old cookbooks.

We are convinced that this integrated framework represents a better way to learn and think about data analysis. Students who understand data analysis through our framework will be able to ask the questions of their data that they want to ask instead of the questions that the designer of some textbook or of some computer statistical package assumed they would want to ask. In the end, students will use theory and their own intelligence to guide their data analysis instead of turning over the analysis to a cookbook set of procedures or to a computer program. Intelligent data analysis is our goal. Rather than applying recipes that have been learned by rote, we wish to enable students to write the recipes they need for their analysis.

A few words are in order concerning our assumptions about the reader and the structure of the book. We assume that the reader is a student or professional researcher in the behavioral or social sciences, business, education, or a related field who wants to analyze data to answer significant theoretical or practical questions in a substantive discipline. We assume that the reader does not want to become a statistician but is more interested in the substantive discipline, with data analysis being a tool rather than an end in itself.

We assume that the reader is reasonably facile with basic algebraic manipulations because almost all of our models for data are expressed algebraically. But we do not assume any other mathematical training such as calculus or matrix algebra. A very important assumption is that the reader has access to a good multiple regression program in a statistical software package. But, importantly, we do not care which one.

The book is appropriate for semester or year-long statistics and data analysis courses emphasizing multiple regression and/or analysis of variance and taught at the upper undergraduate or graduate levels. Previous editions have been successfully used with students from psychology, sociology, anthropology, political science, linguistics, cognitive science, neuroscience, biology, geology, geography, museum sciences, applied mathematics, marketing, management science, engineering, and organizational behavior.

Our assumptions about the reader and our goal of training data analysts instead of statisticians have prompted several decisions about the structure of the book:

1 We present only enough mathematical derivations to provide conceptual clarity. We generally assume that the mathematical statisticians have done their job correctly, and so we use many of their results without proving them ourselves. At the same time, however, we cannot abandon all mathematical details because it is extremely important that the data analyst be able to recognize when the data analysis and particularly the model of the data are inappropriate or in need of modification. The choice of which derivations to include is therefore guided by the goal of training educated data analysts and not mathematical statisticians.
2 We let the computer do the computational work. Most statistics cookbooks present many different formulas for each statistic. The different formulas facilitate hand calculation for different organizations of the raw data. We assume that the reader is interested in the science and not the calculations and will use the most efficient means to do the calculations—the computer. Ironically, the old

hand-calculation formulas can be disastrous when implemented in computer programs because of problems of rounding errors and other computer idiosyncrasies. Neither the computational formulas that are best for computers nor the hand-calculation formulas are very conceptual. Hence, we avoid both computer- and hand-oriented formulas. We present instead formulas that emphasize the concepts but may not be useful for direct computations either by hand or by computer.

3 We try as much as possible to work from the general to the specific so that an integrated framework emerges. Many statistics books begin with details and simple statistics and then slowly build up to more general models, changing concepts and notation frequently along the way. Although we begin with simple models of data and work up, we do so within the context of a consistent overall framework, and we present the simple models using the same concepts and notation that we use with the more complex models presented later. Most importantly, we use the same inferential statistics throughout.

4 We do not try to cover all of statistics. There are many statistical procedures that we have left out either because they are infrequently used or because the same thing can be accomplished with a more general model that we do cover. We provide the data analyst with a limited stock of models and statistical procedures, but the models we do provide are quite general and powerful. The goal is to learn a few powerful techniques very thoroughly.

5 Our framework for data analysis is consistent with what has been termed the regression approach or the general linear model. Historically, the regression approach has often been contrasted, erroneously, with an analysis-of-variance approach. And, in this tradition, regression has been more often associated with surveys, quasi-experiments, and nonexperimental data, while analysis of variance has more frequently been used for randomized, laboratory experiments. These historical associations caused many people to have a false belief that there is a fundamental difference between the two approaches. We adopt the more general regression approach and show how all of analysis of variance can be accomplished within this framework. This has the important advantage of reducing the number of specialized statistical techniques that must be learned for particular data analysis problems. More importantly, we show how the regression approach provides more control in the analysis of experimental data so that we can ask specific, theoretically motivated questions of our data—questions that often differ from the standard questions that the procedures of a traditional analysis of variance presume we want to ask.

6 We had a difficult decision to make about the vocabulary to be used in this book. On the one hand, we are convinced that the traditional names for the various statistical tests perpetuate false distinctions. For that reason, we would have preferred to abandon them in favor of a shorter list of names from our integrated framework. However, doing so would have been irresponsible because many readers of this book will have had an introductory course using the traditional names, will have to talk to colleagues who have not read this book, and will have to read journal reports of data analysis that perpetuate the traditional names. We therefore do both: Consistent terminology and notation are used throughout the book to provide an integrated framework, but we also provide translations, usually at the beginning and end of each chapter, between our terms and their traditional counterparts.

How the Book Is Organized and What Is New This Time Around

The preface or introduction in most statistics textbooks describes alternative orders in which the chapters may be studied, which is precisely what we think is wrong with most statistics textbooks. When each chapter or section is self-contained, students are often befuddled by the abrupt change of topics across chapter boundaries. To avoid these abrupt changes, we have organized the chapters in a specific sequence and used a unifying framework for data analysis throughout. Each chapter generally focuses on a single concept, but the understanding of that concept will depend on the concepts developed in the previous chapters. Thus, we do not think the chapters should be read in any order other than the one in which we present them. This means, of course, that our book cannot be used as a cookbook. But then, we have made clear that we think the cookbook organization of statistical knowledge is misguided.

The chapters have further been organized into three sections. The first section, Section A, discusses the basic logic and mechanics of an inferential statistical test. To facilitate learning, this discussion is oriented around a very simple statistical model, and it assumes that the data are reasonably "well mannered." Section B then discusses more complex models, allowing the researcher to examine more nuanced questions. Finally, Section C discusses situations where the data are "ill mannered" and discusses procedures to handle them in the context of model comparison.

Although readers who use our book to look up specific tests may be frustrated by the sequential organization of our book, we are convinced that this organization represents a better way to learn statistics. It substantially reduces the number of separate concepts that must be mastered and allows progressive accumulation of integrated knowledge about data analysis rather than disconnected amassing of discrete recipes.

Chapter 1 presents an overview of data analysis through the concepts of data, model, and error and integrates them in the equation:

Data = Model + Error

The concept of proportional reduction in error (PRE) is introduced as a tool for sorting our way through alternative models for a given set of data. Chapter 2 considers alternative definitions of error and shows the implications for the model for each of the alternative definitions. While examining the alternative definitions of error, we develop the usual descriptive statistics for location and spread. However, we prefer to view these descriptive statistics in terms of models of data. Chapter 3 introduces a small set of reasonable assumptions about error and uses those assumptions, in conjunction with the concept of sampling distributions of statistics, to choose from among the alternative definitions of error offered in Chapter 2. The sum of squared errors (SSE) is chosen to be our standard definition of error. This chapter also introduces the concept of sampling distributions of statistics. New to this edition, this chapter also contains an intuitive explanation of maximum likelihood estimation. Chapter 4 uses the material of Chapter 3 to make inferences about models and their parameters. The inferential test developed in this chapter is equivalent to a test usually identified as the one-sample t-test, but we present it in the more general form that we use in all later chapters.

In this new edition, we provide more complete coverage of statistical power, effect sizes, and confidence intervals in a separate chapter, Chapter 5. All the necessary concepts for statistical inference used throughout the book are contained in Chapters 2, 3, 4, and 5, although the models that are explored in these chapters are very simple.

The remainder of the book is mostly about building models that vary in their complexity and appropriateness for different kinds of data. Chapter 6 is about models that make conditional predictions of data based on a single continuous variable. Such models are usually referred to as simple regression. This chapter uses the inferential techniques developed in Chapter 4 to ask questions about the simple regression models and their parameters. Chapter 7 extends the concepts of Chapter 6 to models making conditional predictions of data based upon multiple continuous variables. Elsewhere, the material in this chapter is often referred to as multiple regression. An important concept in this chapter is redundancy among predictors. Chapter 8 introduces the important technique of including multiplicative products of continuous predictor variables in models. We then use that technique to develop curvilinear models and to consider the interaction between two or more predictor variables. Chapters 6–8 contain all the essential information for what is generally known as regression analysis.

Chapters 9–12 consider models that use categorical variables to make conditional predictions of data. Such models are often known as analysis of variance. The key link between this material and the previous chapters is provided by Chapter 9, which introduces contrast codes. The use of contrast codes allows all the model building and testing techniques of regression analysis (Chapters 6–8) to be applied without modification to models with categorical predictor variables. Chapter 9 develops these important contrast codes for the case of single categorical predictor variables, often known as one-way analysis of variance. Chapter 10 shows how models using products, developed in Chapter 8, can be applied to the contrast codes of Chapter 9 to produce models of increasing complexity when there are multiple categorical predictor variables. These models are often referred to as analysis of variance for factorial designs. Chapter 11 makes the generalization to models that include both continuous and categorical predictor variables; these models are sometimes known as analysis of covariance and mediational analyses. Also included here are models that include products of continuous and categorical variables, models known elsewhere as moderated regression models.

An important assumption underlying statistical inference with all the models is that each data observation is independent from all other observations. This assumption is unlikely to be true when more than one observation comes from each unit of analysis (e.g., when each person or group provides more than one data observation). Chapters 12 and 13 present an approach for building models in such situations that is entirely consistent with the material in previous chapters. This problem is known in psychology as within-subject or repeated measures analysis of variance, in contrast to the models in Chapters 9–11 that apply to between-subject analysis of variance. More generally, this problem is referred to as nonindependent or correlated errors. Chapter 12 presents the basics of our approach for analyzing data with correlated errors. This approach consists of developing a separate set of contrast codes for the multiple observations from each unit of analysis, using those codes to construct new data variables, and then using the machinery of previous chapters to model those new data variables. Chapter 13 extends that approach to what might be called within-subject regression by including continuous predictor variables that can vary either between or within the nonindependent observations in a dataset. Included here is an introduction to what are called linear mixed models that allow random error variance to accrue from multiple sources rather than a single one. By extension we provide an introduction to what are called multilevel or hierarchical models as well as models for data with crossed random factors. This material has been more fully developed in this edition.

Chapter 14 considers the problems of outliers and other violations of the assumptions about error made in Chapter 3. We use the concepts developed in previous chapters to develop an outlier model that provides a principled method for detecting outliers and mitigating their effects on the model for the remainder of the data. This chapter also considers techniques for the detection and treatment of violations of assumptions underlying the procedures for model building and statistical inference presented in the previous chapters. The effect of all the remediation techniques presented is to transform data or otherwise restructure the problem so that all the machinery previously developed may be appropriately applied.

New to this edition is a fuller development of logistic regression in Chapter 15, permitting the analysis of dichotomous rather than continuous outcome or dependent variables. Although the inferential machinery for these models is rather different, the basic model comparison approach that we have developed has a nice parallel in logistic models.

Also new to this edition are updated exercises that supplement each chapter, datasets for all examples used in the text, R code for all of the analyses that we describe, and PowerPoint slides. These can be found at our supplementary website, www.routledge.com/9781032572086.

In preparing this revision we realize the huge debt that we owe all of our students and teachers over the years. We thank them for the many ways in which they have made us think harder and deeper as teachers and data analysts. Joshua Correll, Abby M. Folberg, and Carey S. Ryan are especially grateful to Chick Judd and Gary H. McClelland for sharing this approach to data analysis with us and for the opportunity to co-author this fourth edition.

Section A

Statistical Machinery

Introduction to Section A

Statistical Machinery

This book is about data analysis. In the social and behavioral sciences we often collect batches of data that we hope will answer questions, test hypotheses, or disprove theories. To perform any of these functions, we must analyze data. This book is divided into three sections, which deal with distinct aspects of data analysis.

Section A describes a fundamental process for using data to draw conclusions. This section introduces the statistical "machinery" that underlies every test described in the book. It also introduces a set of important assumptions about the nature of our data, which common statistical analyses require us to satisfy. Section A is the foundation of the book.

Holding the aforementioned assumptions in mind, Section B introduces a series of increasingly complex models. For instance, we consider models that involve a linear relationship between a single predictor variable and an outcome variable. We also consider models that make predictions for each of three different groups. These relatively simple models are addressed in earlier chapters. Subsequent chapters use the lessons of these earlier chapters to develop increasingly complicated models. At the end of this section, we consider complex models involving multiple groups, non-linear relationships, and complex interactions. These models allow us to test increasingly nuanced and interesting questions.

The final section, Section C, considers data that violate one or more of the set of assumptions we made about our data in Section A. Many of the chapters in this section describe transformations that bring "badly behaved" data in line with our assumptions. These transformations then allow us to use the statistical machinery developed in Section A and the full suite of models developed in Section B to analyze data, even when those data initially seem problematic.

Section A introduces the namesake of this book: "a model comparison approach." In this section, we describe data analysis in terms of a comparison between two models. This process of comparison lies at the heart of a broad range of statistical tests.

Chapter 1 explains the basic logic of model comparison in very general, non-statistical terms.

Chapter 2 introduces a very simple model—one that makes a single prediction for every observation in our dataset. This model provides a context in which we address a variety of fundamental questions. The chapter then considers a variety of strategies we might use to measure a model's ability (or, really, inability) to account for the data.

Chapter 3 evaluates these different strategies of measuring a model's fit using sampling distributions, a powerful tool that will be used throughout the remainder of this section. In this chapter, we settle on one particular strategy that works very well as long as our data satisfy certain assumptions.

Chapter 4 introduces model comparison in statistical terms. It provides the machinery that we use to determine whether one model fits the data appreciably better than another. In this chapter, we will describe what it means to have a "significant" effect.

DOI: 10.4324/9781003438335-2

Chapter 5 addresses the question of statistical power. In brief, we consider a situation in which some effect exists in the real world, and we present a strategy for determining our ability to detect that effect in our data.

Throughout Section A, we will use very simple statistical models to consider the logical elements of data analysis. The goal is to build an understanding of the statistical machinery that we will rely on through the rest of the book, from the simplest analysis to the most sophisticated. Section B will then proceed to develop these more sophisticated models.

1 Introduction to Data Analysis

In this chapter, we present an overview of what data analysis means. This overview is intentionally abstract with few details so that the "big picture" will emerge, offering a general sense of how to analyze quantitative data to answer a research question. This big picture approach will make it easier to comprehend the details that come later.

Overview of Data Analysis

The process of data analysis is represented by the following simple equation:

DATA = MODEL + ERROR

DATA represents the basic scores or observations, usually but not always numerical, that we want to analyze. MODEL is a more compact description or representation of the data. Our data are usually bulky and of a form that is hard to communicate to others. The compact description provided by the model is much easier to communicate, say, in a journal article, and is much easier to think about when trying to understand phenomena, to build theories, and to make predictions. To be a representation of the data, all the models we consider will make a specific prediction for each observation or element in DATA. Models range from the simple (making the same prediction for every observation in DATA) to the complex (making differential predictions conditional on other known attributes of each observation). To be less abstract, let us consider an example. Suppose our data were, for each of 45 countries, the percentage of individuals who used the internet in the year 2021; these data are listed in Figure 1.1. A simple model would predict the same percentage for each country. A more complex model might adjust the prediction for each country according to its type of government (e.g., authoritarian or democratic), service economy, and the educational and income levels of its population. The amount by which we adjust the prediction for a particular attribute (e.g., service economy, that is, percentage of total employees in services) is an unknown *parameter* that must be estimated from the data.

The last part of our basic equation is ERROR, which is simply the amount by which the model fails to represent the data accurately. It is an index of the degree to which the model mispredicts the data observations. We often refer to error as the *residual*—the part that is left over after we have used the model to predict or describe the data. In other words,

ERROR = DATA − MODEL

DOI: 10.4324/9781003438335-3

i	*Country*	*Percentage*
1	ARG	87.15
2	AUS	96.24
3	BLR	86.89
4	BEL	92.79
5	BWA	73.50
6	BRA	80.69
7	BGR	75.27
8	CAN	92.83
9	CHN	73.05
10	COL	73.03
11	HRV	81.25
12	DNK	98.87
13	RCU	76.20
14	EGY	72.06
15	FIN	92.81
16	FRA	86.10
17	DEU	91.43
18	GRC	78.49
19	GTM	50.84
20	HUN	88.64
21	ISL	99.69
22	IDN	62.10
23	IRL	95.17
24	ISR	90.30
25	ITA	74.86
26	JPN	82.91
27	JOR	82.76
28	MYS	96.75
29	MEX	75.63
30	NLD	92.05
31	NIC	57.15
32	NOR	99.00
33	PER	71.11
34	POL	85.37
35	PRT	82.31
36	ROU	83.59
37	RUS	88.21
38	ESP	93.90
39	SWE	88.31
40	CHE	95.57
41	THA	85.27
42	TUR	81.41
43	UKR	79.22
44	GBR	96.68
45	USA	91.75

Figure 1.1 Percentage of individuals using the internet in the year 2021 by country

Source: World Bank World Development Indicators from the International Telecommunication Union (ITU) World Telecommunication/ICT Indicators Database. https://databank.worldbank.org/reports.aspx?source=world-development-indicators#

The goal of data analysis is then clear: We want to build the model to be a good representation of the data by making the error as small as possible. In the unlikely case when ERROR = 0, DATA would be perfectly represented by MODEL.

How do we reduce the error and improve our models? One way is to improve the quality of the data so that the original observations contain less error. This involves better research designs, better data collection procedures, more reliable instruments, etc. We do not say much about such issues in this book but instead leave those problems to texts and courses in experimental design and research methods. Those problems tend to be much more discipline specific than the general problems of data analysis and so are best left to the separate disciplines. Although we often note some implications of data analysis procedures for the wise design of research, we in general assume that the data analyst is confronted with the problem of building the best model for data that have already been collected.

The method available to the data analyst for reducing error and improving models is straightforward and, in the abstract, the same across disciplines. Error can almost always be reduced (never increased) by making the model's predictions conditional on additional information about each observation. This is equivalent to adding parameters to the model and using data to build the best estimates of those parameters. The meaning of "best estimate" is clear: We want to set the parameters of the model to whatever values will make the error the smallest. The estimation of parameters is sometimes referred to as "fitting" the model to the data. In this book, we will introduce a variety of basic models. It is unlikely that any of these models will provide a good fit "off the rack"; instead, the basic model will need to be fitted or tailored to the particular size and bulges of a given data customer. In this chapter, we are purposely vague about how the error is actually measured. We are also vague about how parameters are actually estimated to make the error as small as possible because that would get us into details to which we devote whole chapters later. But for now, the process in the abstract ought to be clear: Add parameters to the model and estimate those parameters so that the model will provide a good fit to the data by making the error as small as possible.

To be a bit less abstract, let us again consider the example of internet use by country. An extremely simple model would be to predict a priori (that is, without first examining the data) that in each country the percentage of individuals who used the internet was 70. This qualifies as a model according to our definition because it makes a prediction for each of the 45 countries. But in this model, there are no parameters to be estimated from the data to provide a good fit by making the error as small as possible. No matter what the data, our model predicts 70. We will introduce some notation so that we have a standard way of talking about the particulars of DATA, MODEL, and ERROR. Let Y_i represent the ith observation in the data; in this example Y_i is simply the percentage of individuals who used the internet for the ith country. Then our basic equation:

DATA = MODEL + ERROR

for this extremely simple model becomes

$$Y_i = 70 + \text{ERROR}$$

We can undoubtedly improve our model and reduce the error by using a model that is still simple but has one parameter: Predict that the percentage is the same in all countries but leave the predicted value as an unspecified parameter to be estimated from the data. For example, the average of all

45 percentages might provide a suitable estimate. We will let β_0 represent the unknown value that is to be estimated so that our slightly more complex, but still simple, model becomes

$$Y_i = \beta_0 + \text{ERROR}$$

It is important to realize that we can never know β_0 for certain; we can only estimate it from the data.

We can make our model yet more complex and reduce the error further by adding more parameters to make *conditional predictions*. For example, people in authoritarian countries may use the internet less than those in more democratic countries. We could implement that in a model that starts with a basic percentage of internet use (β_0) for all countries, which is adjusted upward by a certain amount (β_1) if the country has a democratic government and reduced by that same amount if the country has an authoritarian government. More formally, our basic equation now has a more complex representation, namely

$$Y_i = \beta_0 + \beta_1 + \text{ERROR if the country is democratic}$$
$$Y_i = \beta_0 - \beta_1 + \text{ERROR if the country is authoritarian}$$

Another slightly more complex model would make predictions conditional on a continuous, rather than a categorical, predictor. For example, we might make predictions conditional on the percentage of employees in service industries, presuming that internet use is higher in countries with stronger service economies. We again start with a basic percentage of internet users (β_0) for all countries, which is adjusted upward by a certain amount (β_1) for each percentage point a country's percentage of service employees is above the national average and reduced by the same amount for each percentage point a country's percentage of service employees is below the national average. More formally, letting X_i represent the amount a country's percentage of service employees is above or below the national average:

$$Y_i = \beta_0 + \beta_1 X_i + \text{ERROR}$$

In words, the percentage of service employees is the condition in this model on which we base our differential or conditional prediction of a country's internet use.

We can continue making our model yet more complex by adding parameters to make similar adjustments for gross national product (GNP), educational level, and urban versus rural population, etc. By so doing we will be adding still more implicit hypotheses to the model.

It might appear that the best strategy for the data analyst would be to add as many parameters as possible, but this is not the case. The number of observations in DATA imposes an inherent limit on the number of parameters that may be added to MODEL. At the extreme, we could have separate parameters in our model for each observation and then estimate the value of each such parameter to be identical to the value of its corresponding DATA observation. For example, our prediction might contain statements such as, *if the country is Belgium, then estimate its parameter to be 92.79*, which is the percentage of individuals in Belgium that used the internet. That procedure would clearly reduce the error to zero and provide a perfect fit. But such a model would be uninteresting because it would simply be a duplicate of data and would provide no new insights, no bases for testing our theories, and no ability to make predictions in slightly different circumstances. A paramount goal of science is to provide simple, parsimonious explanations for phenomena. A model with a separate parameter for each observation is certainly not parsimonious. Our ideal model, then, is a compact description of the data and has many fewer parameters than the number of observations in data.

We now have an obvious conflict. The goal of reducing the error and providing the best description of DATA leads us to add parameters to the model. On the other hand, the goal of parsimony and the desire for a compact, simple model leads us to remove parameters from the model. The job of the data analyst is to find the proper balance between these two conflicting objectives. Thus, the ultimate goal is to find the smallest, simplest model that provides an adequate description of the data so that the error is not too large ("too large" will be defined later). In still other words, the data analyst must answer the question of whether it is worthwhile to add yet more parameters to a model.

Returning to the example of internet use, we will want to ask whether the extra complexity of making predictions conditional on the percentage of employees in service industries, type of government, income, educational level, urban versus rural population, etc. is worth the trouble. By so doing, we will simultaneously be asking whether the hypotheses implicit in the more complex models are true. For example, if we decide that conditioning our prediction of internet use on the percentage of service employees is not worthwhile, then we will have effectively rejected the hypothesis that a stronger service economy is related to higher internet use. This is the essence of testing hypotheses.

Although we are still being vague about how to measure the error, we can be more precise about what we mean by: "Are more parameters worthwhile?" We will call the model without the additional parameters the *compact model* and will refer to it as Model C. The alternative, *augmented model*, Model A, includes all the parameters, if any, of Model C plus some additional parameters. The additional parameters of Model A may reduce the error or leave it unchanged; there is no way the additional parameters can increase the error. So it must be that

$$\text{Error(A)} \leq \text{Error(C)}$$

where ERROR(A) and ERROR(C) are the amounts of error when using Models A and C, respectively. The question of whether it is worthwhile to add the extra complexity of Model A now reduces to the question of whether the difference between ERROR(C) and ERROR(A) is big enough to worry about. It is difficult to decide based on the absolute magnitude of the errors. We will therefore usually make relative comparisons. One way to do that is to calculate the *proportional reduction in error* (PRE), which represents the proportion of Model C's error that is reduced or eliminated when we use the more complex Model A to describe DATA. Formally,

$$\text{PRE} = \frac{\text{ERROR}(\text{C}) - \text{ERROR}(\text{A})}{\text{ERROR}(\text{C})}$$

The numerator is simply the difference between the two errors (the amount of error reduced) and the denominator is the amount of error for the compact model with which we started. An equivalent expression is

$$\text{PRE} = 1 - \frac{\text{ERROR}(\text{A})}{\text{ERROR}(\text{C})}$$

If the additional parameters do no good, then ERROR(A) will equal ERROR(C), so PRE = 0. If Model A provides a perfect fit, then ERROR(A) = 0 and (assuming Model C does not also provide a perfect fit) PRE = 1. Thus, values of PRE will be between 0 and 1. The larger the value of PRE, the more it will be worth the cost of increased complexity to add the extra parameters to the model.

The smaller the value of PRE, the more we will want to stick with the simpler, more parsimonious compact model.

For example, ignoring for the moment how we calculate the error, assume that total ERROR = 50 for the simple model that says that internet use is the same in all 45 countries and that ERROR = 30 for the model with the additional parameter for the percentage of service employees. Then, ERROR(C) = 50, ERROR(A) = 30, and

$$\text{PRE} = 1 - \frac{30}{50} = .40$$

That is, increasing the complexity of the model by considering the percentage of service employees would reduce the error by 40%.

Let us review where we are. We have transformed the original problem of the conflicting goals for the model (parsimony and accurate representation of data) into a consideration of the size of PRE for comparing Model C and Model A. Unfortunately, we have still not solved the problem of the conflicting goals for the model because now we must decide whether a given PRE (e.g., the 40% reduction in the previous example) is big enough to warrant the additional parameter(s). The transformation of the original problem has moved us closer to a solution, however, for now we have a PRE index that will be used no matter how we finally decide to measure the error. More importantly, PRE has a simple, intuitive meaning that provides a useful description of the amount of improvement provided by Model A over Model C.

Deciding whether a PRE of, say, 40% is worthwhile involves inferential statistics. An understanding of inferential statistics must await the details of measuring the error, sampling distributions, and other topics that are developed in Chapters 2, 3, and 4. We can, however, now specify two considerations that will be important in inferential statistics. First, we will be much more impressed with a PRE of, say, 40% if it were obtained with the addition of only one parameter instead of with four or five parameters. Hence, our inferential statistics will need to consider the number of extra parameters added to Model C to create Model A. *PRE per parameter added* will be a useful index. Second, we noted that n, the number of observations in DATA, serves as an upper limit to the number of parameters that could be added to the model. We will be more impressed with a given PRE as the difference between the number of parameters that were added and the number of parameters that could have been added becomes greater. Hence, our inferential statistics will consider how many parameters *could have been added* to Model C to create Model A but were not. In other words, we will be more impressed with a PRE of 40% if the number of observations greatly exceeds the number of parameters used in Model A than if the number of observations is only slightly larger than the number of parameters.

The use of PRE to compare compact and augmented models is the key to asking questions of our data. For each question we want to ask of DATA, we will find appropriate Models C and A and compare them using PRE. For example, if we want to know whether the percentage of service employees is useful for predicting the percentage of individuals who use the internet, we would compare a Model C that does not include a parameter for service employees to a Model A that includes all the parameters of Model C plus an additional parameter for the percentage of service employees. If Model C is a simple model (i.e., a single-parameter model that makes a constant prediction for all observations), then we are asking whether the percentage of service employees *by itself* is a useful predictor of internet use. If there are other parameters in Model C, then we are asking whether the percentage of service employees is a useful predictor of internet use *over and above* the other parameters. (We discuss this at length in Chapter 6.) As another example, if we want to ask whether several factors, such as income level, percentage of service employees, educational level, and type of government, are simultaneously useful in predicting internet use, we would use PRE to compare a Model C that did not have parameters for any of those factors to a Model A that did have those parameters in addition to those in Model C.

In the usual language for statistical inference, Model C corresponds to the *null hypothesis* and Model A corresponds to the alternative hypothesis. More precisely, the null hypothesis is that all the parameters included in Model A but not in Model C are zero (hence, the name "null") or equivalently that there is no difference in error between Models A and C. If we reject Model C in favor of Model A, then we reject the null hypothesis in favor of the alternative hypothesis that is implied by the difference between Models C and A. That is, we conclude that it is unreasonable to presume that all the extra parameter values in Model A are zero. We discuss this fully in Chapter 4.

Notation

To facilitate seeing relationships between procedures, we use consistent notation throughout. Y_i represents the *i*th observation from DATA. The first observation is numbered 1 and the last is n, for a total of n observations in DATA. $\hat{Y}_i$ represents the model's prediction of the *i*th observation. Y will always represent the variable that we are trying to predict with our model. Other variables giving information about each observation on which we might base conditional predictions will be represented by X. In other words, X will always be used to represent the predictor(s) of Y. So X_{ij} represents the value of the *j*th predictor variable for the *i*th observation. For example, in the internet use example the 45 percentages would be represented as, $Y_1, Y_2, \ldots Y_{45}$, with $n = 45$. X_{i1} could be 1 if the country has a democratic government, and –1 if the country has an authoritarian government. And we could use X_{i2} to represent service employees, in which case the value of X_{i2} would be the percentage of employees working in service industries for the *i*th country.

For model parameters we will use $\beta_0, \beta_1, \ldots, \beta_j, \ldots, \beta_{p-1}$, for a total of p parameters. Even if we were to know the values of these parameters exactly, we would not expect the model to predict the data exactly. Instead, we expect that some random error will cause the model to predict less than perfectly. We let ε_i represent the unknown amount by which we expect the model to mispredict Y_i. Thus, for the simple model, the basic equation

DATA = MODEL + ERROR

can be expressed in terms of the true parameter β_0 and the error ε_i as

$$Y_i = \beta_0 + \varepsilon_i$$

We can never know the true β parameters (or ε) exactly. Instead, we will have estimates of β that we calculate from the data. These estimates will be labeled, $b_0, \ldots, b_1, \ldots, b_j, \ldots, b_{p-1}$, respectively. We use $\hat{Y}_i$ to represent the prediction for the *i*th observation based on the calculated b values. We then let e_i represent the amount by which the predicted value or $\hat{Y}_i$ mispredicts Y_i; that is,

$$e_i = Y_i - \hat{Y}_i$$

The Greek letters β and ε represent the true but unknowable parameters and the Roman letters b and e represent estimates of those parameters calculated from DATA. For the simple model, the model part of the basic data analysis equation is

MODEL: $\hat{Y}_i = b_0$

and we can express that basic equation in terms of our parameter estimates as either

$$Y_i = \hat{Y}_i + e_i$$

or

$$Y_i = b_0 + e_i$$

Either of the two previous equations are estimates for the basic equation expressed in terms of the unknown parameters as

$$Y_i = \beta_0 + \varepsilon_i$$

The quantity $\beta_j X_{ij}$ tells us how much we should adjust the basic prediction for the *i*th observation based on the *j*th predictor variable. For example, if X_{ij} equals the percentage of service employees in a country, then β_j specifies how much to adjust, upward or downward, depending on the sign of β_j the internet use prediction for a particular country. A more complicated model involving more parameters can be expressed as

$$Y_i = \beta_0 + \beta_1 X_{i1} + \beta_2 X_{i2} + \ldots + \beta_j X_{ij} + \ldots + \beta_{p-1} X_{i,p-1} + \varepsilon_i$$

$\hat{Y}_i$, the MODEL portion of the data analysis equation, is then represented in terms of the parameter estimates as

$$\text{MODEL: } \hat{Y}_i = b_0 + b_1 X_{i1} + b_2 X_{i2} + \ldots + b_j X_{ij} + \ldots + b_{p-1} X_{i,p-1}$$

The equation

DATA = MODEL + ERROR

can again be expressed in two ways:

$$Y_i = \hat{Y}_i + e_i$$

or

$$Y_i = b_0 + b_1 X_{i1} + b_2 X_{i2} + \ldots + b_j X_{ij} + \ldots + b_{p-1} X_{i,p-1} + e_i$$

Note that when β values are used on the right side of the equation the appropriate symbol for the error is always ε_i and when *b* values (i.e., estimates of βs) are used on the right side of the equation the appropriate symbol for the error is always e_i. The reason is that in the first instance the error ε_i is unknown, while in the second instance an estimated value e_i can actually be calculated once $\hat{Y}_i$ is calculated.

We will have to develop a few special symbols here and there, but in general the previous notation is all that is required for all the models we consider in this book.

Summary

The basic equation for data analysis is

DATA = MODEL + ERROR

The data analyst using this equation must resolve two conflicting goals: (a) to add parameters to MODEL so that it is an increasingly better representation of DATA with correspondingly smaller ERROR, and (b) to remove parameters from MODEL so that it will be a simple, parsimonious representation of DATA. Resolving this conflict is equivalent to asking whether the additional parameters are worth it. We use PRE, the index of the proportional reduction in error, to answer this question by comparing appropriately chosen Models C and A. In the traditional language of statistical inference, this is equivalent to comparing a null hypothesis and an alternative hypothesis. The next several chapters provide the information necessary to judge when PRE is large enough to warrant rejecting Model C in favor of Model A.

We use consistent notation throughout the book to specify our statistical models. Y represents the variable that we are trying to predict ("y-hat," that is, $\hat{Y}$, represents a predicted value of Y) and X represents a predictor variable. Lower-case Greek letters represent true, but unknown, characteristics of the population, and Roman letters represent the estimates of those characteristics. Thus, β represents a true, but unknown, population parameter, and b represents an estimate of that parameter, which is calculated from DATA. Similarly, ε_i represents the true, but unknown, error of prediction, and e_i represents an estimate of that error, which is calculated from DATA. The same statistical model might therefore be expressed using unknown population parameters, for example, $Y_i = \beta_0 + \beta_1 X_{i1} + \varepsilon_i$, or using symbols that represent estimates that are calculated from DATA, for example, $Y_i = b_0 + b_1 X_{i1} + e_i$. Finally, we may also express the model in terms of predicted values, for example, $\hat{Y}_i = b_0 + b_1 X_{i1}$.

2 Simple Models

Definitions of Error and Parameter Estimates

In this chapter we consider the very simplest models—models with one or even no parameters. These simple models make the same prediction for all data observations. Such a simple model may not seem very realistic or useful. However, this simple model provides a useful baseline against which we can compare more complicated models, and it will turn out to be more useful than it first appears. For many of the questions we will want to ask about our data, the appropriate Model C will be the simple model with no parameter or one parameter; this will be compared to the more complex Model A. Also, the simple model provides a useful first-cut description of data.

Overview of the Simple Model

Formally, the simplest model is

$$Y_i = B_0 + \varepsilon_i$$

where B_0 is some specified value not based on this particular batch of data (i.e., it is a specific a priori numeric value), and ε_i is the true error or the amount by which Y_i differs from B_0. This simple model in which no parameters are estimated from the data is not frequently used in the social sciences because we seldom have theories sufficiently powerful to make an explicit prediction for a parameter. Such models are much more common in fields such as biology. For example, a medical study measuring human body temperature might reasonably consider the model: except for error, all temperatures are 37°C. In this case, $B_0 = 37$ so the formal model would be

$$Y_i = 37 + \varepsilon_i$$

Although we will sometimes want to consider a specified or hypothesized value of B_0 in order to ask an explicit question about data, it is much more common to consider the equation

$$Y_i = \beta_0 + \varepsilon_i$$

where β_0 is a true parameter that is estimated from the data. Continuing with the medical example, suppose that the body temperatures were all from people who had taken a certain drug. We might suspect that, except for error, they all have the same body temperature, but it is not the usual body temperature of 37°C. We use β_0 to represent whatever the body temperature might be for those who have taken the drug. It is important to realize that β_0 is unknowable; we can only estimate it from the data. In terms of these true values, ε_i is the amount by which Y_i differs from β_0 if we were ever

DOI: 10.4324/9781003438335-4

to know β_0 exactly. We use b_0 to indicate the estimate of β_0 that we derive from the data. Then the predicted value for the *i*th observation is

$$\hat{Y}_i = b_0$$

and

$$\text{DATA} = \text{MODEL} + \text{ERROR}$$

becomes

$$Y_i = b_0 + e_i$$

where e_i is the amount by which our prediction misses the actual observation. Thus, e_i is the estimate of ε_i. The goal of tailoring the model to provide the best fit to the data is equivalent to making the errors

$$e_i = Y_i - b_0$$

as small as possible. We have only one parameter, so this means that we want to find the estimate b_0 for that one parameter β_0 that will minimize the errors. However, we are not really interested in each e_i but in some aggregation of all the individual e_i values. There are many ways to perform this aggregation. In this chapter, we consider some of the different ways of aggregating the separate e_i into a summary measure of the error. Then we show how each choice of a summary measure of the error leads to a different method of calculating b_0 to estimate β_0 so as to provide the best fit of the model to the data, that is, to minimize errors of prediction. Finally, we consider expressions that describe the "typical" error.

Measures of location or *measures of central tendency* are the traditional names for the parameter estimates of β_0 developed from the different definitions of error. These names are appropriate because the parameter estimate in the simple model tells us about the location of a typical observation or about the center of a batch of data. Specific instances include the mode, median, and mean. *Measures of variability* or *measures of spread* are the traditional names for the expressions for typical errors. These names are appropriate because expressions for typical errors tell us how variable the observations are in a batch of data or, equivalently, how far the data spread out from the center. Specific instances include the median absolute deviation and standard deviation. Together, measures of central tendency and spread are known as *descriptive statistics*. However, this suggests a false distinction between these statistics and those to come later. We want to emphasize that the parameter estimates for β_0 in the simple model are no more nor less descriptive than the parameter estimates we will develop for more complicated models. Models and their parameter estimates always provide descriptions of data. Hence, we will generally avoid the phrase "descriptive statistics" and just refer to parameter estimates. The reader should be aware, however, that when other textbooks refer to descriptive statistics they are generally referring to the material in this chapter.

Conceptual Example

Before considering simple models and measures of error more formally, we consider some conceptual examples to help build intuitions so that the subsequent mathematical formulas will be less abstract. Suppose that the data consisted of the five observations 1, 3, 5, 9, 14, representing the number of books read over the summer by each of five elementary school students. These observations are plotted in Figure 2.1.

The simple model makes the same prediction for all five observations. The horizontal line represents the value of that constant prediction. The vertical lines drawn from each observation to the prediction line represent the amount by which the prediction misses the actual data value. In other words, the length of the line is e_i. One way to find the best value for $\hat{Y}$ and, equivalently, the best estimate b_0 for β_0 is to adjust the $\hat{Y}$ line up or down so that the length of the lines is a minimum. (We have dropped the subscript i from $\hat{Y}_i$ because all the predictions are the same for the simple model.) In other words, we can use trial and error to find the best estimate. For example, we might want to try 7 as our estimate. The data, the prediction line for $\hat{Y} = b_0 = 7$, and the errors are graphed in Figure 2.2. Note that the five line lengths are now 6, 4, 2, 2, 7, with a sum of 21.

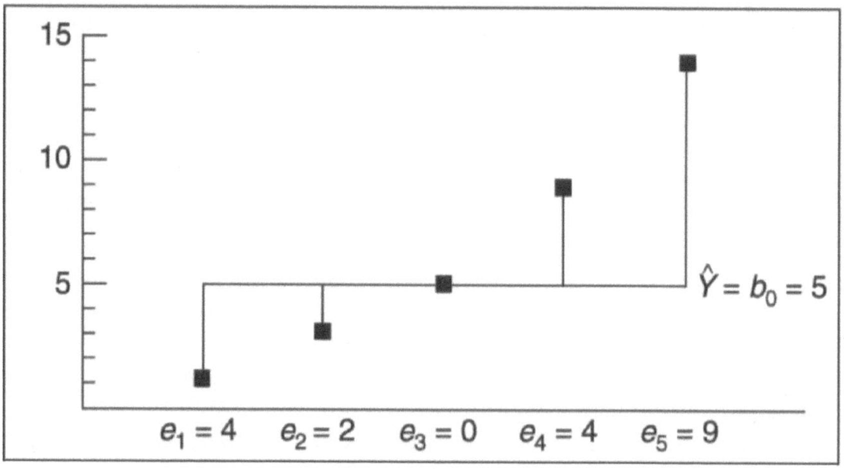

Figure 2.1 Error as the sum of line lengths (estimate is $\hat{Y} = b_0 = 5$)

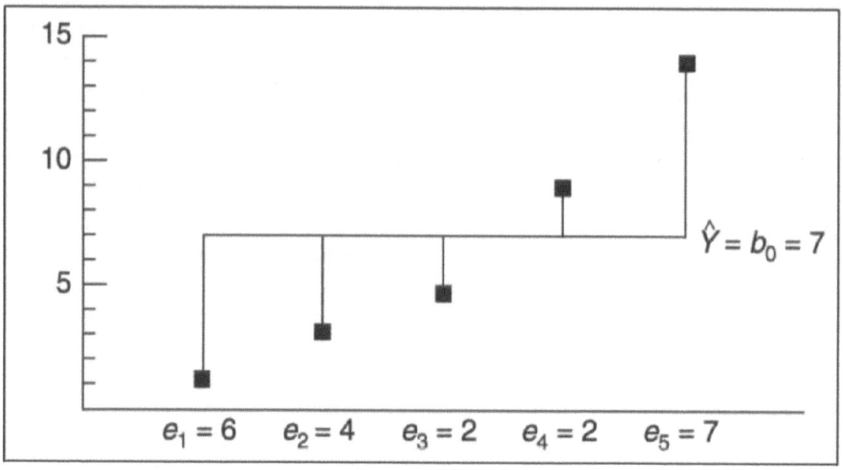

Figure 2.2 Error as the sum of line lengths (estimate is $\hat{Y} = b_0 = 7$)

For an estimate of 5, the line lengths were 4, 2, 0, 4, 9, with a sum of 19. The estimate $b_0 = 5$ thus produces less total error than $b_0 = 7$, so we can eliminate 7, in favor of 5, as an estimate if our goal is to minimize the total error. We can continue to try other estimates until we find the best one. Figure 2.3 shows the sum of the line lengths for different choices of b_0 between 0 and 10. The sum of the line lengths reaches a minimum of 19 when $b_0 = 5$. We would get more total error, a larger sum of line lengths, if we used a value of b_0 that was either lower or higher than 5. Hence, $b_0 = 5$ is the best estimate. Note that 5 is the middle of our five observations; the middle observation in a batch of data that have been sorted from smallest to largest is often called the *median*.

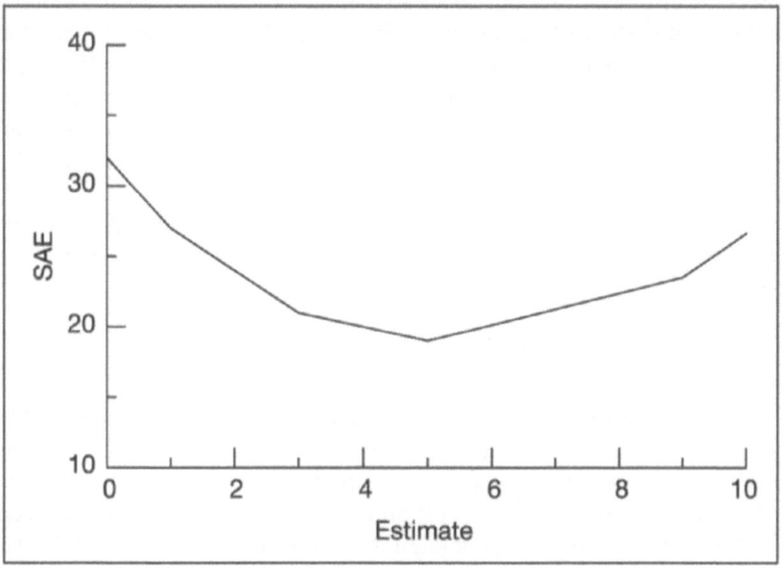

Figure 2.3 Sum of absolute error (SAE) as a function of $\hat{Y}_i$

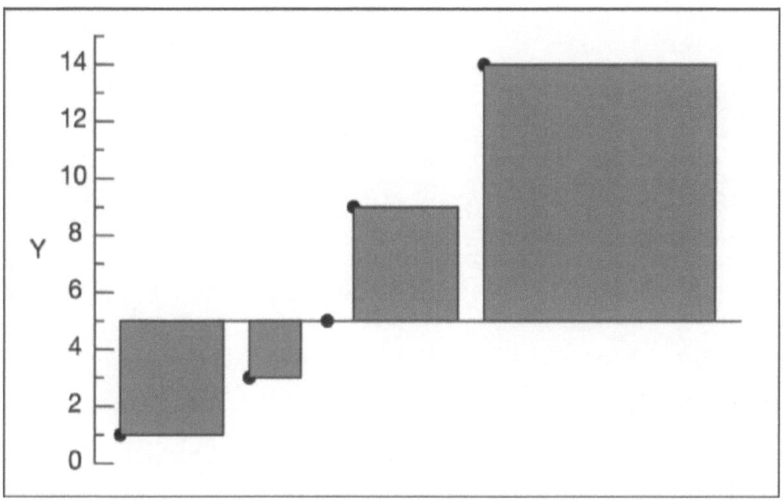

Figure 2.4 Error as the sum of squares (estimate is $\hat{Y}_i = b_0 = 5$)

It is interesting to ask how we might adjust the estimate b_0 if one of the observations were dramatically changed, for example, if the 14 were replaced by 140 so that the five observations were 1, 3, 5, 9, and 140. In this case, the e_5 line would be very long. We could shorten e_5 by one unit if we defined b_0 as 6 instead of 5, but that change would lengthen e_1, e_2, and e_3 by one unit each—yielding an increase in total error. Thus the minimum total error still occurs when $b_0 = 5$! The middle or median observation is the best estimate even though one of the observations has been increased by a factor of 10.

The sum of the error line lengths as an aggregate summary index of error is not always reasonable. For example, the simple sum implies that several small errors (e.g., four errors of length 1) are equivalent to one large error (e.g., one error of length 4). Instead, we may want to charge a higher penalty in the error index for big errors so that an error of 4 counts more than four errors of 1. One way to accomplish this is to square the line lengths before summing them. For example, $4^2 = 16$ adds a lot more to the error sum than $1^2 + 1^2 + 1^2 + 1^2 = 4$. Figure 2.4 depicts the original set of five observations with this new definition of error; each e_i is now represented by a square, the length of whose side is determined by the distance between the observation and the horizontal line representing the constant prediction $\hat{Y}_i$ of the simple model. The aggregate error is simply the sum of those squares.

Again, we can use trial and error to find a value of $\hat{Y}_i$ and b_0 that makes that sum of squares as small as possible. Let us again consider the possible estimates of 5 and 7. For $b_0 = 5$, the five areas of the squares are

$$4^2 = 16,\ 2^2 = 4,\ 0^2 = 0,\ 4^2 = 16, \text{ and } 9^2 = 81$$

and the sum of those squares is 117. For $b_0 = 7$, the five areas are 36, 16, 4, 4, 49, and the sum of squares is 109. So $b_0 = 5$, which was the best estimate when using line lengths, is no longer the best estimate when we use squares, because $b_0 = 7$ produces a smaller sum of squares or smaller error. Figure 2.5 shows the sum of squares for different possible values of b_0 between 0 and 10. The best value for b_0 is about 6.4 with a minimum sum of squares of about 107. The estimates of 5 and 7 are not bad—sums of squares of 117 and 109, respectively—but clearly inferior to

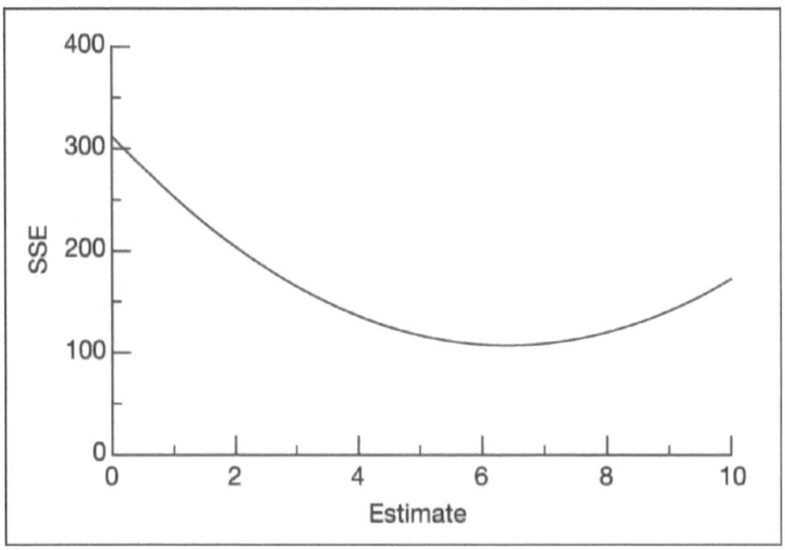

Figure 2.5 Sum of squared errors (SSE) as a function of $\hat{Y}_i$

the optimum estimate of 6.4. Although not obvious, we will prove later that the best estimate, that is, the estimate that minimizes the sum of squared errors, is the arithmetic average or *mean* of the observations. For the five observations,

$$\frac{1+3+5+9+14}{5} = \frac{32}{5} = 6.4$$

which produces b_0, the best estimate of β_0.

Consider what would happen to the estimate b_0 if one of the observations were dramatically changed: say, the 14 were replaced by 140. The minimum sum of squares no longer occurs when $b_0 = 6.4$; instead, the minimum now occurs when b_0 is a whopping 31.6, which is again the average of the five observations. But note that although that estimate is the best when error is defined as the sum of squared errors, it is not very good, with a total sum of squares of about 14,723.

Before formalizing these examples in equations, let us summarize the concepts we have introduced. First, the best estimate b_0 for a simple, one-parameter model is the constant prediction $\hat{Y}$, which minimizes the sum of the errors. Although we will generally have better ways to estimate the parameter than trial and error, it is important to realize that the best estimate could be found by trial and error until we could no longer make the error any smaller. In fact, computer programs sometimes use precisely this strategy. Second, the choice of a method for summarizing or expressing the error—the lengths of the error lines or their squares in the previous examples—affects the best estimate b_0. We will soon see that there are many other plausible ways to define error. Third, if total error is the sum of the line lengths, then the median of the observations provides the best estimate b_0, where the median is simply the middle observation. Fourth, if total error is the sum of the squared line lengths, then the best estimate b_0 is the arithmetic average or mean of the observations. Fifth, the median does not change when an extreme observation is made more extreme, but the mean can change dramatically. Sixth, many times in this book we will encounter the phrase "sum of squares"; it is useful to realize that it can indeed be represented geometrically as a literal summation of squares.

Formalities for Simple Models

As always, we begin with the basic equation for data analysis:

DATA = MODEL + ERROR

For simple models, MODEL states that all observations in DATA are essentially the same, so

$$Y_i = \beta_0 + \varepsilon_i$$

where β_0 represents the true, but unknown, parameter and ε_i represents the individual error disturbances around the unknown parameter. Then b_0 is the estimate of that parameter based on the data at hand. So, the actual model used for predicting each Y_i becomes

MODEL: $\hat{Y}_i = b_0$

The basic data analysis equation can then be written as

$$Y_i = \hat{Y}_i + e_i \text{ or } Y_i = b_0 + e_i$$

The error or residual associated with each observation is then simply the difference between the data and the model prediction, or

$$e_i = Y_i - \hat{Y}_i$$

Our problem is how to select a single value for b_0 to represent all of the data. Clearly, we want b_0 to be in the "center" of the data so that it will be generally close to all of the observations. Hence, estimates of b_0 are often called *measures of central tendency*. But we need to be much more precise about defining the center and what we mean by "close." As always, the key is to make e_i as small as possible. Instead of looking at each individual e_i, we need to consider ways of aggregating the separate e_i values into a summary measure of the total error. Once we have done that, it is a simple procedure to choose a b_0 that will make the summary measure of error as small as possible. We now turn to a consideration of possible summary measures.

Count of Errors (CE)

One possibility is simply to count the number of times Y_i does not equal $\hat{Y}_i$. This ignores the size of the individual errors and only counts whether an error occurred. Formally,

$$\text{ERROR} = \sum_{i=1}^{n} I(e_i) = \sum_{i=1}^{n} I\left(Y_i - \hat{Y}_i\right) = \sum_{i=1}^{n} I\left(Y_i - b_0\right)$$

where $I(e_i) = 1$ if $e_i \neq 0$ and $I(e_i) = 0$ if $e_i = 0$. Functions such as $I(\ )$ *are* often called *indicator* functions and are simply a fancy way of representing whether or not something is to be included in an equation.

Sum of Errors (SE)

In order not to lose information about the size of the e_i, we might add all the e_i values so that

$$\text{ERROR} = \sum_{i=1}^{n} e_i = \sum_{i=1}^{n} \left(Y_i - \hat{Y}_i\right) = \sum_{i=1}^{n} \left(Y_i - b_0\right)$$

But this is unsatisfactory because it allows positive and negative errors to cancel one another. For example, suppose that b_0 underestimated one observation by 1000 and overestimated another observation by the same amount so that $e_1 = 1000$ and $e_2 = -1000$. Adding those two errors would produce zero, incorrectly implying that there was no error. We have no reason to be more interested in overestimates than underestimates, so one solution is to ignore the positive and negative signs of the e_i.

Sum of Absolute Errors (SAE)

One way to remove the signs is to sum the absolute values of the errors:

$$\text{ERROR} = \sum_{i=1}^{n} |e_i| = \sum_{i=1}^{n} \left|Y_i - \hat{Y}_i\right| = \sum_{i=1}^{n} \left|Y_i - b_0\right|$$

The sum of absolute errors is the formal equivalent of summing the line lengths in Figure 2.1. As we have noted, it may not always be desirable to treat one big error (e.g., an error of 4) the same as the equivalent amount of small errors (e.g., four errors of 1). The conceptual example therefore suggests the next measure of error.

Sum of Squared Errors (SSE)

Another way to remove the signs from the e_i is to square each one before summing them:

$$\text{ERROR} = \sum_{i=1}^{n} e_i^2 = \sum_{i=1}^{n} \left(Y_i - \hat{Y}_i\right)^2 = \sum_{i=1}^{n} \left(Y_i - b_0\right)^2$$

The sum of squared errors is the formal equivalent of adding up the error squares in Figure 2.4. Besides removing the signs, the squaring has the additional effect of making large errors more important.

Estimators of β_0

As demonstrated by the conceptual examples presented earlier in this chapter, the choice of a method for aggregating the e_i influences the estimate b_0. It should not be surprising, therefore, that for each definition of aggregate error presented earlier there is a different way of calculating b_0 from the data. We could use trial and error for each definition of error, trying different values of b_0 until we found the one that gave the minimum value for error. However, it turns out that for each of the earlier definitions of error we can calculate b_0 from the data so that b_0 is guaranteed to produce the minimum possible value for error. While we will almost always use the calculation method, it is important to remember that the definition of the best estimate of β_0 is the value of b_0 that produces the least error. We list in Figure 2.6 the definition of b_0 for each definition of error considered so far.

Proof That the Mean Minimizes SSE

We now present a formal proof that the mean $\bar{Y}$ does indeed produce the smallest possible value of SSE. Although we generally avoid proofs, we think it is important to understand that the choice of the mean as an estimator is not arbitrary; instead, the choice of SSE as an aggregate measure of ERROR also dictates the choice of the mean as the best estimator. Similar proofs can be given to show that the indicated estimator minimizes the corresponding definition of error in the list in Figure 2.6.

Let us begin by assuming that $\hat{Y}$, the best estimator for reducing the sum of squared errors, is something other than

$$\bar{Y} = \frac{\sum_{i=1}^{n} Y_i}{n}$$

Error definition	*b_0, estimator of β_0*
Count of errors	Mode = most frequent value of Y_i
Sum of absolute errors	Median = middle observation of all the Y_i
Sum of squared errors	Mean = average of all the Y_i

Figure 2.6 Estimators for each definition of error

(Note that here $\hat{Y}$ does not have an i subscript because we make the same prediction for all observations in the simple model.) Our strategy is to proceed with that assumption until we reach a point where we can see that it is a bad assumption. At that point we will know that the only reasonable assumption is $\hat{Y} = \bar{Y}$.

The sum of squared errors is

$$\text{SSE} = \sum_{i=1}^{n} \left(Y_i - \hat{Y}\right)^2$$

Obviously, $\bar{Y} - \bar{Y} = 0$; we can add zero within the parentheses without changing the sum of squared errors. That is

$$\text{SSE} = \sum_{i=1}^{n} \left(Y_i - \hat{Y} + \bar{Y} - \bar{Y}\right)^2$$

Rearranging the terms, we get

$$\text{SSE} = \sum_{i=1}^{n} \left[\left(Y_i - \bar{Y}\right) + \left(\bar{Y} - \hat{Y}\right)\right]^2$$

Squaring gives

$$\text{SSE} = \sum_{i=1}^{n} \left[\left(Y_i - \bar{Y}\right)^2 + 2\left(Y_i - \bar{Y}\right)\left(\bar{Y} - \hat{Y}\right) + \left(\bar{Y} - \hat{Y}\right)^2\right]$$

Breaking the sums apart yields

$$\text{SSE} = \sum_{i=1}^{n} \left(Y_i - \bar{Y}\right)^2 + \sum_{i=1}^{n} \left[2\left(Y_i - \bar{Y}\right)\left(\bar{Y} - \hat{Y}\right)\right] + \sum_{i=1}^{n} \left(\bar{Y} - \hat{Y}\right)^2$$

The last term contains no subscripts so the summation is equivalent to adding up the same quantity n times; hence, the summation sign can be replaced by multiplication by n. Similarly, the quantities without subscripts in the middle term can be taken outside the summation sign to give

$$\text{SSE} = \sum_{i=1}^{n} \left(Y_i - \bar{Y}\right)^2 + 2\left(\bar{Y} - \hat{Y}\right)\sum_{i=1}^{n} \left(Y_i - \bar{Y}\right) + n\left(\bar{Y} - \hat{Y}\right)^2$$

Now let us concentrate on the middle term. Note that

$$\begin{aligned}
\sum_{i=1}^{n} \left(Y_i - \bar{Y}\right) &= \sum_{i=1}^{n} Y_i - n\bar{Y} \\
&= \sum_{i=1}^{n} Y_i - n\left(\frac{\sum_{i=1}^{n} Y_i}{n}\right) \\
&= \sum_{i=1}^{n} Y_i - \sum_{i=1}^{n} Y_i \\
&= 0
\end{aligned}$$

Hence, the middle term of SSE includes a multiplication by zero, which eliminates that term. We are left with

$$\text{SSE} = \sum_{i=1}^{n}\left(Y_i - \bar{Y}\right)^2 + n\left(\bar{Y} - \hat{Y}\right)^2$$

We want SSE to be as small as possible. We have no freedom in the first term: Y_i and $\bar{Y}$ are whatever they happen to be. We do, however, have freedom to choose $\hat{Y}_i$ in the second term to make SSE as small as possible. Clearly, $n(\bar{Y} - \hat{Y})^2$ is positive, so it is making SSE larger. But if we let $\hat{Y} = \bar{Y}$, then $n(\bar{Y} - \hat{Y})^2 = 0$ and we are no longer adding anything extra to SSE. For any estimate of $\hat{Y}$ other than $\bar{Y}$, we will be making SSE larger. Hence, SSE is as small as possible when $\hat{Y} = \bar{Y}$, and the minimum is

$$\text{SSE} = \sum_{i=1}^{n}\left(Y_i - \bar{Y}\right)^2$$

Any other choice for $\hat{Y}$ would produce a larger SSE; thus, the mean is the best estimator for reducing SSE in the simple model.

Describing Error

If the goal is simply to describe a batch of data, then we can apply the simple model and use one or all of the measures of central tendency—mode, median, and mean—as descriptive statistics. When doing so, it is also useful to present a description of the typical error. Reporting the total error (e.g., SAE or SSE) is not desirable because the total depends heavily on the number of observations. For example, aggregate error based on 50 observations is likely to be larger than aggregate error based on 15 observations, even if the typical errors in the former case are smaller. For each measure of central tendency there is a corresponding customary index of the typical error. We consider each next.

Modal Error

When we use the count of the errors (CE) as the aggregate index of error, there can only be two values for e_i: either $e_i = 0$ when $Y_i = \hat{Y}$ or $e_i = 1$ when $Y_i \neq \hat{Y}$. The typical error is simply the more frequent or modal error. Modal error is seldom used, but we have presented it for completeness.

Median Absolute Deviation

When we use the sum of absolute errors (SAE) as the aggregate index of error, it is customary to use the median absolute error or deviation from the prediction to represent the typical error. To find the median absolute deviation, simply sort the $|e_i|$ into ascending order and find the middle one.

Standard Deviation

When we use the sum of squared errors (SSE) as the aggregate index of error, the index is somewhat more complex. We will be making extensive use of SSE throughout this book. To avoid having to introduce more general formulas later, we present the more general formula now and then show how it applies in the case of simple models. In a general model with p parameters, those p parameters have been used to reduce the error. In principle, the maximum number of parameters we could have is n—one parameter for each observation—in which case the error would equal zero. Thus, there are $n - p$ potential parameters remaining that could be used to reduce the remaining error. A useful index of error is then the remaining error per remaining potential parameter. This index has the name *mean squared error* (MSE) and is given by

$$\text{MSE} = \frac{\text{SSE}}{n-p} = \frac{\sum_{i=1}^{n}\left(Y_i - \hat{Y}_i\right)^2}{n-p}$$

For the simple model considered in this chapter there is only the one parameter β_0 to be estimated, so $p = 1$ and the estimate of Y_i is $b_0 = \bar{Y}$, the mean value. For the simple model, MSE has the special name *variance* and is commonly represented by s^2, that is

$$\text{Variance} = s^2 = \text{MSE} = \frac{\text{SSE}}{n-1} = \frac{\sum_{i=1}^{n}\left(Y_i - \bar{Y}\right)^2}{n-1}$$

MSE represents the typical *squared* error; to express the typical error in the units in which the original data were recorded, it is useful to take the square root of MSE, which is often referred to, especially on computer printouts, as *ROOT MSE*. For the simple model, the square root of the variance or the MSE has the special name *standard deviation* and is given by

$$\text{Standard deviation} = s = \sqrt{\text{MSE}} = \sqrt{\frac{\sum_{i=1}^{n}\left(Y_i - \bar{Y}\right)^2}{n-1}}$$

Another index sometimes used when SSE is used as the aggregate index of error is the *coefficient of variation*. It is common for the size of the standard deviation to be proportional to the size of the mean. For example, if $\bar{Y} = 10{,}000$, we would expect the typical error or standard deviation to be much larger than when $\bar{Y} = 10$. Although this need not be the case, it usually is true. To remove the effect of the overall magnitude of the data from the description of the error, the coefficient of variation expresses the size of the standard deviation as a proportion of the mean, that is

$$\textit{Coefficient of variation} = CV = \frac{s}{\bar{Y}}$$

An Example

We will use the percentage of individuals who used the internet in the year 2021 by country as an example to illustrate the simple model and the descriptors of central tendency and error. To facilitate finding the mode and the median, we have rearranged the data of Figure 1.1 in Figure 2.7 in order of increasing percentages. Our goal is to fit to these data the simple model that has just one parameter. Thus, the basic data analysis equation is

$$Y_i = \beta_0 + \varepsilon_i$$

and we want to fit the model to the data by finding the estimate b_0 for β_0 that minimizes error—the e_i in the equation

$$Y_i = b_0 + e_i$$

How we find the estimate b_0 depends on which definition of aggregate error we adopt. If CE is adopted as the criterion, then the best estimate for β_0 is the mode. To find the mode, we simply

i	*Country*	*Percentage*	*Rank*
19	GTM	50.84	1
31	NIC	57.15	2
22	IDN	62.10	3
33	PER	71.11	4
14	EGY	72.06	5
10	COL	73.03	6
9	CHN	73.05	7
5	BWA	73.50	8
25	ITA	74.86	9
7	BGR	75.27	10
29	MEX	75.63	11
13	RCU	76.20	12
18	GRC	78.49	13
43	UKR	79.22	14
6	BRA	80.69	15
11	HRV	81.25	16
42	TUR	81.41	17
35	PRT	82.31	18
27	JOR	82.76	19
26	JPN	82.91	20
36	ROU	83.59	21
41	THA	85.27	22
34	POL	85.37	23
16	FRA	86.10	24
3	BLR	86.89	25
1	ARG	87.15	26
37	RUS	88.21	27
39	SWE	88.31	28
20	HUN	88.64	29
24	ISR	90.30	30
17	DEU	91.43	31
45	USA	91.75	32
30	NLD	92.05	33
4	BEL	92.79	34
15	FIN	92.81	35
8	CAN	92.83	36
38	ESP	93.90	37
23	IRL	95.17	38
40	CHE	95.57	39
2	AUS	96.24	40
44	GBR	96.68	41
28	MYS	96.75	42
12	DNK	98.87	43
32	NOR	99.00	44
21	ISL	99.69	45

Figure 2.7 Percentage of individuals using the internet in the year 2021 by country (sorted by percentage)

observe which percentage is the most frequent. For these data, each value occurs only once so there really is no mode. It is frequently the case with continuous variables that there is no mode, or at least not a single mode, so the mode is usually not as useful as either the median or the mean for such data. If we were to round the data to the nearest whole number, there would be only one mode of 93. Using a mode of 93 to predict the rounded data, the prediction would be accurate 3 times and incorrect 42 times.

If SAE is adopted as the criterion, then the best estimate for β_0 is the median. There are 45 observations, so the middle observation is the 23rd observation. (If there is an even number of observations then there will be two middle values, which can be averaged to obtain the median.) So, in this case, the best estimate is 85.37. The middle columns of Figure 2.8 present the prediction

			Median			*Mean*		
i	*Country*	*Percentage*	$\hat{Y}_i$	e_i	$\lvert e_i \rvert$	$\hat{Y}_i$	e_i	e_i^2
1	ARG	87.15	85.37	1.78	1.78	83.982	3.168	10.036
2	AUS	96.24	85.37	10.87	10.87	83.982	12.258	150.259
3	BLR	86.89	85.37	1.52	1.52	83.982	2.908	8.456
4	BEL	92.79	85.37	7.42	7.42	83.982	8.808	77.581
5	BWA	73.50	85.37	-11.87	11.87	83.982	-10.482	109.872
6	BRA	80.69	85.37	-4.68	4.68	83.982	-3.292	10.837
7	BGR	75.27	85.37	-10.10	10.10	83.982	-8.712	75.899
8	CAN	92.83	85.37	7.46	7.46	83.982	8.848	78.287
9	CHN	73.05	85.37	-12.32	12.32	83.982	-10.932	119.509
10	COL	73.03	85.37	-12.34	12.34	83.982	-10.952	119.946
11	HRV	81.25	85.37	-4.12	4.12	83.982	-2.732	7.464
12	DNK	98.87	85.37	13.50	13.50	83.982	14.888	221.653
13	RCU	76.20	85.37	-9.17	9.17	83.982	-7.782	60.560
14	EGY	72.06	85.37	-13.31	13.31	83.982	-11.922	142.134
15	FIN	92.81	85.37	7.44	7.44	83.982	8.828	77.934
16	FRA	86.10	85.37	0.73	0.73	83.982	2.118	4.486
17	DEU	91.43	85.37	6.06	6.06	83.982	7.448	55.473
18	GRC	78.49	85.37	-6.88	6.88	83.982	-5.492	30.162
19	GTM	50.84	85.37	-34.53	34.53	83.982	-33.142	1098.392
20	HUN	88.64	85.37	3.27	3.27	83.982	4.658	21.697
21	ISL	99.69	85.37	14.32	14.32	83.982	15.708	246.741
22	IDN	62.10	85.37	-23.27	23.27	83.982	-21.882	478.822
23	IRL	95.17	85.37	9.80	9.80	83.982	11.188	125.171
24	ISR	90.30	85.37	4.93	4.93	83.982	6.318	39.917
25	ITA	74.86	85.37	-10.51	10.51	83.982	-9.122	83.211
26	JPN	82.91	85.37	-2.46	2.46	83.982	-1.072	1.149
27	JOR	82.76	83.25	-2.61	2.61	83.982	-1.222	1.493
28	MYS	96.75	85.37	11.38	11.38	83.982	12.768	163.022
29	MEX	75.63	85.37	-9.74	9.74	83.982	-8.352	69.756
30	NLD	92.05	85.37	6.68	6.68	83.982	8.068	65.093
31	NIC	57.15	85.37	-28.22	28.22	83.982	-26.832	719.956
32	NOR	99.00	85.37	13.63	13.63	83.982	15.018	225.540
33	PER	71.11	85.37	-14.26	14.26	83.982	-12.872	165.688
34	POL	85.37	85.37	0.00	0.00	83.982	1.388	1.927
35	PRT	82.31	85.37	-3.06	3.06	83.982	-1.672	2.796
36	ROU	83.59	85.37	-1.78	1.78	83.982	-0.392	0.154
37	RUS	88.21	85.37	2.84	2.84	83.982	4.228	17.876
38	ESP	93.90	85.37	8.53	8.53	83.982	9.918	98.367
39	SWE	88.31	85.37	2.94	2.94	83.982	4.328	18.732
40	CHE	95.57	85.37	10.20	10.20	83.982	11.588	134.282
41	THA	85.27	85.37	-0.10	0.10	83.982	1.288	1.659
42	TUR	81.41	85.37	-3.96	3.96	83.982	-2.572	6.615
43	UKR	79.22	85.37	-6.15	6.15	83.982	-4.762	22.677
44	GBR	96.68	85.37	11.31	11.31	83.982	12.698	161.239
45	USA	91.75	85.37	6.38	6.38	83.982	7.768	60.342
Sum		3779.20	3841.65	-62.45	388.43	3779.19	.000	5392.860

Figure 2.8 Predictions and errors using the median and mean to estimate b_0 in the simple model

$\hat{Y}_i$, the error $e_i = Y_i - \hat{Y}_i$, and the absolute error based on the median as the estimate b_0. In this case, total error = 388.43. Any other estimate for β_0 would produce a larger value for SAE. Although it is not obvious from Figure 2.8, the 23rd largest absolute error is 7.42, so the median absolute deviation or MAD = 7.42.

If SSE is adopted as the criterion, then the best estimate for β_0 is the mean. The average of the 45 observations gives 83.982 as the estimate b_0. The last set of columns in Figure 2.8 gives the values of $\hat{Y}$ (or b_0 in the case of the simple model), $e_i = Y - \hat{Y}$, and e^2_i. Note that the sum of the errors equals zero exactly and, necessarily, that the sum of the data observations equals the sum of the predictions. This is characteristic of predictions based on minimizing the SSE. The actual SSE equals 5392.860. Again, any other estimate for β_0 would produce a larger SSE. The variance or, more generally, the MSE equals

$$s^2 = \frac{\text{SSE}}{n-1} = \frac{5392.860}{44} = 122.565$$

and the standard deviation or root-mean-squared error equals

$$s = \sqrt{MSE} = \sqrt{122.565} = 11.071$$

Finally, CV is given by

$$\frac{s}{\bar{Y}} = \frac{11.071}{83.982} = .132$$

Note that in this example the three estimates of β_0 (using the three definitions of error) are similar: 93, 85.37, and 83.982. It is often the case for "well-behaved" data that the three estimates are very similar, but there is no guarantee that data will be well-behaved and that the three estimates will be similar. Later we will see that a major discrepancy between the three estimates, especially between the median and the mean, should alert us to special problems in the analysis of such data. Note also in this example that the median absolute deviation and the standard deviation produce different estimates for the typical error—7.42 and 11.071, respectively. This is not surprising given the different definitions of error used.

Summary

In terms of the basic data analysis equation

$$\text{DATA} = \text{MODEL} + \text{ERROR}$$

the simple model with one parameter is expressed as

$$Y_i = \beta_0 + \varepsilon_i$$

Fitting the model to the data consists of finding the estimator b_0 of β_0 that makes the errors e_i as small as possible in the equation

$$Y_i = b_0 + e_i$$

To fit the simple model to data, we must first define how the individual error terms e_i are to be aggregated into a summary index of error. Once we have chosen an aggregate index of error, we can find, by trial and error if necessary, the best estimate b_0 of β_0 that minimizes error. Important definitions of aggregate error are: (a) the count of errors, (b) the sum of absolute errors, and

(c) the sum of squared errors. For each of these definitions, there is a different, well-defined best estimator for β_0 that can be found by calculation rather than by trial and error. These estimators are, respectively: (a) the mode—the most frequent value of Y_i, (b) the median—the middle value of all the Y_i, and (c) the mean—the arithmetic average of all the Y_i. Also, for each of these three definitions of aggregate error there is an expression for representing the "typical" value for e_i. These expressions are, respectively: (a) the modal error, (b) the median absolute deviation, and (c) the standard deviation. Collectively, these best estimators and these expressions for the typical error are known as *descriptive statistics* because they provide a first-cut description of a batch of data. We prefer to view them simply as estimators for the simple model using different definitions of error.

In Chapter 3, we will choose one of the three definitions of error to be our standard definition. We will make this choice based on reasoned principles. However, the reader should anticipate that we will choose the sum of squared errors to be our standard definition of aggregate error.

3 Simple Models

Models of Error and Sampling Distributions

In Chapter 2 we considered alternative aggregate summary measures for error and saw how different choices of an aggregate error index led us to different estimates for the single parameter of the simple model. Now we need to make a choice of which aggregate index of error we will generally use. To do so, we must state our assumptions about the behavior of the errors e_i. We need to have a model for error. In this chapter, we consider a reasonable model of error that is often appropriate and show how that model of error directs us to a certain aggregate index. In later chapters, we show how to check the appropriateness of the chosen model for error and then consider what remedies to apply in our data analysis when the model for error is inappropriate.

In this chapter, our focus is on the error term ε_i, which should be included in the full statement of our models. For the simple model, the complete statement is

MODEL: $Y_i = \beta_0 + \varepsilon_i$

This model says that were it not for random perturbations represented by the ε_i, all the Y_i would equal β_0 exactly. What are these random perturbations? They are anything that causes the observed datum Y_i to deviate from the true value β_0. For the internet use example, the availability of equipment may have been unusually low in some countries; other countries may have had either unusually good or unusually poor recordkeeping procedures; still others may have had unusually high or low numbers of individuals whose work involved internet use. Any of these or countless other factors may have caused the observed percentages of individuals using the internet to deviate from the single value β_0 predicted by the simple model.

You might object that a factor such as the availability of equipment ought to be included in the model of internet use rather than in the error. Indeed, the factors previously listed as causes of error may be important predictors that ought to be included in the model. However, if we are using a simple model with only one parameter, then all those factors are included in the error because those factors cause the observed Y_i to deviate from the single value β_0 predicted by the model. Later we will consider more complex models that will allow us to move such factors from the error to the model. To the extent that moving a factor from the error to the model reduces the error, we will have improved our model and its predictions. How to do that must await subsequent chapters; for now, our problem is what assumptions we can reasonably make about the nature of the error.

Sampling Distributions

As was shown in Chapter 2, for each aggregate index of error there is an associated estimator of β_0 in the simple model. For example, for the sum of absolute errors the median is the best estimator, and for the sum of squared errors the mean is the best estimator. Part of our strategy for selecting

DOI: 10.4324/9781003438335-5

an aggregate index of error will be to examine the performance of various estimators under some reasonable assumptions about the behavior of the errors ε_i. Next we develop a metaphorical representation of the simple model, which will help our intuitions about error and provide a context for examining the relative performance of the median and the mean.

Metaphorical Representation of Error

Suppose that Nature has a bag containing many tickets and on each ticket is written a particular value for ε_i. Figure 3.1 is a list of 100 values for ε_i that might be on the tickets.

To determine the value for a particular observation Y_i, Nature reaches into her bag of tickets and *randomly* draws one ticket; she then adds the number on the ticket, the value of the error ε_i, to the true value β_0. Let us assume, for example, that $\beta_0 = 50$ so that the simple model becomes

MODEL: $Y_i = 50 + \varepsilon_i$

Suppose the following 20 values were on tickets randomly drawn from a bag containing the tickets of Figure 3.1:

−9	28	−17	22	35
−12	−23	1	10	11
6	−24	−19	−26	−31
−4	8	20	−14	−2

Then the first datum would equal 50 + (−9) = 41 because −9 is on the first error ticket drawn. The second datum would equal 50 + 28 = 78 and so on. The 20 observed values would be

41	78	33	72	85
38	27	51	60	61
56	26	31	24	19
46	58	70	36	48

Note that for the moment we are ignoring the issue of how it is determined which numbers are to be written on the tickets representing the errors. This is a very important issue to which we return later. Had a different set of error tickets been drawn we would have obtained a different set of data.

−11	6	1	7	14	0	−1	9	1	2
32	15	12	−5	4	7	−6	−22	1	−2
20	11	29	−14	−8	−13	−7	−25	−3	−4
−19	8	−4	0	−15	17	−23	−8	6	17
3	−5	−15	11	2	−14	−20	4	31	24
2	−28	−31	12	−5	10	29	−10	19	−5
−6	5	0	−4	−5	−2	−12	13	4	−23
25	−19	−11	−22	3	4	3	28	10	−26
−9	35	7	−10	22	−17	10	5	−15	20
−5	−24	−4	−2	12	−26	10	−7	−11	−24

Figure 3.1 100 hypothetical values of ε_i in Nature's bag of tickets

Statisticians often use the word *sample* to refer to a batch of data that could have been different if, in terms of our metaphor, a different set of error tickets had been drawn. Thus, the aforementioned 20 data observations represent a sample of the possible data values that could have been obtained. The mean and median for this particular sample of 20 observations are 48 and 47, respectively. Note that neither the mean nor the median exactly equals the actual value of β_0 specified in the model as 50 for this example. Had a different sample been drawn, the mean and median would likely have been different. For some samples the mean and median would be closer to the true value, and for others they would be farther away. If the mean and median are to be good estimates of β_0, we would hope that on average they would be close to β_0. For this example, we can perform a simulation experiment to determine likely values for the mean and median for samples of size 20. On each simulation round we randomly select another sample of 20 error tickets, add 50 to each one, and then calculate the mean and median of the resulting sample. If we perform many, many simulation rounds, generating many, many different samples and computing the mean and median in each, then we can treat each mean or median as a datum itself. We can then construct a frequency polygon for these data. Such frequency distributions are known as *sampling distributions* because they show the distribution of a particular statistic based on many, many samples.

Properties of Estimators

Sampling distributions provide all the information necessary for us to evaluate the relative performance of various estimators. Next we examine the performance of the median and mean with respect to three desirable properties of an estimator: unbiased, efficient, and consistent. Each concept is illustrated and defined in terms of Figure 3.2, which depicts the sampling distributions for both the mean and the median based on an infinite number of samples of size 20 from the bag of error tickets in Figure 3.1.

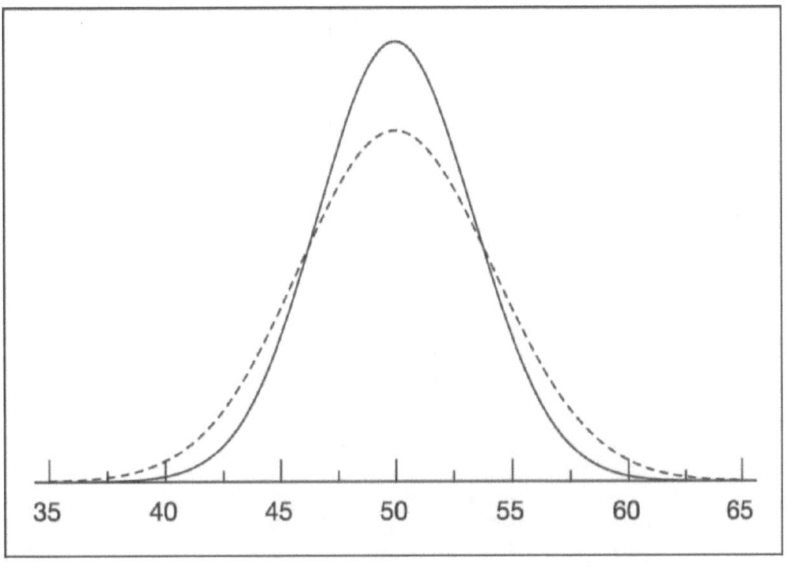

Figure 3.2 Sampling distributions for the mean (solid line) and median (dashed line) for $n = 20$

Unbiasedness

Reassuringly, both distributions have peaks at 50, indicating that 50 is the most likely value for both the mean and the median. If the peak and the average value from the sampling distribution for an estimator or statistic equal the specified value for the parameter (50 in this case), then the estimator is said to be *unbiased.*

Efficiency

For these sampling distributions, there are very few instances for which the mean or the median is very far away from 50; the bulk of the observed values are between about 45 and 55. Note, however, that the mean appears to be a slightly better estimator than the median because there are somewhat fewer values of the median near 50 and there are somewhat more values of the median in the "tails" of the distribution far away from 50. For this example, the mean is said to be more efficient than the median because with the same number of observations the mean provides more precision.

Consistency

It is interesting to ask what the sampling distributions would be like if the sample sizes were larger. For example, what if the mean and median were based on samples of size 100 instead of samples of size 20? Figure 3.3 shows the sampling distributions for the mean and the median based on an infinite number of samples of size 100.

Compared with Figure 3.2, it is clear that the sampling distributions based on samples of size 100 are much narrower than those based on samples of size 20. The narrowness indicates that a greater proportion of the observed values of the mean and median were near 50 and that even fewer observed values were far away from 50. If the sampling distribution of an estimator or statistic becomes narrower as the sample size increases, then it is said to be a *consistent* estimator. Note also

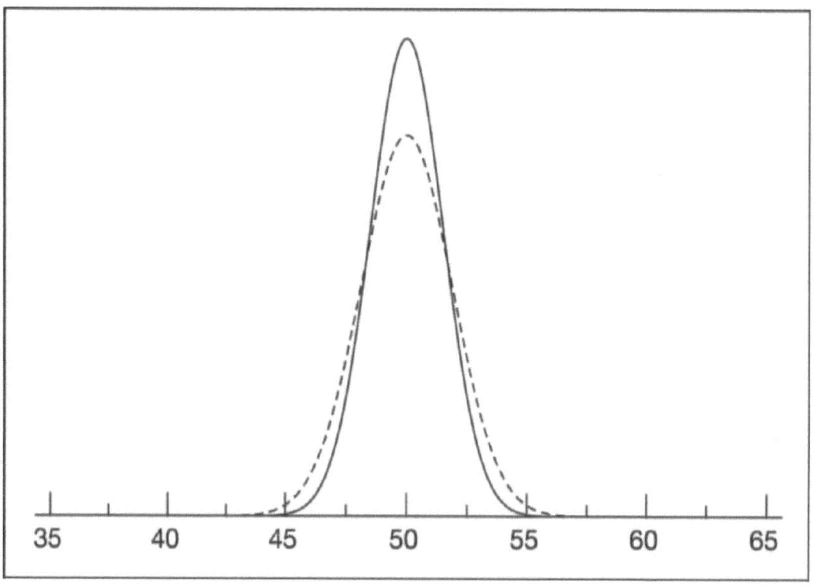

Figure 3.3 Sampling distributions for the mean (solid line) and median (dashed line) for $n = 100$

that again the mean is slightly more efficient than the median because its sampling distribution is a little narrower.

Normal Distribution of Errors

Unbiasedness, consistency, and efficiency are obviously desirable attributes for an estimator of β_0 in the simple model. Both the mean and the median and many other estimators are unbiased and consistent, so those properties offer no basis for choosing an estimator and its corresponding index of aggregate error. The efficiency of estimators does differ, and in this example the mean was slightly more efficient than the median. Hence, our example suggests that choosing the sum of squared errors as our aggregate index of error and using its associated estimator (i.e., the mean) would serve us well. But that conclusion depends on the particular distribution of error tickets in Figure 3.1. Had there been some other distribution of error tickets in Nature's bag, we might have found the median to be more efficient. Thus, we must make an assumption, or state our model, about the contents of Nature's bag of error tickets.

We will assume that the errors have a *normal distribution*. That is, we assume that a frequency distribution for the error tickets would have the general shape of the normal curve depicted in Figure 3.4. This is an idealized representation that would only be obtained from a bag containing an infinite number of tickets. The 100 error tickets of Figure 3.1 were, in effect, randomly selected from a bag containing an infinite number of normally distributed errors.

There are many other possible distributions of errors that we might have assumed, but there are good reasons for assuming the normal distribution. First, there is an abundance of empirical data indicating that the distribution of errors is often approximately normal. That is, in fact, how it acquired the name "normal" distribution. Second, as we shall see in Chapter 14, it is often possible to transform a non-normal distribution of errors into a nearly normal distribution. Third, and most importantly, there is a reasonable model of the errors that causes us to expect them to be distributed normally. We turn now to a description of that model.

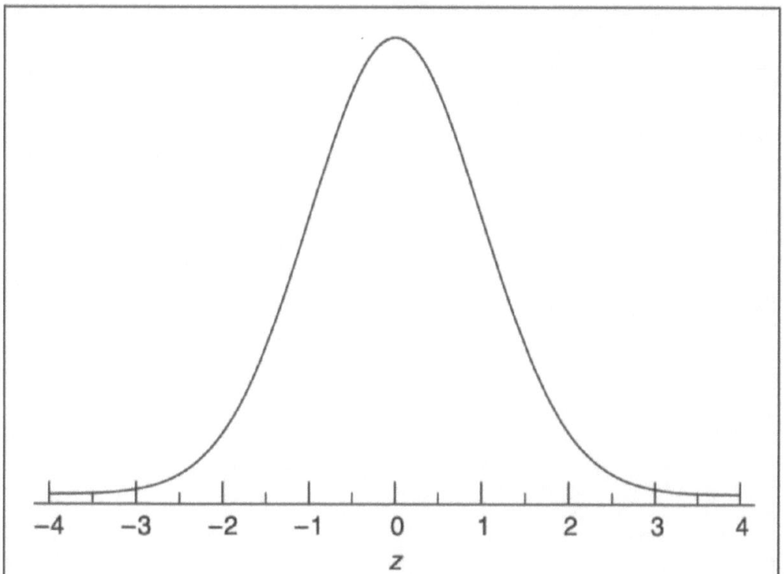

Figure 3.4 Normal probability curve

Previously we mentioned a number of possible components of the errors for the internet access data. It is unlikely that in any particular case only a single error component is perturbing the observed value away from the parameter value specified in the simple model. For example, it might be that internet use was altered because of unusual installation processes *and* unusual record keeping *and* publicity about internet security problems. Some of the factors will push the observed value above the model value, and others will push it below the model value. The observed difference between the observed and model values, $Y_i - \hat{Y}_i$, will actually be determined by the *sum* of the individual error components. The central limit theorem, a well-known theorem in mathematical statistics, shows that the distribution of the sum or average of a number of components will approximate to a normal distribution no matter what the original distribution of the individual components. Thus, if we assume that the actual error is the sum of a number of individual errors, then it is reasonable to expect that the distribution of the errors will approximate to a normal distribution. The central limit theorem also shows that the greater the number of components in the sum or average, the better the approximation to a normal distribution.

The central limit theorem is of such importance that a demonstration of the concept is worthwhile. Suppose that the distribution of the individual error components is such that any value between –0.5 and +0.5 is just as likely as any other. This distribution is known as the *uniform distribution* because each possible value has the same probability. As Figure 3.5 illustrates, the uniform distribution is very different from a normal distribution.

Suppose that each value on an error ticket is the sum of 12 small individual error components randomly selected from the uniform distribution. Figure 3.6 shows the frequency distribution for 10,000 such error tickets. Clearly, even though the distribution for the individual components was not normal, the distribution for the sum closely approximates to a normal distribution. Thus, if it is reasonable to assume that the observed error is actually the sum of a lot of small random error components, then it is reasonable to assume that the distribution of the errors is approximately normal.

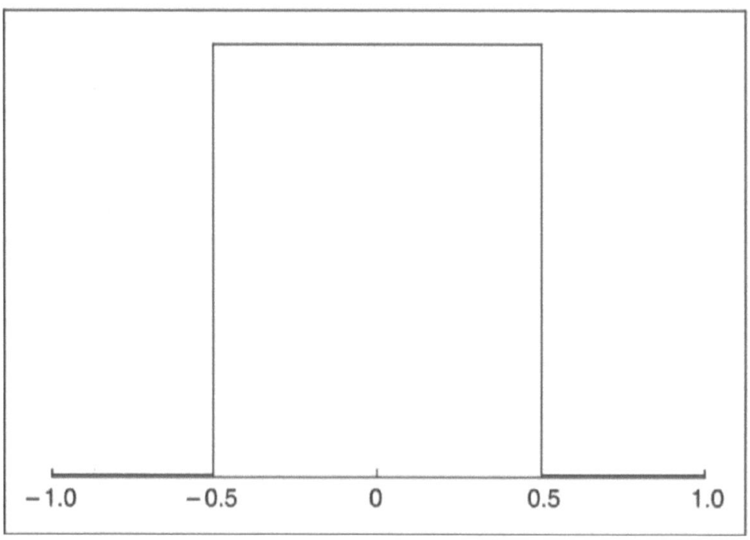

Figure 3.5 Uniform distribution

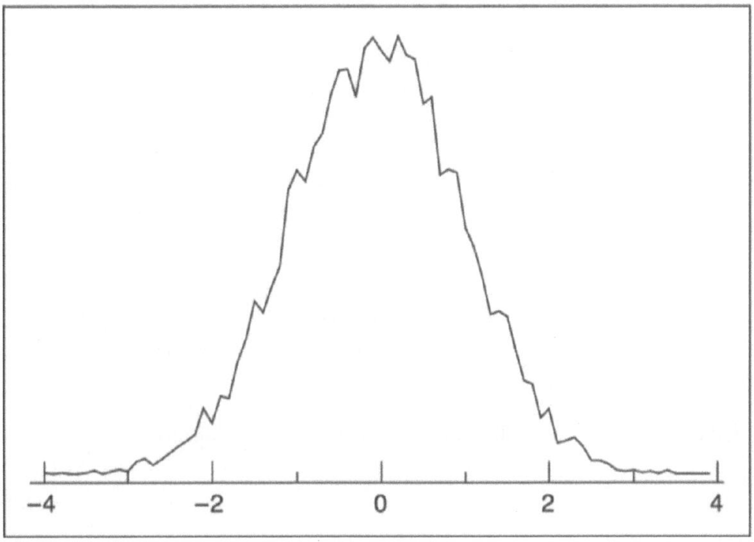

Figure 3.6 Frequency distribution for the sum of 12 uniform errors

Formal Specification of the Normal Distribution

Fortunately, there are precise mathematical statements of the normal distribution, so we do not have to depend on simulations of drawing tickets from a bag, as we used in our metaphor earlier. The curve for the relative frequency of the normal distribution represented in Figure 3.4 is given by

$$f(x) = \frac{1}{\sigma\sqrt{2\pi}} e^{-(x-\mu)^2/2\sigma^2}$$

This equation has two parameters: μ, the expected value or mean of the distribution, and σ^2, the variance of the distribution. Different values of μ simply shift the curve of Figure 3.4 back and forth along the horizontal axis. Figure 3.7 shows two normal curves that have the same variance ($\sigma^2 = 1$) but different means ($\mu_1 = 0, \mu_2 = 3$). The variance parameter σ^2 determines the spread of the normal curve. The larger σ^2 is, the wider the distribution. Figure 3.8 shows three normal curves that each have a mean of $\mu = 0$ but increasing variances of 0.25, 1, and 4.

All normal distributions have the same shape, so it is common to use the distribution with $\mu = 0$ and $\sigma^2 = 1$ as a standard reference distribution. Figure 3.9 gives the probability that a value x randomly sampled from the normal distribution will be less than the tabled values of z ranging between –3 and +3. For example, the probability for $z = 1.0$ equals .841, which indicates that 84.1% of the time a randomly sampled value from this distribution will be less than 1.0. Note that the probability for $z = 0$, the mean for this particular distribution, equals .5, indicating that we would expect 50% of the observations to be below zero and 50% of the observations to be above zero. Tables of the normal distribution are available in many statistical books, and most computer statistical packages provide functions for easily calculating the required probabilities.

To find the expected proportion of observations between any two particular values, we simply subtract their associated probabilities. For example, the proportion of observations between $z = 0$ and $z = 1$ is .841 –.500 = .341. Figure 3.10 shows the expected proportion of observations between various landmarks.

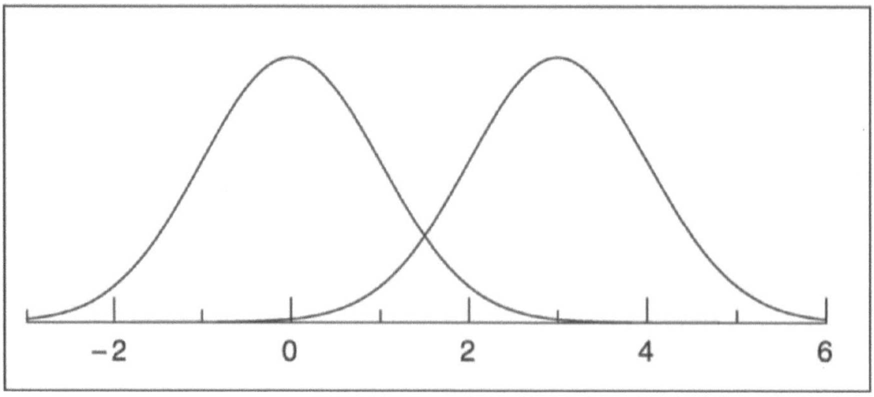

Figure 3.7 Normal probability curves with same variance and different means

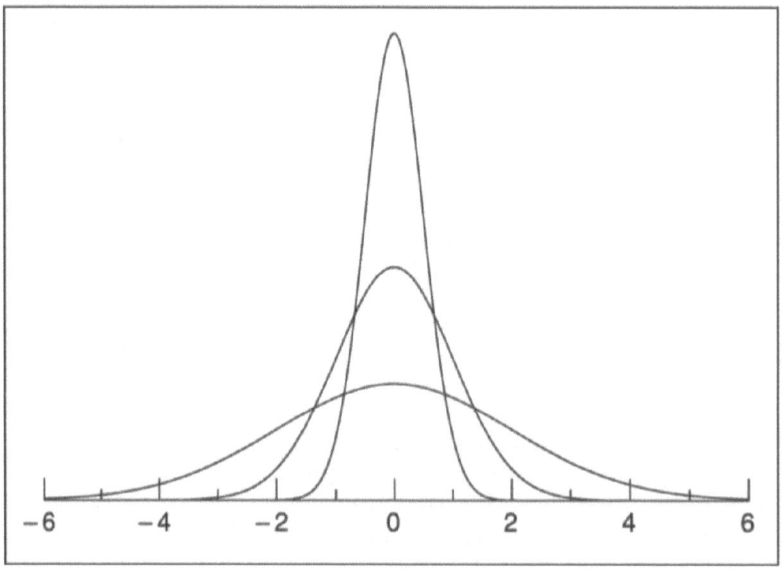

Figure 3.8 Normal probability curves with same mean but different variances

We will not be using the z table of Figure 3.9 very much; instead, we will be using similar tables for some special distributions. For most purposes, it is sufficient to remember that approximately 68% of the observations from a normal distribution fall between –1 and +1, approximately 95% fall between –2 and +2, and almost all (approximately 99.7%) fall between –3 and +3. The tails of the normal distribution theoretically extend to $+\infty$ and $-\infty$; however, observations more extreme than 3 are infrequent, and observations more extreme than 4 are very rare.

Normal distributions with means other than 0 and/or variances other than 1 are easily compared with the reference distribution in Figure 3.9. Suppose that Y is an observation from a normal distribution with mean μ and variance σ^2. Subtracting μ from Y will shift the distribution along the horizontal axis, so the mean for the constructed variable $Y - \mu$ equals 0. Although it is not obvious, it can be shown that dividing $Y - \mu$ by σ will change the spread of the distribution so that it coincides with our reference distribution. (Note that we divide by σ, not σ^2.) The division by σ

z	$p(x < z)$	z	$p(x < z)$
–3.0	.001	0.0	.500
–2.9	.002	0.1	.540
–2.8	.003	0.2	.579
–2.7	.003	0.3	.618
–2.6	.005	0.4	.655
–2.5	.006	0.5	.691
–2.4	.008	0.6	.726
–2.3	.011	0.7	.758
–2.2	.014	0.8	.788
–2.1	.018	0.9	.816
–2.0	.023	1.0	.841
–1.9	.029	1.1	.864
–1.8	.036	1.2	.885
–1.7	.045	1.3	.903
–1.6	.055	1.4	.919
–1.5	.067	1.5	.933
–1.4	.081	1.6	.945
–1.3	.097	1.7	.955
–1.2	.115	1.8	.964
–1.1	.136	1.9	.971
–1.0	.159	2.0	.977
–0.9	.184	2.1	.982
–0.8	.212	2.2	.986
–0.7	.242	2.3	.989
–0.6	.274	2.4	.992
–0.5	.309	2.5	.994
–0.4	.345	2.6	.995
–0.3	.382	2.7	.997
–0.2	.421	2.8	.997
–0.1	.460	2.9	.998
0.0	.500	3.0	.999

Figure 3.9 Cumulative probabilities for the normal distribution for $\mu = 0$ and $\sigma^2 = 1$

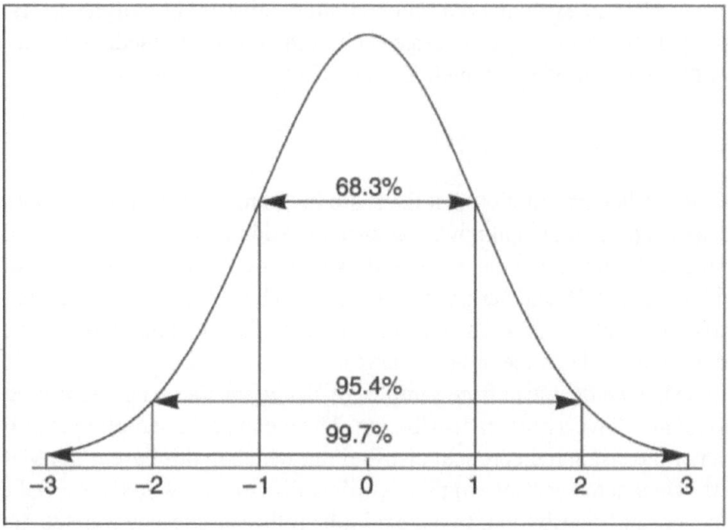

Figure 3.10 Standard normal probability curve ($\mu = 0$, $\sigma^2 = 1$)

rescales the original variable so that it is measured in units of the standard deviation. Thus, if Y is from a normal distribution with mean μ and variance σ^2, then z, defined as

$$z = \frac{Y - \mu}{\sigma}$$

has a normal distribution with mean 0 and variance 1. Values calculated according to this formula are known as *z scores* or, because of their use with the standard reference distribution, *standard normal deviates*. "Deviates" refers to the size of the deviation from the mean, measured in units of the standard deviation. For example, intelligence test scores (IQ scores) are scaled so that $\mu = 100$ and $\sigma = 15$. To find the proportion of individuals we would expect to have IQ scores that are less than or equal to 127, given that IQ scores are normally distributed, we would first compute the corresponding z score. In this case,

$$z = \frac{127 - 100}{15} = 1.8$$

That is, an IQ score of 127 is 1.8 units (i.e., 1.8 standard deviations) above the mean (of 100). The associated cumulative probability in Figure 3.9 for $z = 1.8$ is .964. So we would expect about 96% of the population to have IQ scores lower than 127.

We can also convert z scores back to the original scale. The Y score that corresponds to a particular z is given by

$$Y = \mu + z\sigma$$

In the same IQ example, if we wanted to know what IQ score corresponded to $z = 1.6$ (i.e., a score 1.6 standard deviations above the mean), we would calculate

$$100 + 1.6(15) = 124$$

Sampling Distributions Revisited

Now that we have made an explicit assumption about the distribution of the errors, we can reconsider the sampling distributions for estimators such as the mean and median. Let us again return to our example of the simple model in which $\beta_0 = 50$. That is, our model is

$$Y_i = 50 + \varepsilon_i$$

We now make the explicit assumption that the ε_i are randomly sampled from a normal distribution with $\mu = 0$ and some constant but unknown variance σ^2. Adding 50 to each error value simply translates the normal distribution along the horizontal axis, so the Y_i data values will also have a normal distribution with mean $\mu = 50$ and variance σ^2. In order to be able to draw a picture of this distribution, we have assumed that $\sigma^2 = 225$. Then the distribution of the Y_i values would have the theoretical curve depicted in Figure 3.11 as the broader curve.

If we were to calculate the mean from samples of size n, the theoretical sampling distribution of such means would also have a normal distribution. The mean of the sampling distribution of means equals the mean of the original theoretical distribution, which in this case is 50. Mathematical statisticians have shown that the variance of this sampling distribution would be σ^2/n if the error tickets are drawn from a normal distribution. In our example, if the sample size were 20, then the variance

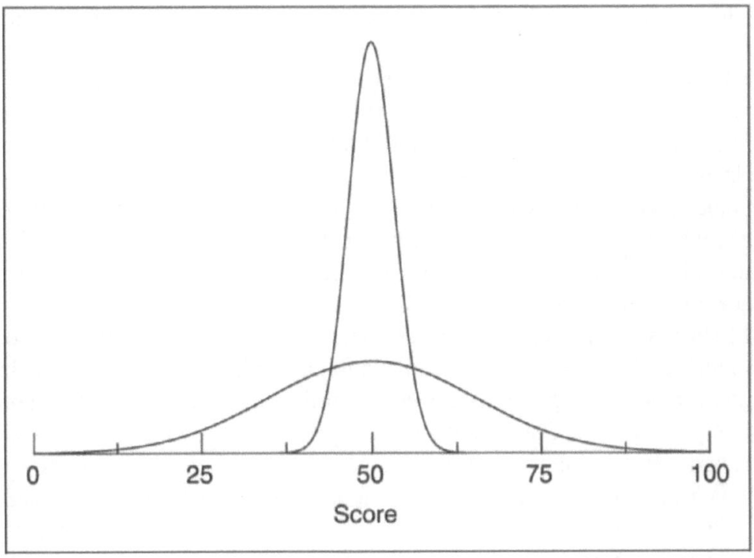

Figure 3.11 Normal distribution of observations and sampling distribution of the mean for $n = 20$

of the sampling distribution for the means would equal 225/20 = 11.25. If we were to collect many, many samples of size 20 and calculate the mean for each sample, the sampling distribution of those means would approximate the theoretical sampling distribution also depicted in Figure 3.11 as the narrower curve. Note that, as was the case when we simulated the sampling distribution, the spread for the mean is very much smaller than the original spread. If we were to collect larger samples, the sampling distribution for the mean would be even narrower because we would be dividing σ^2 by a larger value in order to determine its variance.

Mathematical statisticians have also shown that the variance for the sampling distribution of the median based on samples of size n is approximately equal to

$$\frac{\pi\sigma^2}{2n}$$

if the error tickets are drawn from a normal distribution. This is the same as the variance for the mean except that it is multiplied by $\pi/2$ (which equals approximately 1.57), so the variance for the median must be larger. That is, the median is less efficient when error values have a normal distribution. In particular, the variance for the sampling distribution of the mean is about $2/\pi = 64\%$ of the variance for the sampling distribution of the median. Another way to say this is that for a sample of size 100 using the sample median has about the same efficiency as using the mean with a sample of size 64. As our earlier simulation suggested, but did not prove, the mean is clearly preferable to the median if the distribution of errors is normal. However, for some non-normal distributions of errors the median can actually be more efficient than the mean.

Choice of an Estimator and Index of Error

Now that we have made an explicit assumption about the distribution of the errors, we can choose the estimator and its corresponding definition of aggregate error. We want an estimator that is unbiased, efficient, and consistent.

It can be shown mathematically that if the distribution of errors is normal, then the mean is a more efficient estimator than any other unbiased estimator. That is, the mean is more efficient than the median, the mode, and any other statistic whose sampling distribution would be centered at the parameter value specified in the simple model. The mean is also a consistent estimator. Hence, we will choose it as the best estimator. We also will adopt its corresponding definition of aggregate error, which is the sum of squared errors (SSE).

There are other good reasons for selecting SSE and the mean besides the aforementioned theoretical reasons, which rely on the assumption that the errors ε_i have a normal distribution. One reason is practicality. By far the greatest number of statistical procedures have been developed for SSE. Consequently, there are very good computer procedures for accomplishing the statistical tests based on SSE. There are some very practical reasons for the ubiquity of statistical procedures based on SSE. It is much easier to work with squares in mathematical proofs than with absolute values. Calculating an average is also generally easier than ordering all of the data to find the middle one. More importantly, for complex models with many parameters, the only way to estimate the best values for the parameters using most other aggregate indices of error, such as the sum of absolute errors (SAE), is brute force or iterative search algorithms. In contrast, when using SSE, we will almost always be able to use explicit formulas for calculating the best parameter estimates. Finally, the parameters based on SAE are not necessarily unique. That is, sometimes many different values for the parameters all produce the same minimum value of SAE. This is never a problem with SSE.

Other Distributions of Errors

It is important to remember that our choice of the mean and SSE is highly dependent on our assumption that the errors are normally distributed. If the errors have some other distribution, then choosing the mean and SSE might not be appropriate. Figure 3.12 shows another possible distribution for the errors. Asymmetric distributions such as this one are *skewed* distributions. Note that in this case very large positive errors are more likely than very large negative errors. There are many types of data that can produce errors with a skewed distribution. For example, in studies measuring reaction times the smallest scores are obviously constrained to be above zero, but there is no constraint, other than the experimenter's patience, on how large the scores can be. This will tend to produce a scattering of observations very far above the mean and many observations just below the mean, resulting in skewed distributions of the errors as in Figure 3.12. In Chapter 14, we discuss procedures for detecting and transforming skewed and other non-normal distributions of the errors.

One of the most troubling possible error distributions is one that has the general shape of the normal distribution except that it has "thicker" tails. That is, extreme, especially very extreme, errors are more likely to occur than would be expected from the normal distribution. These extreme errors produce extreme observations that can have a very deleterious effect on the mean. As we demonstrated in Chapter 2, one wild observation can cause a dramatic change in the mean. The problem is that SSE pays a lot of attention to extreme observations in order to reduce the really big errors, so estimates are pulled powerfully in the direction of the extreme observations, so much so that the mean may not be a very good estimate of central tendency when there are a few extreme observations. The median is much more resistant to the effects of these wild observations or outliers, but we have to give up a lot of efficiency to get this robustness. In general, we will stick with the mean and SSE, but we must remember that they are not resistant to outliers. To make sure that a few extreme observations are not seriously distorting our results, we will have to check carefully for outliers each time we use statistical procedures based on SSE. A number of useful follow-up procedures have been developed precisely for this purpose, and we will examine them in Chapter 14.

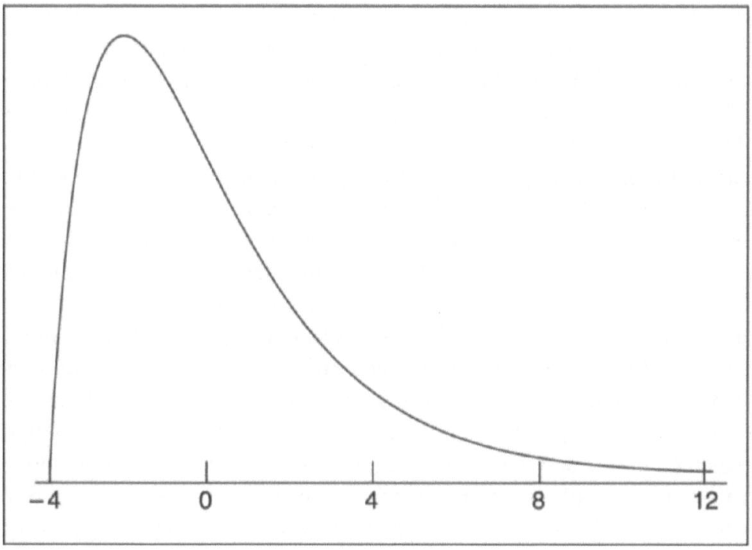

Figure 3.12 Example of a skewed distribution

Other Criteria for Estimators

We have adopted unbiasedness, efficiency, and consistency as reasonable criteria for an estimator. These criteria have been adopted by many statisticians. However, not all statisticians agree on the desirability of these criteria or, especially, on their relative desirability. For example, some statisticians argue that we should be willing to accept a little bit of bias in order to achieve greater efficiency. For some of the complex models that we consider later there are biased estimators that may have greater efficiency than the unbiased estimators we present. Other statisticians argue that the mean and SSE are so susceptible to outlier problems that estimators that are somewhat less efficient but more robust should be preferred. In general, our approach is to stick with the mean and SSE while always checking assiduously for potential outliers that may be distorting the results.

Other Assumptions About Errors

In the preceding discussion of our model of errors and its metaphorical representation of Nature drawing error tickets from a bag, we made several implicit assumptions. We need to make these assumptions explicit so that we will understand what assumptions have been made and we can be alert to potential violations of these assumptions. We consider in turn three standard assumptions: that the ε_i are independent of one another, that they are identically distributed, and that their mean is zero. These assumptions will be especially important in considering the sampling distributions for statistics developed later, particularly in Chapter 4.

Independence

The formal assumption is that each error is independent of the value of any other error. In terms of our metaphor, this means that each error ticket is selected randomly from the bag and then replaced in the bag before the next ticket is drawn. A violation of independence occurs when the value on one error ticket allows us to make a reasonable guess about the value on another error ticket. In terms of the metaphor, it would be a violation of independence if error tickets came in pairs attached to one another so that, if we drew one, we would automatically draw the other. As a practical example, suppose that our data consisted of internet use rates for each country for each of several years. If a given country's internet usage rate were above the mean or median in one year, then it would likely be above the mean or median in other years. Thus, when we use the simple model, all the errors for that country are likely to be positive and probably of about the same size. Thus, knowing the error for that country in one year allows us to make a reasonable guess about the errors for other years. As a result, the errors would not be independent. This is an example of *time-series* data—the same units are observed or measured across a number of different time periods. Time-series data seldom have errors that satisfy independence.

As a similar example, suppose that we have two attitude scores from each subject: one score measured before the subject reads a persuasive communication and one measured after the subject reads the communication. This is known as a *within-subject* design. Again, it is probable that someone who is below the mean or median on the first measure (the "pretest") will also be below the mean or median on the second measure ("the post-test"). Thus, if the simple model were used, the errors for each subject would likely be either both positive or both negative, so independence would once again be violated. As a final example, consider a survey that asks married couples what proportion of various household tasks they perform. For a given couple, if one partner claims to perform a small proportion of a particular task, then it is likely that the other partner would claim to perform a large proportion of that particular task. If we tried to use a single simple model for all the data, it is likely that independence would be violated within couples. Knowing that the error for one partner is negative (i.e., the proportion claimed is below the mean or median) would lead us to expect that the error for the other partner would be positive (i.e., the proportion claimed is above the mean or median). Note that this dependence, unlike that in the other examples, is a *negative dependence* because the signs of the errors for each partner are likely to be opposite.

Unlike the assumption about the normal distribution of errors, which is very robust, the assumption that the errors are independent of each other is not at all robust. That is, our statistical inference will be seriously affected if the independence assumption is incorrect. Thus, we must be especially alert for possible dependencies among the errors. If the independence assumption is violated, then all is not lost. In Chapters 12 and 13, we show how to analyze such data so that, after suitable transformations and rearrangements, independence is satisfied.

Identically Distributed

Another very important assumption is that all the errors are sampled from the *same* distribution. In terms of our metaphor, this assumption would be violated if Nature had *two* bags of error tickets: one bag containing error values of a reasonable size and another bag containing error tickets with much larger values. If Nature draws mostly from the first bag, then most of the resulting data values will be of a reasonable size. But if Nature occasionally draws an error ticket from the second bag, we will observe several "wild" data values that are very different from the other observations. Such data values are known as *outliers*. In other words, a violation of this assumption occurs if most of the errors are sampled from a normal distribution with variance σ^2, and a few errors are sampled from a normal distribution with a variance very different from σ^2. Although it would technically be a problem if the second variance were much smaller than the first one, the greater problem occurs when the second variance is very much larger than the first.

A slightly different type of violation of this assumption occurs, for example, in experiments when the manipulation itself causes the error variance to change. In general, there is a problem whenever the size of the variance depends on the model's predicted value, $\hat{Y}$. In this context, the assumption of identical distributions is equivalent to assuming that the variances are equal, so it is sometimes called the *homogeneity of variance* assumption or sometimes *homoscedasticity*. Its violation is *heterogeneity of variance* or *heteroscedasticity*.

The identical distribution assumption is fairly robust. The variances must be very different before there are important problems for the data analysis. The most serious problems arise from outliers—a few observations that have errors sampled from a distribution with a very much greater variance. In Chapter 14, we consider procedures for detecting outliers and consider what to do about them. We also develop techniques for detecting heterogeneity of variance and the use of transformations to reduce the heterogeneity to acceptable limits.

Errors Are Unbiased

The final important assumption is that the errors are unbiased or, equivalently, that the expected value or mean of the distribution from which the errors are sampled equals zero. That is, our assumption is that, on average, the errors do not change the model parameters. If, instead, the errors were not random but systematic, such that errors averaged, say, below zero, then the systematic bias would appear in the model rather than in the error. Consider, for example, internet use data. Suppose that internet use was not properly documented for some individuals. Then any estimate of the true internet use would necessarily be an underestimate. In effect, the systematic underreporting of error would be represented in the model and not in the error. As data analysts we cannot do much to detect a systematic bias in the error by examining the data. Hence, it is incumbent on the collector of the data to ensure that the errors are indeed random with mean zero.

Depending on the substantive domain, ensuring that the errors are random involves aspects of experimental design, construction of survey questionnaires, interviewing procedures, instrument reliability, and so forth. We do not address such issues because they are highly domain-specific. However, if those issues have not been addressed in the actual collection of the data, then the use of the statistical procedures described in this book may be very misleading. As data analysts we offer no remedies for the problems of systematic error. Hence, we will always assume that errors are sampled from a distribution that has a mean of zero.

Estimator for σ^2

A final issue we need to consider is an estimator of σ^2 (see the discussion in Chapter 2). We know that the mean is an estimator of the true but unknown population mean μ or β_0. It turns out that the mean squared error (MSE) is an unbiased estimator of the true but unknown population error variance σ^2. MSE is an index of the extent to which data deviate from the values predicted by the model. For the simple model that predicts the mean ($Y_i = b_0 + e_i$), MSE has the special name *variance* and special symbol s^2. Thus, for the simple model that predicts the mean ($Y_i = b_0 + e_i$), we have the formulas

$$MSE = Variance = s^2 = \frac{SSE}{n-1} = \frac{\sum_{i=1}^{n}\left(Y_i - \hat{Y}_i\right)^2}{n-1} = \frac{\sum_{i=1}^{n}\left(Y_i - \bar{Y}\right)^2}{n-1}$$

which estimates the population error variance σ^2. For more complex models with p parameters

$$MSE = \frac{SSE}{n-p} = \frac{\sum_{i=1}^{n}\left(Y_i - \hat{Y}_i\right)^2}{n-p}$$

which also estimates σ^2.

Note that MSE is an index of the *squared* deviations of the data from the values that are predicted by the model. A more easily interpreted index is the square root of MSE or root-mean-squared error (RMSE). RMSE estimates the population error standard deviation σ. For the simple model that predicts the mean, RMSE has the special name *standard deviation* and the special symbol s. Thus, the mean $\hat{Y}$, the variance s^2, and the standard deviation s calculated from a sample of data provide estimates of the parameters μ, σ^2, and σ that characterize a normal distribution.

Summary

As a result of random perturbations in data, sample statistics, such as the mean and the median, are unlikely to equal the true population parameters that they estimate; in other words, different samples will usually have different means (and medians) even when the samples are from the same population.

Sampling distributions show the distributions of statistics based on a very large number of samples of a particular size. So, for example, we could draw a thousand samples, each of size n, from the same population, calculate the mean of each sample, and then graph the set of 1000 means. The mean of this sampling distribution would equal the mean of the original theoretical distribution, but the sampling distribution would be narrower. Sampling distributions become narrower as the size of the samples from which the statistic is calculated becomes larger. The variance of the sampling distribution of the mean is equal to σ^2/n.

A comparison of the sampling distributions for the mean and the median indicates that both are unbiased and consistent estimators of the true population mean and median. However, the mean is generally more efficient than the median. For this reason, as well as practical considerations, we use the mean to estimate β_0 in the simple model and its associated aggregate index of error (the sum of squared errors) to estimate σ^2. However, the conclusion that the mean is more efficient than the median relies on the assumption that errors are normally distributed. If the distribution of errors is skewed or contains unusual values, the mean may not be a good estimate of central tendency.

All normal distributions have the same shape and are thus often described with reference to a standard normal distribution in which $\mu = 0$ and $\sigma^2 = 1$. Normal distributions with other means and/or variances can be compared with the standard normal distribution by computing z scores or standard normal deviates. A z score reflects an observation's distance from the mean in standard deviation units. Approximately 68% of the observations in a normal distribution fall between z scores of –1 and +1, approximately 95% fall between z scores of –2 and +2, and nearly all (about 99.7%) fall between z scores of –3 and +3.

In addition to being normally distributed, errors are assumed to be independent of one another, identically distributed, and unbiased (i.e., have a mean of zero). Data analysts cannot determine whether errors are unbiased. Determining whether the normality, independence, and identical distribution assumptions have been violated is discussed in subsequent chapters, as are potential remedies.

Just as the mean is an estimator of the true but unknown population mean μ or β_0, the mean squared error is an unbiased estimator of the true but unknown population error variance σ^2, and the RMSE or square root of MSE is an unbiased estimator of the population error standard deviation σ. In the case of the simple model that predicts the mean, the MSE is known as the *variance* s^2 and the RMSE is known as the *standard deviation* s.

4 Simple Models

Statistical Inferences About Parameter Estimates

In Chapter 2 we considered various alternatives for how we should measure aggregate error in the equation

DATA = MODEL + ERROR

Although the sums of both the absolute errors and the squared errors seem to be reasonable alternatives, we decided in Chapter 3, for reasons of efficiency, practicality, and tradition, to define total error as the sum of the squared errors. In the simplest models, where we are estimating the single parameter β_0, the choice of the sum of squared errors as the definition of error implies that the best estimate is the sample mean (as we proved in Chapter 2).

In this chapter, we develop procedures for asking questions or testing hypotheses about simple models. Defining and answering interesting questions is the purpose of data analysis. We first consider the logic of answering questions about data for the case of the simplest models because it is easy to focus on the logic when the models are simple and because the logic generalizes to more complex models. The specific statistical test presented in this chapter is equivalent to the one-sample *t*-test. We do not derive this test in terms of the *t* statistic; we prefer instead to construct this test using concepts and procedures that are identical to those required for the more complex models we will consider later.

The generic problem is that we have a batch of data for which we have calculated b_0, the mean, as an estimate of β_0. Our question is whether β_0 is equal to some specific value. For example, we might want to know whether the body temperatures of a group of patients administered a therapeutic drug differed from the normal body temperature of 37°C. We will let B_0 equal the a priori value specified in our question, 37. The statement

$$\beta_0 = B_0$$

represents our *hypothesis* about the true value of β_0. Such statements are often called *null hypotheses*. The calculated value of b_0 will almost never exactly equal B_0, the hypothesized value of β_0. That is, the compact model,

$$\text{MODEL C: } Y_i = B_0 + \varepsilon_i$$

in which no parameters are estimated will almost always produce a larger error than the augmented model,

$$\text{MODEL A: } Y_i = \beta_0 + \varepsilon_i$$

DOI: 10.4324/9781003438335-6

in which β_0 is estimated by b_0, the mean of the batch of data. We will calculate PRE, the proportional reduction in error index developed in Chapter 1, to see how much better the predictions of Model A are than those of Model C. The original question then becomes not whether Model A is better than Model C, but whether Model A is better *enough* than Model C that we should reject the hypothesis that $\beta_0 = B_0$. Deciding what value of PRE is "better enough" is the essence of statistical inference and is the focus of this chapter.

To be less abstract, we will consider a detailed example. Suppose that 20 tickets were available for a lottery that had a single prize of $1000. How much would individuals be willing to pay for a 1-in-20 or 5% chance of winning the $1000 prize? The expected value of a ticket in this particular lottery would be $1000/20 tickets or $50. One might hypothesize that people would focus on the magnitude of the prize (i.e., the $1000 payoff) and thus be willing to pay more than the expected value of a ticket (i.e., $50). But one might also hypothesize that people would focus on the likelihood of losing whatever amount they paid and thus be generally willing to pay less than the $50 expected value of a ticket. Formally, we are comparing these two models:

MODEL A: $\text{Bid}_i = \beta_0 + \varepsilon_i$
MODEL C: $\text{Bid}_i = 50 + \varepsilon_i$

Suppose that 20 individuals participated in our hypothetical lottery by submitting the following bids to buy a ticket:

41	50	51	28	29
24	82	37	42	37
45	50	37	22	52
25	53	29	65	51

These data are displayed in Figure 4.1 in what is called a stem and leaf plot, where the left column of numbers indicates the left-most digit of each bid (the values of ten) and the numbers to the right of this column indicate the second digit of each bid (i.e., there were six bids in the twenties: 22, 24, 25, 28, 29, 29).

The average bid is $42.5, which is obviously less than the expected value of $50. However, we want to know whether these bids are meaningfully different from $50 or whether their mean is below $50 simply as a result of random variation in the data. In other words, even if the true state of the world were such that people generally are willing to pay the expected value of $50, it is unlikely that the average bid in our sample would equal $50 *exactly*. So we need to determine whether

2	245899
3	777
4	125
5	001123
6	5
7	
8	2
Stem width = 10.00	

Figure 4.1 Stem and leaf plot for the 20 lottery bids

the average bid of 42.5 that we obtained is different *enough* from 50 that we would be willing to conclude that, on average, people are willing to pay less than the expected value. In this example, β_0 represents the *true* typical amount that people would be willing to pay for a lottery ticket. We do not know, nor can we ever know, exactly what this *true* value is. The *hypothesized* value for β_0 is B_0, and in this example it equals 50, the expected value of a ticket in our lottery. Note that $50 was not estimated from our data. Rather, $50 is an a priori hypothesized value: It is a specific value that was determined *before* looking at the data. The *estimated* value for β_0 is b_0, and in this case it equals 42.5, the mean or average of the 20 bids, because the mean minimizes the sum of squared errors. In other words, for the compact model (which represents the null hypothesis), the prediction is given by

MODEL C: $\hat{Y}_i = 50$

and for the augmented model (in which β_0 is estimated from the data), the prediction is given by

MODEL A: $\hat{Y}_i = 42.5$

So our question is whether the predictions of Model A are better enough to infer that Model C is unreasonable.

To answer our question, we calculate PRE. To do so, we first need to calculate the error for each model. The necessary calculations are displayed in Figure 4.2. For the compact model, $\hat{Y}_{iC} = 50$, so the sum of squared errors from the compact model, SSE(C), is given as

$$\text{SSE}(\text{C}) = \sum_{i=1}^{n} \left(Y_i - \hat{Y}_{iC}\right)^2 = \sum_{i=1}^{n} (Y_i - 50)^2 = 5392$$

The squared error using the compact model for each bidder is listed in the third column of Figure 4.2, along with the sum of 5392 for SSE(C). For the augmented model, $\hat{Y}_{iA} = 42.5$, so

$$\text{SSE}(\text{A}) = \sum_{i=1}^{n} \left(Y_i - \hat{Y}_{iA}\right)^2 = \sum_{i=1}^{20} (Y_i - 42.5)^2 = 4267$$

The squared error using the augmented model for each bidder is listed in the fourth column, along with its sum of 4267 for SSE(A). Then the proportional reduction in error using Model A instead of Model C is given by

$$\frac{\text{SSE}(\text{C}) - \text{SSE}(\text{A})}{\text{SSE}(\text{C})} = \frac{5392 - 4267}{5392} = .209$$

That is, Model A using b_0, the *estimated* value of β_0, has 20.9% less error than Model C using B_0, the *hypothesized* value of β_0. Later in this chapter, we will determine whether 20.9% less error is enough to warrant rejecting Model C ($50) in favor of Model A.

We note in passing that for Model A one observation (bidder 7) is responsible for a substantial proportion of the total SSE. Although the presentation of formal procedures for investigating outliers must wait until Chapter 14, large errors associated with a few observations should make us suspect the presence of outliers. Remember that SSE and its associated estimators, such as the mean, are not resistant to outliers.

		Squared errors	
	Bid	*Compact*	*Augmented*
Bidder number	Y_i	$(Y_i - B_0)^2$	$(Y_i - b_0)^2$
1	41	81	2.25
2	50	0	56.25
3	51	1	72.25
4	28	484	210.25
5	29	441	182.25
6	24	676	342.25
7	82	1024	1560.25
8	37	169	30.25
9	42	64	0.25
10	37	169	30.25
11	45	25	6.25
12	50	0	56.25
13	37	169	30.25
14	22	784	420.25
15	52	4	90.25
16	25	625	306.25
17	53	9	110.25
18	29	441	182.25
19	65	225	506.25
20	51	1	72.25
Sum	850.00	5392.00	4267.00
Mean	42.50		

Figure 4.2 Lottery bids and error calculations for 20 bidders in a hypothetical lottery

Decomposition of SSE

At this point it is useful to introduce a table that summarizes our analysis so far. The sum of squares reduced (SSR) is defined as

$$\text{SSR} = \text{SSE(C)} - \text{SSE(A)}$$

and represents the amount of error that is reduced by using Model A instead of Model C. Rearranging the equation, we see that

$$\text{SSE(C)} = \text{SSR} + \text{SSE(A)}$$

In other words, the original error SSE(C) can be decomposed into two components: (a) the reduction in error due to Model A (i.e., SSR) and (b) the error remaining from Model A (i.e., SSE[A]). It is common to summarize the results of an analysis in a table having separate rows for SSR, SSE(A), and SSE(C). Figure 4.3 provides the generic layout for such tables, which are referred to as analysis of variance or ANOVA source tables because they analyze (i.e., separate or partition) the original variance or error into component parts. Figure 4.4 presents the ANOVA summary table for our example. Note that the SS (sum of squares) for the total line, which represents SSE(C), is

Source	*SS*	*PRE*
Reduction using Model A	SSR	SSR/SSE(C)
Error for Model A	SSE(A)	
Total	SSE(C)	

Figure 4.3 Generic ANOVA layout

Source	*SS*	*PRE*
Reduction (using β_0)	1125	.209
Error (using β_0)	4267	
Total (error using B_0)	5392	

Figure 4.4 ANOVA summary table for lottery example

indeed the sum of SSR and SSE(A); for our example, 5392 = 1125 + 4267. PRE is readily obtained from the SS column using the formula

$$\text{PRE} = \frac{\text{SSR}}{\text{SSE}(\text{C})}$$

We will use these tables, which give the decomposition of the sums of squares, as the basic summary for our statistical tests. Later, we will add several other useful columns to such tables.

SSR is understood and often easily calculated as the difference between SSE(C) and SSE(A). However, there is another representation for SSR, which provides additional insight for the comparison of Models C and A. It can be shown (we essentially did it in Chapter 2 for the case of the simple model; the more general proof does not provide useful insights so we omit it) that

(Equation 4.1)

$$\text{SSR} = \sum_{i=1}^{n}\left(\hat{Y}_{iC} - \hat{Y}_{iA}\right)^2$$

where $\hat{Y}_{iC}$ and $\hat{Y}_{iA}$ are, respectively, the predictions for the *i*th observation using Model C and Model A. This formula will be useful later for calculating certain SSRs that are not automatically provided by computer programs. More important are the insights it provides. For a fixed SSE(C), the larger SSR is, the larger PRE is, and the larger the improvement provided by using Model A instead of Model C. This formula shows that SSR is small when Models C and A generate similar predictions for each observation. In the extreme case when Models C and A are identical (i.e., they produce the same predictions), then SSR = 0 and PRE = 0. Conversely, SSR will be large to the extent that Models C and A generate *different* predictions. Thus, SSR is a direct measure of the difference between Models C and A and PRE = SSR/SSE(C) is a proportional measure of that difference.

Equation 4.1 is useful for calculating the SSR for the simple models considered in this chapter. Although we generally avoid multiple computational formulas, we present this one because many computer programs do not conveniently provide the necessary information for computing PRE in our terms. We will illustrate the use of Equation 4.1 by computing the SSR for the lottery example.

The value predicted by MODEL A is $\hat{Y}_{iA} = \bar{Y} = 42.5$ and the value predicted by Model C is the hypothesized value $\hat{Y}_{iC} = B_0 = 50$. So according to Equation 4.1,

$$\text{SSR} = \sum_{i=1}^{n} \left(\hat{Y}_{iC} - \hat{Y}_{iA}\right)^2 = \sum_{i=1}^{20} (50 - 42.5)^2 = \sum_{i=1}^{20} (7.5)^2 = 20(56.25) = 1125$$

That is, SSR equals the constant $7.5^2 = 56.25$ summed 20 times. Thus, SSR = 20(56.25) = 1125, which is the same value that we obtained by calculating SSR directly as SSE(C) – SSE(A) in Figure 4.3. Thus, for simple models, by comparing a Model A that estimates one parameter with a Model C that estimates none, the following formula is often handy:

(Equation 4.2)

$$\text{SSR} = \sum_{i=1}^{n} \left(B_0 - \bar{Y}\right)^2 = n\left(B_0 - \bar{Y}\right)^2$$

We will have many occasions to use Equation 4.2.

Sampling Distribution of PRE

It might seem that a difference in parameter estimates of 50–42.5 = 7.5 and PRE = 20.9% are "large enough" to infer that Model C is unreasonable relative to Model A. Unfortunately, such a conclusion may not be warranted statistically, and it is important to understand why. To gain this understanding, we need to focus on the error term ε_i that is included in the full statement of Model C:

MODEL C: $Y_i = B_0 + \varepsilon_i$

As noted before, this model says that, were it not for random perturbations represented by ε_i, all the Y_i values would equal B_0 exactly. In Chapter 3 we made the assumption that the ε_i values are all sampled randomly and independently from a normal distribution with mean 0 and variance σ^2. We also saw in Chapter 3 that the exact value for the mean calculated from a sample of size n would depend on the particular sample of errors. Sometimes the calculated mean would be above the true value B_0, and sometimes it would be below. That is, there would be a sampling distribution for the mean. If Model C were correct, sometimes the sample mean would be somewhat higher than 50 and other times it would be somewhat lower, but it would seldom equal 50 exactly. For example, we would most likely have obtained a different mean if the bids had been gathered before lunch or the day before or a day later, because the pattern of random perturbations would have been different.

Similar to the sampling distribution for the mean, there is also a sampling distribution for PRE. Just as b_0, calculated from the data, is the estimate of the unknown true parameter β_0, so too PRE, calculated from the data, is the estimate of the unknown true proportional reduction in error η^2. For the moment, let us assume that Model C is correct (i.e., that $\beta_0 = B_0$) and consider what the sampling distribution for PRE would be. In other words, we begin by assuming that the null hypothesis is true. In these terms, $\eta^2 = 0$ is equivalent to Model C being correct and to Model A making absolutely no improvement relative to Model C. We saw in Chapter 3 that b_0 has a sampling distribution, so even if Model C is correct, we would not expect our estimate b_0 to equal B_0 exactly. But we know that b_0 produces the smallest possible sum of squared errors, so SSE(A), using b_0, will almost always be at least a little smaller than SSE(C), using B_0. For example, in the lottery data the mean will seldom equal 50 exactly, even if the true parameter value were 50, and thus the SSE calculated using the sample mean will almost always be a little less than SSE(C), and SSE(A) can never exceed SSE(C) (see the proof in Chapter 2). Hence, even if the true proportional reduction in error $\eta^2 = 0$ (as it must when Model C is correct), the calculated PRE will usually be at least a little greater than zero (and never less than zero). PRE is therefore a biased estimator of η^2 because PRE overestimates the true value of η^2. We will return to this issue of the bias in PRE later. For now, the important point is that we should not expect the calculated PRE to equal zero even when Model A makes no improvement on Model C. Thus, we cannot base our decision about the validity of Model C simply on whether PRE = 0.

We cannot use PRE $\neq 0$ as a criterion for rejecting Model C, but we can consider the sampling distribution of PRE to determine whether the calculated value of PRE is a *likely* value, *assuming that Model C is correct*. If the calculated value of PRE is a likely value, then we ought not to reject Model C. Neither should we reject its equivalent null hypothesis that $\beta_0 = B_0$. On the other hand, if the calculated value of PRE is an unlikely value when Model C is assumed to be true, then we ought to reject Model C in favor of Model A. In terms of our example, we need to consider the sampling distribution of PRE to determine for this case whether PRE = .209 is a likely value, assuming that Model C is correct (i.e., that $\beta_0 = 50$). If it is a likely value, then we ought not to reject Model C and its equivalent hypothesis that $\beta_0 = B_0 = 50$; there would be no evidence that the lottery bids are different from what would be expected if the null hypothesis were true. If PRE = .209 is an unexpected value, then we ought to reject Model C and its hypothesis, and we should instead choose Model A and its estimate that $\beta_0 = b_0 = 42.5$; in other words, we would conclude that the lottery bids were significantly lower than the expected value of $50 for a ticket.

We could develop the sampling distribution for PRE in this example using the same simulation strategy we used in the previous chapter. That is, we could put error tickets of the appropriate size into a bag and then do many simulation rounds, in each of which we would randomly select 20 error tickets, add 50 to each, and calculate PRE for that sample. The only problem with this strategy is that we do not know what size error tickets to place in the bag to be sampled. In other words, we do not know the variance σ^2 of the normal distribution of errors. However, as was noted in the previous chapter, the mean squared error provides an estimate of σ^2. In particular, for Model A with one parameter the estimate is

$$s^2 = \frac{\text{SSE}}{n-1} = \frac{\sum_{i=1}^{n}\left(Y_i - \bar{Y}\right)^2}{n-1} = \frac{4267}{19} = 224.58$$

We could therefore conduct the simulation by sampling error values from a normal distribution with mean 0 and variance 224.58. The 100 error tickets in Figure 3.1 were sampled from such a normal distribution, so they could be used as the bag of error tickets for the simulation.

As an example of a simulation round, suppose that the following 20 error tickets were drawn from Figure 4.1.

17	31	−2	−6	−17
−4	28	29	17	1
−12	−6	3	−25	2
24	15	−20	4	−3

These error terms when added to the value of $B_0 = 50$ of Model C yield the following 20 scores.

67	81	48	44	33
46	78	79	67	51
38	44	53	25	52
74	65	30	54	47

The mean of the resulting 20 scores is 53.8 and SSE(C), using 50 as the model prediction for all the observations, and SSE(A), using 53.8 as the model prediction, are easily calculated to be 5554 and 5265.2, respectively. Thus,

$$\text{PRE} = \frac{5554 - 5265.2}{5554} = .052$$

Then a new simulation round with a new sample of error values would produce a different mean and PRE. These simulation rounds could conceptually be repeated until there were enough PRE values to make a sampling distribution for PRE.

Alas, the simulation strategy previously outlined for generating the sampling distribution of PRE will not work because $s^2 = 224.58$ is only an *estimate* of the true variance of the errors σ^2. Just as it is unlikely that $\bar{Y} = b_0$ (the calculated mean for a sample) will equal B_0 exactly, it is also unlikely that the calculated variance s^2 will equal σ^2 exactly. In other words, we are uncertain about the exact size of the error tickets that should be placed in the bag.

Also, it would be tedious if we had to do a new simulation round for each data analysis. Fortunately, this is not necessary because mathematical statisticians have specified the sampling distribution for PRE based on the assumptions we made in Chapter 3 about the behavior of the errors ε_i. Even though we will not actually do simulations for generating a sampling distribution, it is important to remember that the mathematical formula for that distribution is derived from the assumption that the error values are randomly sampled from a distribution with mean 0 and variance σ^2 and that the sampling could be represented by drawing error tickets from a bag.

Figure 4.5 provides a tabular description of the sampling distribution of PRE for the case of samples of size 20, again assuming the validity of Model C. That is, if Model C were correct (in our case, $\beta_0 = 50$), and if we compared Model C ($\hat{Y} = 50$) with Model A ($\hat{Y} = b_0 = \bar{Y}$), then Figure 4.5 presents the proportional frequency and cumulative proportion for each range of PRE. The proportional frequencies are plotted in Figure 4.6. As we would expect, the sampling distribution in Figure 4.6 shows that values of PRE near zero are the most likely. It also shows that values of PRE greater than .2 are infrequent.

The cumulative proportions are generally more useful than the proportions for individual ranges. The cumulative proportion is simply the total proportion for the range of PREs from zero to the value of interest. For example, to find the cumulative proportion for the range from 0 to .03, we simply add the proportions for the three component ranges: 0 to .01, .01 to .02, and .02 to .03. For this range, .334 + .125 + .088 = .547. That is, 54.7% of the PRE values are less than or equal to .03. The cumulative proportions are displayed in the last column of Figure 4.5 and graphed in Figure 4.7. We can see, for example, from both the table and the graph of the cumulative proportions that PRE is less than .1 about 84% of the time.

We can now ask whether a value for PRE of .209 is likely if Model C is correct. From the table in Figure 4.5 we see that approximately 96% of the time PRE would be less than .209. Or, in other words, a PRE as large as .209 would be obtained only about 4% of the time if Model C were correct. We can finally answer our question: It is unlikely (less than 5% chance) that we would have obtained a PRE this large had Model C been correct. We can therefore reasonably reject Model C in favor of Model A. This is equivalent to rejecting the null hypothesis that $\beta_0 = B_0 = 50$.

PRE range	*Proportion*	*Cumulative proportion*
.00–.01	.334	.334
.01–.02	.125	.459
.02–.03	.088	.547
.03–.04	.068	.615
.04–.05	.055	.670
.05–.06	.045	.716
.06–.07	.038	.754
.07–.08	.032	.786
.08–.09	.028	.814
.09–.10	.024	.838
.10–.11	.021	.858
.11–.12	.018	.876
.12–.13	.016	.892
.13–.14	.014	.905
.14–.15	.012	.917
.15–.16	.010	.928
.16–.17	.009	.937
.17–.18	.008	.945
.18–.19	.007	.952
.19–.20	.006	.958
.20–.21	.005	.963
.21–.22	.005	.968
.22–.23	.004	.972
.23–.24	.004	.976
.24–.25	.003	.979
.25–.26	.003	.982
.26–.27	.002	.984
.27–.28	.002	.986
.28–.29	.002	.988
.29–.30	.002	.990
.30–.31	.001	.991
.31–.32	.001	.993
.32–.33	.001	.994
.33–.34	.001	.995
.34–.35	.001	.995
.35–.36	.001	.996
.36–.37	.001	.997
.37–.38	.001	.997
.38–.39	.000	.998
.39–.40	.000	.998

Figure 4.5 Tabular description of the sampling distribution of PRE for testing the simple model with 20 observations

Substantively, the data indicate that participants were willing to pay less than the expected value of the lottery tickets.

Critical Values

From the mathematical equations describing the sampling distribution for PRE, we can determine for our example data that, if Model C were correct (and thus $\eta^2 = 0$), we would expect 95% of the simulated values of PRE to be below the precise value of .187. This sampling distribution is plotted

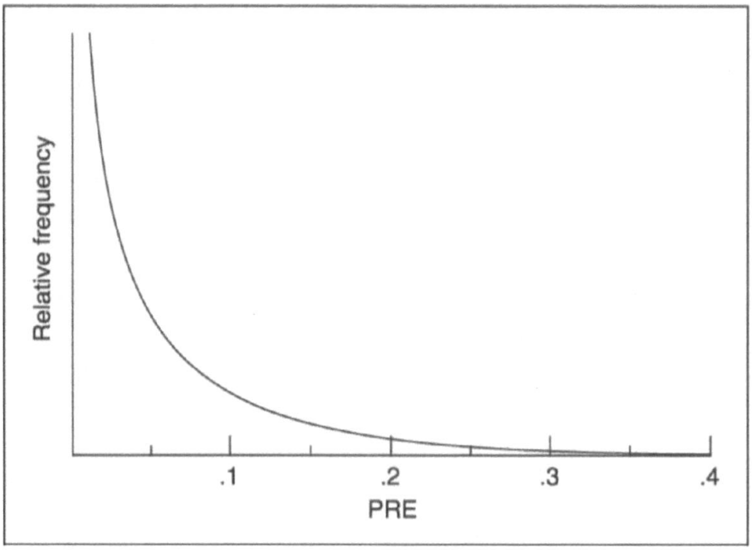

Figure 4.6 Sampling distribution of PRE for testing the simple model with 20 observations

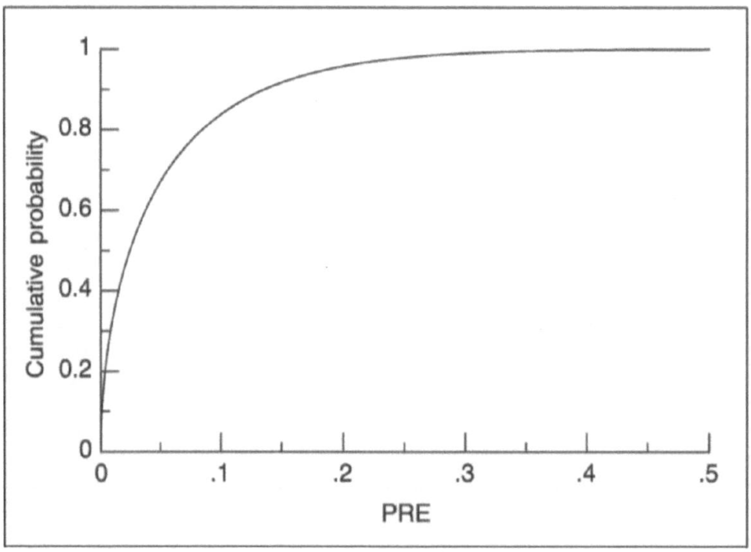

Figure 4.7 Cumulative proportions for n – PA = 19 (PA = number of parameters for Model A)

in Figure 4.8. In the social sciences it is customary to consider a value of PRE to be surprising if it occurs by chance less than 5% of the time when the null hypothesis is true. Thus, .187 is the *critical value* for this example; any value of PRE > .187 causes us to reject Model C. Using the equations, for any number of observations we can calculate the value of PRE for which we would expect 95% of PRE values to be below if Model C were correct. Figure 4.9 gives the 95% (and 99%) critical values for selected numbers of observations.

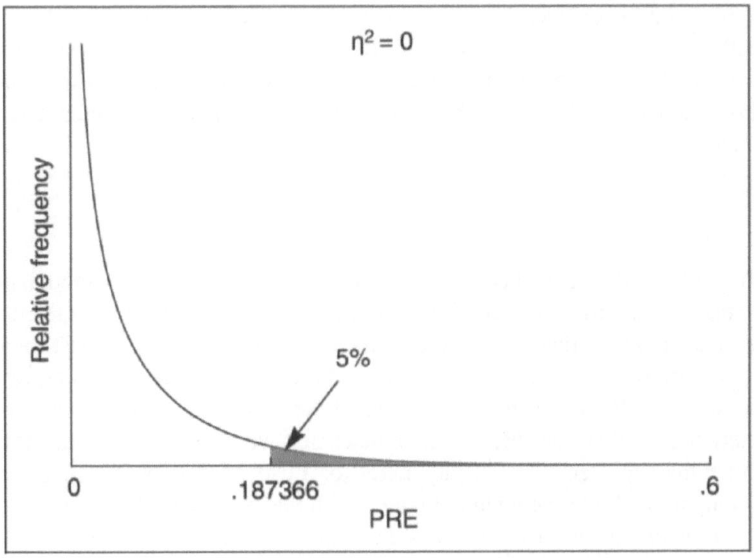

Figure 4.8 Distribution of PRE for n – PA = 19, assuming that $\eta^2 = 0$ (PA = number of parameters for Model A)

Statistic *F*

Figures of the critical values for PRE are rare in statistics books. Much more common, largely for historical reasons, are tables of *F*, a statistic closely related to PRE. As we will see, *F* is a simple function of PRE, so if we know PRE, the number of observations, and the number of parameters in Models C and A, then we also know *F* and vice versa. By re-expressing PRE, *F* also provides additional insights about the proportion of error reduced. We therefore turn to the motivation for calculating *F* and then consider its sampling distribution.

The two reasons for calculating *F* are (a) to examine the proportional reduction in error *per additional parameter* added to the model and (b) to compare the proportion of error that was reduced (PRE) with the proportion of error that remains (1 – PRE). In the context of the simple models that we are considering in this chapter, PRE is obtained by the addition of only one parameter. But later we will want to consider the improvement produced by models that add more than one parameter. To avoid having to present different formulas as the models become more complex, we will present the general definition of *F* here. The key idea is that a given PRE, let us say .35, is more impressive when obtained by the addition of a single parameter than when it is obtained by the addition of several parameters. So we want to consider *PRE per parameter*. That is, we divide PRE by the number of additional parameters used in Model A (parameters that are not used in Model C). We will let PA and PC represent the number of parameters for Model A and Model C, respectively. Then, the number of additional parameters is simply PA – PC. Hence, *F* is based on the quantity

$$\frac{\text{PRE}}{\text{PA} - \text{PC}}$$

which is simply the proportional reduction in error per additional parameter. For the simple models of this chapter, there are no parameters to be estimated for Model C and only one for Model A, so PC = 0, PA = 1, and PA – PC = 1.

Similarly, we need to consider the remaining proportion of the error, 1 – PRE, in terms of the number of additional parameters that *could* be added to reduce it. As noted in Chapter 1, the most parameters we can have is one for each observation. If there are n observations, and we have already used PA parameters in Model A, then at most we could add n – PA parameters to some more complex model. So

$$\frac{1-\text{PRE}}{n-\text{PA}}$$

is the proportion of remaining error per parameter that *could* be added to the model. In other words, this is the average remaining error per additional parameter. If we added a parameter to the model at random, even a parameter that was not really useful, we would expect at least some reduction in error. The proportion of remaining error per parameter or the average remaining error tells us the value of PRE to expect for a worthless parameter. If the parameter or parameters added in Model A are genuinely useful, then the PRE per parameter that we actually obtain ought to be substantially larger than the expected PRE per parameter for a useless, randomly selected parameter. An easy way to compare PRE per parameter obtained with the expected PRE per parameter is to compute the ratio of the two quantities; this gives the definition of F as

$$F = \frac{\text{PRE}/(\text{PA}-\text{PC})}{(1-\text{PRE})/(n-\text{PA})} \tag{4.3}$$

We can think of the numerator of F as indicating the average proportional reduction in error per parameter added, and the denominator as the average proportional reduction in error that could be obtained by adding all possible remaining parameters. For Model A to be significantly better than Model C, we want the average error reduction for the parameters added to be much greater than the average error reduction we could get by adding the remainder of the possible parameters. Hence, if F is about 1, then we are doing no better than we could expect on average. Values of F near 1 suggest that we should not reject the simpler Model C. Values of F much larger than 1 imply that the average PRE per parameter added in Model A is much greater than the average that could be obtained by adding still more parameters. In that case, we would want to reject Model C (and its implicit hypothesis) in favor of Model A. For the example of Figure 4.2 where PRE = .209 and $n = 20$,

$$F = \frac{\text{PRE}/(\text{PA}-\text{PC})}{(1-\text{PRE})/(n-\text{PA})} = \frac{.209/(1-0)}{.791/(20-1)} = \frac{.209}{.0416} = 5.02$$

In other words, we obtained a 20.9% reduction in error per parameter by using Model A, and the further reduction we could get by adding all the additional parameters is only 4.16% per parameter. Their ratio of 5.02 suggests that adding to Model C the one specific additional parameter β_0, which is estimated by the mean, yields a substantially better (about five times better) PRE than we could expect by randomly adding a parameter from the remaining ones. In other words, the increased complexity of Model A is probably worth it. But again, we need to consider the sampling distribution of F to determine whether a value of 5.02 is indeed "surprising."

Again, mathematical equations exist for describing the sampling distribution of F, given the assumptions about error discussed in Chapter 3. If we assume that the errors ε_i are independently, identically, and normally distributed, then F has what is known as an *F distribution*. The 95% and 99% critical values for F for testing simple models are listed in Figure 4.9 next to their corresponding values of PRE. F and PRE are redundant in the sense that one exceeds its critical value if and only if the other one exceeds its critical value. For the example, PRE = .209 exceeds its critical value of .187, and so necessarily the corresponding $F = 5.02$ exceeds its critical value of 4.38. Thus, either PRE or F leads us to reject Model C and its implicit hypothesis that $\beta_0 = B_0 = 50$. Note that for most reasonable numbers of observations the 95% critical value for F is about 4 when PA – PC = 1. If we ignore the fractional part of F, then a useful rule of thumb that reduces the need to consult statistical tables frequently is to reject Model C in favor of Model A whenever F is greater than 5. If F is between 4 and 5, then you will probably have to look it up in the table, and if it is below 4, then we will likely retain Model C unless the number of observations is extremely large. Critical values of PRE and F for

	95%		*99%*	
n – PA	*PRE*	*F*	*PRE*	*F*
1	.994	161.45	1.000	4052.18
2	.903	18.51	.980	98.50
3	.771	10.13	.919	34.12
4	.658	7.71	.841	21.20
5	.569	6.61	.765	16.26
6	.499	5.99	.696	13.75
7	.444	5.59	.636	12.25
8	.399	5.32	.585	11.26
9	.362	5.12	.540	10.56
10	.332	4.97	.501	10.04
11	.306	4.84	.467	9.65
12	.283	4.75	.437	9.33
13	.264	4.67	.411	9.07
14	.247	4.60	.388	8.86
15	.232	4.54	.367	8.68
16	.219	4.49	.348	8.53
17	.208	4.45	.331	8.40
18	.197	4.41	.315	8.29
19	.187	4.38	.301	8.19
20	.179	4.35	.288	8.10
22	.164	4.30	.265	7.95
24	.151	4.26	.246	7.82
26	.140	4.23	.229	7.72
28	.130	4.20	.214	7.64
30	.122	4.17	.201	7.56
35	.105	4.12	.175	7.42
40	.093	4.09	.155	7.31
45	.083	4.06	.138	7.23
50	.075	4.03	.125	7.17
55	.068	4.02	.115	7.12
60	.063	4.00	.106	7.08
80	.047	3.96	.080	6.96
100	.038	3.94	.065	6.90
150	.025	3.90	.043	6.81
200	.019	3.89	.033	6.76
500	.008	3.86	.013	6.69
∞		3.84		6.63

Figure 4.9 Critical values (95% and 99%) for PRE and F for testing models that differ by one parameter (PA = number of parameters for Model A)

testing the more complex Model A that differs from Model C by more than one parameter are listed in the Appendix as a function of PA – PC, often called the "numerator degrees of freedom" because that term appears in the numerator of the formula for F (Equation 4.3), and n – PA, often called the "denominator degrees of freedom" because it appears in the denominator of the formula for F.

We can also rearrange Equation 4.3 to express PRE as a function of F. This formula is often helpful:

$$\text{PRE} = \frac{F}{F + \left[(n - \text{PA})/(\text{PA} - \text{PC})\right]}$$

It is useful to add the degrees of freedom (df) and F to the basic summary table we began earlier. Figure 4.10 presents such a table for our example. It is our policy to avoid multiple computational formulas for the same quantity and instead to present only one conceptual formula. However, we must deviate from that policy for F in this instance because F is traditionally calculated by an equivalent but different formula based on Figure 4.10. Figures constructed using the alternative formula for F are ubiquitous, so the reader has no choice but to learn this alternative in addition to the conceptual formula for F previously presented. The alternative formula for F is

$$F = \frac{\text{SSR}/(\text{PA} - \text{PC})}{\text{SSE}(\text{A})/(n - \text{PA})} = \frac{\text{MSR}}{\text{MSE}}$$

For our example, this yields

$$F = \frac{\text{SSR}/(\text{PA} - \text{PC})}{\text{SSE}(\text{A})/(n - \text{PA})} = \frac{1125/(1-0)}{4267/(20-1)} = \frac{1125}{224.58} = 5.01$$

This agrees with our previous calculation except for a small rounding error. MSR represents the *mean squares reduced*, and MSE represents the *mean square error*. To facilitate this calculation, we usually add an "MS" column to the ANOVA summary table. The final column, labeled "p," gives the probability of obtaining a PRE and F that is that large or larger if $\eta^2 = 0$. In this case, PRE and F exceed the 95% critical values so the probability of getting a PRE or F as large as or larger than the values we calculated if Model C were correct is less than .05. With the additional columns, Figure 4.11 provides a detailed summary of our analysis of the lottery bids.

Statistical Decisions

We have now defined the essence of statistical inference: If PRE and F exceed their respective critical values then the simpler Model C is rejected in favor of the more complex Model A. We

Source	*SS*	*df*	*MS*	*F*	*PRE*	*p*
Reduce, Model A	SSR	PA – PC	$MSR = \frac{SSR}{PA - PC}$	$\frac{MSR}{MSE}$	$\frac{SSR}{SSE(C)}$	
Error for Model A	SSE(A)	n – PA	$MSE = \frac{SSE(A)}{n - PA}$			
Total	SSE(C)	n – PC				

Figure 4.10 Analysis of variance summary table: decomposition of sums of squares

Source	*SS*	*df*	*MS*	*F*	*PRE*	*p*
Reduce, Model A	1125	1	1125.00	5.01	.209	.04
Error for Model A	4267	19	224.58			
Total	5392	20				

Figure 4.11 ANOVA summary table for lottery example

now have a rule for resolving the inherent tension in data analysis between reducing the error as much as possible and keeping the model of the data as parsimonious as possible. It is important to recognize, however, that statistical inference is probabilistic and therefore not infallible. That is, if Model C is actually the correct model, then 5% of the time we will obtain values of PRE and F that exceed their 95% critical values. Our rule is to reject Model C if those statistics exceed their 95% critical values, so in such instances we will have made a mistake in rejecting Model C. There is no way to avoid making occasional mistakes of that type. By adopting a 95% critical value, we are implicitly accepting that, when we have data for which Model C is correct, we are willing to risk a 5% chance of incorrectly rejecting it. Mistakes of this type are known as *Type I* errors. The choice of 5% as an acceptable rate of Type I errors is inherently arbitrary. If we want to be more cautious, we could choose a rate of 1%, or if we are willing to risk more Type I errors then we might choose a rate of 10%.

We also can commit a *Type II error*. A Type II error occurs when Model C is incorrect but the obtained values of PRE and F still do not exceed their critical values. Thus, a Type II error occurs when we fail to reject Model C when we should. In this case, Model C is incorrect, and Model A is actually better, but we are unlucky in terms of the error tickets drawn and therefore miss seeing the difference. Figure 4.12 summarizes the statistical decision that confronts us and defines the ways in which both the right and wrong decisions can be made. Statistical inference can be viewed as a game with Nature. Nature determines whether Model C is correct or incorrect. The goal of the data analyst is to guess which is the case. The data analyst uses the data to make an informed guess. Specifically, if PRE and F exceed their critical values, then the analyst's guess is, "reject Model C"; otherwise, the guess is, "do not reject Model C." If Nature has determined that Model C is correct, then, using a 95% critical value, we will decide correctly 95% of the time and incorrectly (i.e., make a Type I error) 5% of the time. The chance of making a Type I error is often labeled α and referred to as the *significance level*.

On the other hand, if Nature has determined that Model C is incorrect, then we will decide correctly the proportion of times that PRE and F exceed their critical values, and we will decide incorrectly (i.e., make a Type II error) the proportion of times that PRE and F fall below their critical values. The chance of making a Type II error is, unfortunately, often labeled β, which should not be confused with β_0, β_1, etc., which we use to represent parameters in a model. The proportion of times the correct decision is made when Nature has determined that Model C is incorrect, $1 - \beta$, is often referred to as the *power* of a statistical test. Power will be considered further in Chapter 5.

Equivalence to the *t*-Test

In this section we demonstrate the equivalence between the statistical test for the simple model developed in this chapter and the traditional *one-sample t-test* presented in most statistics textbooks. We do so to allow readers with exposure to traditional textbooks to make the comparison between approaches and because *t*-tests are frequently reported in scientific work.

	True state of Nature	
Statistical decision	*Model C correct*	*Model C incorrect*
"Reject Model C"	Type I error	Correct decision
"Do not reject Model C"	Correct decision	Type II error

Figure 4.12 The statistical decision and the two types of errors

The one-sample *t-test* answers whether the mean of a set of observations equals a particular value specified by the null hypothesis. The formula for the one-sample *t-test* is

$$t_{n-1} = \frac{\sqrt{n}\left(\bar{Y} - B_0\right)}{s}$$

where n is the number of observations, $\bar{Y}$ is the calculated mean, B_0 is the value specified by the null hypothesis, and s is the standard deviation of the set of observations. With the appropriate assumptions about Y—the same assumptions that we made about the distribution and independence of the ε_i—calculated values of t can be compared to critical values of *Student's t-distribution.* Tables of this distribution are available in many statistics textbooks. However, separate tables are not really needed because squaring a t value with $n - 1$ degrees of freedom yields an F with 1 and $n - 1$ degrees of freedom. Thus, the F tables in the Appendix may readily be used.

To show the equivalence between F as presented in this chapter and the usual t-test, we begin with the definition of F for the simple model; that is

$$F_{1,n-1} = \frac{\text{PRE}/1}{(1-\text{PRE})/(n-1)}$$

We know that PRE = SSR/SSE(C), so we can show that

$$1-\text{PRE} = 1 - \frac{\text{SSR}}{\text{SSE}(\text{C})} = \frac{\text{SSE}(\text{C}) - \text{SSR}}{\text{SSE}(\text{C})} = \frac{\text{SSE}(\text{A})}{\text{SSE}(\text{C})}$$

Substituting these values into the definition for F yields

$$F_{1,n-1} = \frac{\text{SSR}/\text{SSE}(\text{C})}{\left[\text{SSE}(\text{A})/\text{SSE}(\text{C})\right]/(n-1)} = \frac{\text{SSR}}{\text{SSE}(\text{A})/(n-1)}$$

But from Equation 4.2 we know that for the simple model SSR can be replaced with $n(B_0 - \bar{Y})^2$, and from Chapter 2 we know that SSE(A)/(n – 1) is s^2, the variance of the set of observations. Substituting these values yields

$$F_{1,n-1} = \frac{n\left(B_0 - \bar{Y}\right)^2}{s^2}$$

Taking the square root of this last equation gives the final result of

$$\sqrt{F_{1,n-1}} = t_{n-1} = \frac{\sqrt{n}\left(\bar{Y} - B_0\right)}{s}$$

We provide the aforementioned derivation not to present yet another computational formula but to show that our model comparison approach to statistical inference is statistically identical to the traditional approach. The use of PRE and F for comparing models is nothing more than a repackaging of the traditional approach. However, this repackaging has the important consequence of making it easier to generalize to more complicated models and data analysis questions. Throughout the remainder of the book, we will use PRE and F just as we did in this chapter for statistical inference. In contrast, the traditional *t-test* does not generalize nearly so easily. Also, even though the *t-test* must produce exactly the same conclusion with respect to the null hypothesis, it does not provide a measure of the magnitude of the result. In our model comparison approach, PRE provides a useful measure of the magnitude.

An Example

In this section we illustrate the techniques of this chapter using the internet use data that were presented in Figure 1.1. Suppose that marketing researchers had projected that 90% of individuals would use the internet by the year 2021. Was the marketing researchers' projection overly optimistic? The question is equivalent to comparing the following two models:

MODEL A: $Y_i = \beta_0 + \varepsilon_i$
MODEL C: $Y_i = 90 + \varepsilon_i$

We know that the mean is $Y = 83.982$ (see Chapter 2), so the estimated Model A is

$$\hat{Y}_i = 83.982$$

For the statistical inference we need to calculate PRE and F from SSE(A) and SSE(C). We also know that the variance or the MSE is 122.565 (see Chapter 2). Since MSE = SSE(A)/(n – 1), we can easily obtain SSE(A) by multiplying MSE by n – 1; thus, SSE(A) = 122.565(44) = 5392.86; this value is also given in Figure 2.10. We can compute SSR using

$$\text{SSR} = n(B_0 - \bar{Y})^2 = 45(90\text{–}83.982)^2 = 1629.735$$

Then we can compute SSE(C) from

$$\text{SSE(C)} = \text{SSE(A)} + \text{SSR} = 5392.86 + 1629.735 = 7022.595$$

The computations of PRE and F are then

$$\text{PRE} = \frac{\text{SSR}}{\text{SSE}(\text{C})} = \frac{1629.735}{7022.595} = .232$$

and

$$F_{1,44} = \frac{\text{PRE}/1}{(1-\text{PRE})/(n-1)} = \frac{.232}{.768/44} = 13.29$$

From the tables in the Appendix, the critical values for PRE and F (for $\alpha = .05$) are, respectively, about .085 and 4.06. The obtained values clearly exceed the critical values, so we reject Model C in favor of Model A. Thus, the 23.2% reduction in error obtained by using the estimate $b_0 = 83.982$ instead of the null hypothesis value of $B_0 = 90$ is statistically significant. We therefore conclude that the percentage of individuals who used the internet was significantly lower than the marketing researchers' projection. Figure 4.13 provides a summary of our analysis.

We might summarize our results for a journal article as follows:

On average, across countries the percentage of individuals who used the internet in the year 2021 ($M = 83.98$) was significantly lower than the projected value of 90%, PRE = .232, $F(1, 44) = 13.29$, $p < .05$.

Summary

In Chapter 1 we noted that the equation

DATA = MODEL + ERROR

implies an inherent tension in data analysis between reducing error as much as possible and keeping the model as simple or parsimonious as possible. Whenever we consider adding an additional parameter to the model so that it will fit the data better and thereby reduce error, we must ask whether the additional complexity of the model is worth it. In this chapter we have developed inferential machinery for determining whether the additional complexity is worth it.

To decide whether the benefits of the additional parameters in Model A outweigh the benefits of the parsimony and simplicity of Model C, we first calculate SSE(A) and SSE(C), respectively, the sum of squared errors for the augmented model (which incorporates the additional parameters) and the compact model (which does not include those parameters). The sum of squares reduced, SSR, is simply the difference between them:

$$\text{SSR} = \text{SSE(C)} - \text{SSE(A)}$$

Then we calculate the proportional reduction in error attributable to the additional parameters, which is given by

$$\text{PRE} = \frac{\text{SSE}(\text{C}) - \text{SSE}(\text{A})}{\text{SSE}(\text{C})} = \frac{\text{SSR}}{\text{SSE}(\text{C})}$$

Source	*SS*	*df*	*MS*	*F*	*PRE*	*p*
Reduce, Model A	1629.735	1	1629.735	13.29	.232	<.05
Error for Model A	5392.86	44	122.565			
Total	7022.595	45				

Figure 4.13 ANOVA summary table for internet example

A related statistic is the ratio of the proportional reduction in error per parameter added to the potential proportional reduction in error per remaining unused parameter, which is given by

$$F = \frac{\text{PRE}/(\text{PA}-\text{PC})}{(1-\text{PRE})/(n-\text{PA})}$$

We then compare the calculated values of PRE and F to the distribution of values we would expect *if* Model C, the compact model, *were true*. If the calculated values of PRE and F would have been unlikely if Model C were true, then we reject Model C and conclude that the extra complexity of Model A is worth it. On the other hand, if the calculated values are ones that might reasonably have been obtained if Model C were true, then we do not reject Model C and without further evidence we would not accept the additional complexity of Model A.

This inferential machinery is merely a guide for decision making. It is not infallible. There are two kinds of errors that we can make. A Type I error occurs when Model C is in fact correct but, by chance, we happen to get unusual values of PRE and F and so reject Model C. The probability of a Type I error is α and defines how unusual PRE and F must be before we reject Model C. A Type II error occurs when Model C is in fact false or inferior to Model A but, by chance, we happen to get values of PRE and F that are not unusual and so fail to reject Model C. We generally select α, the probability of a Type I error, according to convention. Decisions about power and the probability of a Type II error will be discussed in the next chapter.

5 Statistical Power

Power, Effect Sizes, and Confidence Intervals

In this new edition, we provide more complete coverage of statistical power, effect sizes, and confidence intervals in a separate chapter, Chapter 5. All the necessary concepts for statistical inference used throughout the book are contained in Chapters 2, 3, 4, and 5, although the models that are explored in these chapters are very simple.

In recent years, researchers have become increasingly concerned with questions of replication and power, not just statistical significance. This chapter is dedicated to these topics. Following Chapter 4, we consider power in the context of the simplest model comparison: a test of a single parameter β_0. But again, as in Chapter 4, the logic we develop here generalizes to more complex model comparisons that are discussed in later chapters. As we briefly mentioned, power is defined as $1 - \beta$, where β is the likelihood of a Type II error. Power thus refers to the likelihood that a statistical test leads the researcher to reject Model C *when Nature has determined that Model C is incorrect*. Critically, whereas the logic developed in Chapter 4 operates from the assumption that Model C is correct, consideration of power requires the researcher to assume a different perspective: that some alternative model is correct. In addition, whereas the statistical machinery in Chapter 4 may easily be implemented after data collection is complete, questions of power are best addressed prior to data collection.

Considering power when designing the study and taking steps to maximize it is important for the simple reason that doing so increases the chances of detecting an effect if it exists. Unfortunately, low power is a problem that plagues data analysis far more frequently than is commonly realized. One estimate suggests that a typical study in the field of psychology has power of only .35 (Bakker et al., 2012). This means that, assuming that Model C is incorrect, only 35 times out of 100 would the researcher reject the null hypothesis. Low power creates serious difficulties. For example, consider the plight of researchers trying to evaluate the effectiveness of an innovative educational curriculum or a new therapy for a serious illness. If the power of detecting the effect is only .36, there is a high risk of concluding that the new curriculum or therapy is not effective even when it actually is. When one considers the time and money that may be invested in research—not to mention the potential of the findings to benefit people and advance science—it generally makes little sense to design and conduct studies that have such a low chance of finding effects even when the effects really exist.

The goal of a power analysis conducted before the collection of data is not to obtain an exact calculation of the statistical power but to obtain an indication of whether there is much hope for detecting the effect you want to find given a particular research design and sample size. If a study would not have much chance of distinguishing between Model C and Model A for a large effect (e.g., $\eta^2 = .3$), then there is little if any reason for conducting the study.

Unfortunately, a meaningful consideration of power requires the researcher to specify a specific hypothesized alternative to Model C, which poses a number of practical challenges that are

DOI: 10.4324/9781003438335-7

discussed further. Because of these challenges, power analysis is often performed in ways that severely undermine its utility. In this chapter, we discuss how to estimate power and how to improve power. We also discuss related concepts of confidence intervals and effect sizes.

Estimating Statistical Power

To determine the chances of a Type II error or to determine the power, we need to know the sampling distribution for PRE and F, *assuming that Model C is incorrect.* We cannot determine such a sampling distribution in general because to say that Model C is incorrect is to say only that the true proportional reduction in error η^2 is greater than zero. However, using the equations provided by mathematical statisticians, we can derive the sampling distributions for specific values of η^2 we might want to consider. Consider our lottery example (described in Chapter 4) in which 20 individuals submitted bids to buy a ticket. Figure 5.1 displays plots of the sampling distribution for PRE for PA – PC = 1 and n – PA = 19 assuming progressively greater values of *true* PRE, that is, η^2. Note that in this chapter we must frequently distinguish between PRE, which is a statistic computed from DATA, and the true value of PRE, which is an unknown population parameter and is typically denoted as η^2 (eta squared). If $\eta^2 = .1$, the probability of obtaining a PRE greater than the critical value and thus rejecting Model C is 29% as compared with only 5% if Model C is correct and $\eta^2 = 0$. Figure 5.2 displays the cumulative probability distributions for other observed values of PRE if we were to assume that η^2 were equal to 0, .05, .1, .3, .5, and .75, but again only for the very particular conditions of our example: PA – PC = 1 and n – PA = 19. Different sampling distributions would be obtained for other combinations of PA – PC and n – PA. The column for $\eta^2 = 0$ corresponds exactly to the cumulative probability distribution of Figure 5.1 because it represents the null hypothesis. Each entry in Figure 5.2 is the probability that PRE calculated from the data would be less than the value of PRE specified for that row. For example, the value of .08 in the row for PRE = .30 and the column for $\eta^2 = .50$ means that if η^2 (the true PRE) were .5, then the probability of obtaining a PRE (calculated from the data) of .3 or lower is 8%.

We can use the cumulative probability distributions of PRE for different values of η^2 to perform "what if" analyses. For example, we can ask, "*what* would the chances of making a Type II error be *if* $\eta^2 = .3$ also assuming PA – PC = 1 and n – PA = 19?" To answer this question, we first must decide what chance of a Type I error we are willing to risk. If we adopt the customary value in the social sciences of $\alpha = .05$, then, as before, we select the critical value of PRE so that the calculated value of PRE has only a .05 probability of exceeding that critical value if $\eta^2 = 0$. In this case, we see from Figure 5.2 that the probability of obtaining a PRE less than or equal to .20 equals .96, so only 4% of the observed values of PRE should be greater than .20 if $\eta^2 = 0$. Remember that $\eta^2 = 0$ implies that Model C and Model A are identical, so our decision rule will be to reject Model C in favor of Model A if PRE > .20. Now we can use the column for $\eta^2 = .3$ to answer our "what if" question. The entry in that column for the row for PRE = .20 reveals that the probability that the calculated PRE will be less than .20 is .25. That is, even if $\eta^2 = .3$ (i.e., there is a real difference between Model C and Model A), there is still a 25% chance that we will obtain a value of PRE below the critical value and hence will not reject Model C. In other words, the probability of making a Type II error is .25. Conversely, 1 – .25 = .75 is the probability of obtaining PRE > .20 and hence rejecting Model C in favor of MODEL A if $\eta^2 = .3$. In other words, the power (probability of not making a Type II error) is .75.

We can do "what if" analysis for other values of η^2. For example, if $\eta^2 = .05$, which implies a small difference between Models C and A, then power equals 1 – .86 = .14. That is, the chances of obtaining a PRE large enough to reject Model C in favor of Model A would be only 14% for this small difference between the two models. On the other hand, if $\eta^2 = .75$, which implies a very large

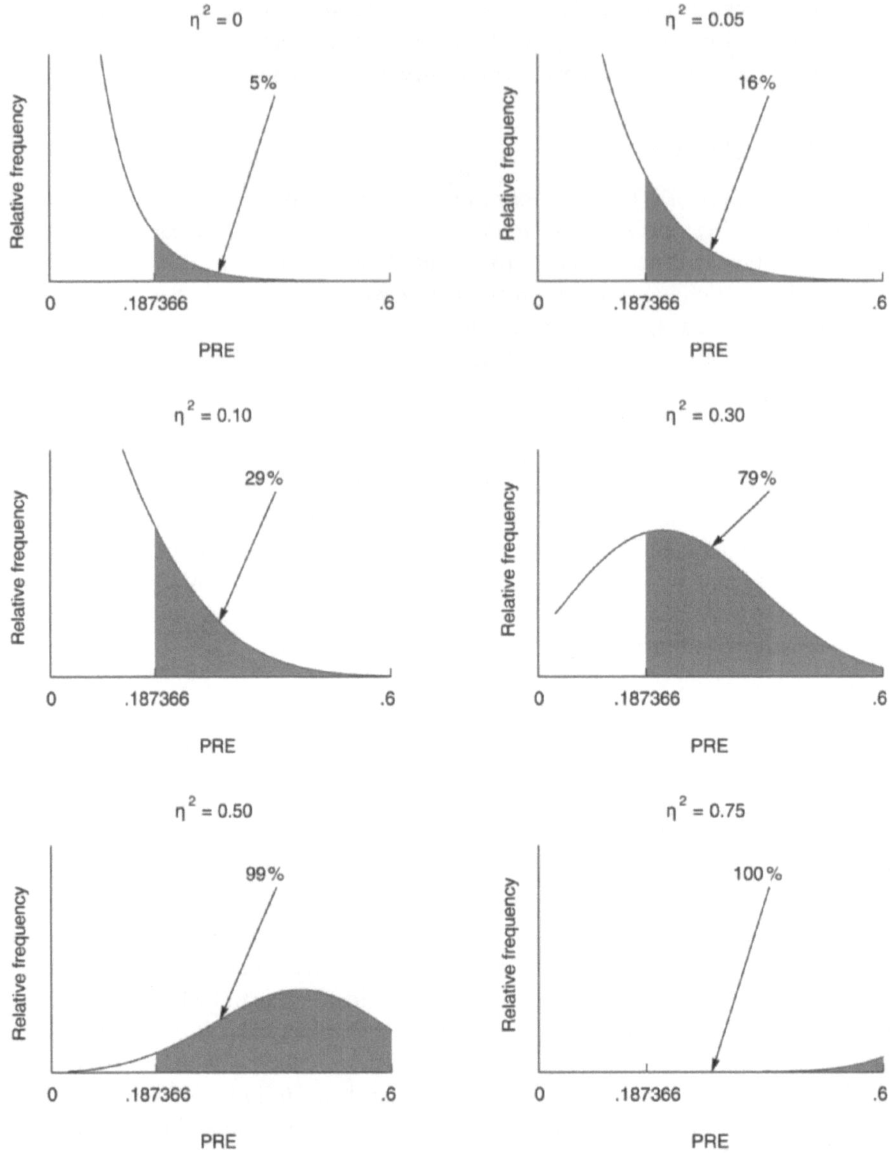

Figure 5.1 Distributions of PRE for PA – PC = 1 and n – PA = 19, assuming various values for η^2

difference between Models C and A, then power equals 1 – .00 = 1. That is, for this large difference we would be virtually certain to obtain a PRE large enough to reject Model C in favor of Model A.

The cumulative sampling distributions are unwieldy, and only a few of the numbers are actually needed for the "what if" analyses, so a "power table" that only gives the power probabilities for specified levels of η^2 is more useful. Figure 5.3 gives the power probabilities for selected values of η^2 and n when PC = 0, PA = 1, and α = .05. We can use this table to do the same "what if" analyses that we did earlier, as well as many others. For our example problem we simply use the row for n = 20: for η^2 = .05, power = .16; and for η^2 = .3, power = .79. (The small differences from the

	True PRE, η2					
PRE	*0*	*.05*	*.1*	*.3*	*.5*	*.75*
.00	.00	.00	.00	.00	.00	.00
.05	.67	.47	.31	.03	.00	.00
.10	.84	.65	.49	.08	.00	.00
.15	.92	.78	.63	.16	.01	.00
.20	.96	.86	.74	.25	.02	.00
.25	.98	.92	.83	.37	.04	.00
.30	.99	.95	.89	.49	.08	.00
.35	1.00	.97	.94	.61	.14	.00
.40	1.00	.99	.96	.72	.23	.00
.45	1.00	.99	.98	.81	.35	.00
.50	1.00	1.00	.99	.88	.48	.00
.55	1.00	1.00	1.00	.93	.62	.01
.60	1.00	1.00	1.00	.97	.75	.04
.65	1.00	1.00	1.00	.99	.86	.11
.70	1.00	1.00	1.00	1.00	.94	.25
.75	1.00	1.00	1.00	1.00	.98	.47
.80	1.00	1.00	1.00	1.00	.99	.73
.85	1.00	1.00	1.00	1.00	1.00	.92
.90	1.00	1.00	1.00	1.00	1.00	.99
.95	1.00	1.00	1.00	1.00	1.00	1.00

Figure 5.2 Cumulative sampling distributions of PRE for various η^2 when PA – PC = 1 and n – PA = 19

			Prob(PRE > critical value)							
	Critical values		*True PRE, η2*							
n	F	PRE	0	.01	.03	.05	.075	.1	.2	.3
2	161.45	.994	.05	.05	.05	.05	.05	.06	.06	.07
3	18.51	.903	.05	.05	.05	.06	.06	.07	.08	.11
4	10.13	.771	.05	.05	.06	.06	.07	.08	.11	.15
5	7.71	.658	.05	.05	.06	.07	.08	.09	.14	.21
6	6.61	.569	.05	.05	.06	.07	.09	.10	.17	.26
7	5.99	.499	.05	.06	.07	.08	.10	.12	.20	.31
8	5.59	.444	.05	.06	.07	.09	.11	.13	.23	.36
9	5.32	.399	.05	.06	.08	.09	.12	.14	.26	.41
10	5.12	.362	.05	.06	.08	.10	.13	.16	.29	.46
11	4.96	.332	.05	.06	.08	.11	.14	.17	.32	.50
12	4.84	.306	.05	.06	.09	.11	.15	.18	.35	.54
13	4.75	.283	.05	.06	.09	.12	.16	.20	.38	.58
14	4.67	.264	.05	.06	.09	.13	.17	.21	.41	.62
15	4.60	.247	.05	.07	.10	.13	.18	.23	.44	.66
16	4.54	.232	.05	.07	.10	.14	.19	.24	.46	.69
17	4.49	.219	.05	.07	.10	.14	.20	.25	.49	.72
18	4.45	.208	.05	.07	.11	.15	.21	.27	.52	.74
19	4.41	.197	.05	.07	.11	.16	.22	.28	.54	.77

Figure 5.3 Power table for η^2 = .05 when PC = 0 and PA = 1

			Prob(PRE > critical value)							
	Critical values		*True PRE, $\eta 2$*							
20	4.38	.187	.05	.07	.12	.16	.23	.29	.56	.79
22	4.32	.171	.05	.07	.12	.18	.25	.32	.61	.83
24	4.28	.157	.05	.08	.13	.19	.27	.35	.65	.87
26	4.24	.145	.05	.08	.14	.20	.29	.37	.69	.89
28	4.21	.135	.05	.08	.15	.22	.31	.40	.72	.92
30	4.18	.126	.05	.08	.15	.23	.33	.42	.75	.93
35	4.13	.108	.05	.09	.17	.26	.37	.48	.82	.96
40	4.09	.095	.05	.10	.19	.29	.42	.54	.87	.98
45	4.06	.085	.05	.10	.21	.32	.46	.59	.91	.99
50	4.04	.076	.05	.11	.23	.36	.51	.64	.93	**
55	4.02	.069	.05	.11	.25	.39	.55	.68	.95	**
60	4.00	.064	.05	.12	.27	.42	.58	.72	.97	**
80	3.96	.048	.05	.14	.34	.53	.71	.84	.99	**
100	3.94	.038	.05	.17	.41	.62	.81	.91	**	**
150	3.90	.026	.05	.23	.57	.80	.93	.98	**	**
200	3.89	.019	.05	.29	.70	.90	.98	**	**	**
500	3.86	.008	.05	.61	.98	**	**	**	**	**

** Power > .995

Figure 5.3 (Continued)

aforementioned power calculations are due to using the more precise critical value of .187 for PRE instead of .20.)

The power table allows us to ask another kind of "what if" question: "What would the power be if the sample size were increased?" For example, how much would the power increase if the lottery bids were evaluated with 30 bidders instead of 20? With this larger sample, if $\eta^2 = .05$, then power = .23, which is better than .16 for 20 bidders, but still not very good. If $\eta^2 = .3$, then power = .93, which is much higher than .79 for 20 bidders and gives an excellent chance of rejecting Model C in favor of Model A. Note that for $n = 50$ we are virtually certain of rejecting Model C whenever the true PRE η^2 is equal to or greater than .3.

Many researchers fail to ask "what if" power questions before they collect their data. The consequence is often that the study that has virtually no chance of rejecting Model C even if the idea that motivated the study is correct. With power tables such as Figure 5.3 (and preferably, software programs, such as R, G*Power, and SAS, which have accessible procedures for calculating power), asking "what if" power questions is computationally simple, and we strongly encourage taking the time for these analyses to ensure that the study has at least some chance of rejecting Model C. But power analysis does require answers to a few key questions.

One natural question is how high should statistical power be? Cohen (1977) suggested that power should be at least .8. However, the ultimate decision depends on how willing the researcher is to risk not finding a significant result even when the ideas motivating the study are correct. For example, the European Organization for Nuclear Research, known as CERN, spent 10 years and roughly $4.5 billion to construct the Large Hadron Collider to test for evidence of the Higgs boson (and other elusive particles). CERN had reason to require power higher than .80. Even a 20% chance of missing the Higgs boson would seem far too high in light of the massive investment. By the same logic, a researcher might tolerate power somewhat lower than .80 for a low-cost exploratory pilot study.

To answer "what if" power questions, we also need to know what values of true PRE or η^2 are appropriate for those "what if" questions. This question is surprisingly tricky. There are multiple strategies that researchers use to obtain a value for η^2, including using Cohen's (1977) suggested values for "small," "medium," and "large" effects; obtaining values of PRE from similar existing research; and computing expectations for PRE from guesses about the parameter values. Some of these strategies may be more reasonable than others. We consider each next. We also offer a method of last resort in cases where it was not possible to conduct a power analysis prior to data collection.

The first commonly used method for specifying η^2 is to use the values defined by Cohen (1977) as "small," "medium," and "large." We do not recommend this approach (see Correll et al., 2020). The suggested values are arbitrary, based only on Cohen's intuitions. They are also inconsistent, because Cohen defined the sizes differently for simple regression, multiple regression, *t*-tests, and analysis of variance. For example, a "large" effect for a *t*-test is defined as $\eta^2 = .14$; a "large" effect for a regression is defined as $\eta^2 = .25$. We will see in the remainder of this book that these tests are often equivalent, so defining "small," "medium," and "large" differently for each test is potentially very problematic. It can seriously compromise power analysis. In truth, Cohen did not recommend this approach either. He offered it as a last resort, and he repeatedly urged researchers to choose a better method. We will therefore consider some other options.

One potential alternative is to use the results of existing studies that are similar to the proposed work. For example, based on a literature review, we might conclude that important effects (such as the effect of the lottery bids) produce PREs greater than or equal to .1. Accordingly, we could use the $\eta^2 = .1$ column of the power table. If we wanted to ensure that power > .8, then going down the column we find that the first power > .8 requires a sample size between 60 and 80, probably about 73, which is a much greater number of participants than we included in our test of whether people were willing to pay significantly less than the expected value of a lottery ticket.

In using the results of past studies to select an appropriate η^2 for the "what if" power analysis, we must remember that calculated values of PRE are biased because on average they overestimate η^2. The following simple formula can be used to remove the bias from PRE:

$$\text{Unbiased estimate of } \eta^2 = 1 - (1 - \text{PRE})\left[\frac{n - \text{PC}}{n - \text{PA}}\right]$$

For our example in which we calculated PRE = .209, the unbiased estimate of η^2 equals

$$1 - (1 - .209)\left[\frac{20 - 0}{20 - 1}\right] = 1 - .791\left[\frac{20}{19}\right] = .167$$

Thus, although the value of PRE calculated from the data is .209, for planning further research our best unbiased guess for the true value of η^2 is only .167. The correction for bias has more of an effect for small values of n – PA than for large values. In essence, the adjustment corrects for the ability of least squares to capitalize on chance for small sample sizes. These unbiased estimates are thus sometimes referred to as "adjusted" values.

It is also important to acknowledge that estimates of η^2 derived from published data may be inflated because of publication bias. If journals only publish significant effects (which may have larger PREs on average than PREs that go unpublished), then smaller PREs will be excluded from our calculations when we try to estimate η^2. It is also problematic to use the results of a single study to estimate power because the value obtained in that particular study may be a poor estimate of the true effect η^2. It is helpful to have multiple estimates.

One solution is to base our guesses about η^2 on meta-analyses. A meta-analysis combines the results of many studies within a specific interest area (like risk tolerance) or within a broad area (like cognitive neuroscience). Though still susceptible to publication bias, one virtue of this approach is that a meta-analysis typically combines PREs of individual studies in a principled manner to develop a general estimate of η^2, which can serve as a reasonable best guess in power analyses. A second important advantage is that a meta-analysis provides some sense of the *variability* of potential values of η^2. Afterall, while we may view the meta-analytic average effect size as our best guess for η^2, that guess could be wrong. What if the true value were somewhat higher than we expected? More problematically for power, what if the true value were lower? Based on a meta-analysis, a researcher can estimate power not just for the most likely value of η^2 but also for a few relatively low values and a few relatively high values. Estimating power for a series of plausible values of η^2 yields a power curve, which shows how power changes as η^2 increases. It conveys a sense of the likelihood of finding a significant effect if the true effect is a little smaller or a little bigger than expected.

A third approach to finding appropriate values of η^2 to use in "what if" power analyses involves guesses about the parameter values and variances. Even with existing data, it is often difficult to generate reasonable expectations about the parameter values and variances. Thus, this third approach is generally less useful than the second. We present this approach partly to be complete, but also for several conceptual reasons. First, the derivation of this approach provides useful insights about the meaning of PRE and η^2. Second, it can be very neat when research has practical implications. Third, it requires describing in detail the data that one expects to obtain, and such an exercise can often be useful for identifying flawed research designs.

We begin with our familiar definition of PRE:

$$\text{PRE} = \frac{\text{SSE}(\text{C}) - \text{SSE}(\text{A})}{\text{SSE}(\text{C})} = \frac{\text{SSR}}{\text{SSE}(\text{C})}$$

We have noted before that SSE(C) = SSE(A) + SSR (i.e., the error for the compact model includes all the error of the augmented model plus the error that was reduced by the addition of the extra parameters in the augmented model). Hence, substituting for SSE(C) yields

$$\text{PRE} = \frac{\text{SSR}}{\text{SSE}(\text{A}) + \text{SSR}} = \frac{1}{\text{SSE}(\text{A})/\text{SSR} + 1}$$

To obtain a definition of the true proportional reduction in error η^2, we simply estimate SSE(A) and SSR, using not the data but the parameter values of B_0 and β_0 that are of interest.

For example, we previously noted that

$$\text{SSR} = \sum_{i=1}^{n} \left(\hat{Y}_{iC} - \hat{Y}_{iA}\right)^2$$

If we thought that the effect of the lottery bids would be to decrease the mean bid from 50 to 40, then $\hat{Y}_{iC} = B_0 = 50$ represents the null hypothesis and $\hat{Y}_{iA} = \beta_0 = 40$ represents an alternative hypothesis that we want to test. For that situation we would *expect*

$$\text{SSR} = \sum_{i=1}^{20} (50 - 40)^2 = \sum_{i=1}^{20} 10^2 = \sum_{i=1}^{20} 100 = 2000$$

In other words, the SSR that we expect is simply 100 added up 20 times (once for each bidder). We saw in Chapter 2 that SSE/(n – PA) was an estimate of the variance σ^2. If we use our expected value of $\beta_0 = 40$ to calculate SSE(A), then we are not using data to estimate any parameters, so PA = 0. Hence, SSE(A)/n = σ^2, so the value of SSE(A) that we *expect* to obtain is

$$\text{SSE(A)} = n\sigma^2$$

Thus, if we have a reasonable guess or expectation for the variance, then we can easily calculate the value of SSE(A) that we would expect. Having good intuitions about what variance to expect is usually quite difficult. Good guesses for the variance often depend on existing data or previous experience with the particular measure for Y. For our example, suppose that past data for the lottery lead us to expect that σ^2 is about 400; then, we would expect

$$\text{SSE(A)} = n\sigma^2 = (20)400 = 8000$$

We now can return to our formula for PRE to calculate the value that we expect for the true proportional reduction in error η^2 (given our guesses for B_0, β_0, and σ^2).

$$\text{Expected}\,\eta^2 = \frac{1}{\text{SSE}(\text{A})/\text{SSR}+1} = \frac{1}{8000/2000+1} = \frac{1}{4+1} = .2$$

In other words, $\eta^2 = .2$ corresponds to our guesses about B_0, β_0, and σ^2. We now can use the power tables to find the power we would have for comparing Model C, which predicts that the mean bid will be 50, against Model A, which predicts that the mean bid will be 40, when the variance is about 400. Using Figure 5.3, we find that for 20 observations the probability of rejecting Model C (i.e., deciding that the lottery bids were less than the expected value of a ticket) is only about .56 even if we think that it will on average be \$10 less than the expected value of \$50. This means that the researcher has only a little more than a 50/50 chance of *deciding* that lottery bids are lower than the expected value of a ticket even when they really are lower by \$10. Using the power table, we can see that testing our hypothesis with twice the number of persons would increase the power substantially to .87. In this case, it would seem advisable to test our hypothesis with a larger sample size rather than with a sample size that offered little hope of finding the effect. We can similarly calculate the power for other "what if" values of B_0, β_0, and σ^2 that we might want to consider.

One interesting application of this approach is to consider the smallest effect that would be of practical interest. For example, imagine a company that is considering modifying an existing product. The cost of that modification is \$1 per unit, and the company wants to know if the benefit of the change is worth the cost. Will customers pay more for the improved product, and in particular, will they pay *enough* more to justify the \$1 cost increase? To answer this question, the company might plan a study in which they ask customers how much they would pay for the new product. They might plan a simple model comparison in which Model A predicts the mean of the data β_0 and Model C predicts the cost of the old product B_0. When they perform a power analysis, they might specify a value of \$1.00 for $\hat{Y}_{iA} - \hat{Y}_{iC}$. If the magnitude of the difference is not at least as great as the cost, the company has no need to detect the effect even if it exists. Using the aforementioned approach, with an estimate of σ^2, they could compute the *smallest effect size of interest* (Lakens, 2014) and evaluate their power to detect that effect.

Finally, we consider an option of last resort. In the absence of a power analysis conducted prior to data collection, what can a researcher do? It does not make sense to conduct a post-hoc power analysis, estimating power based on the study's actual sample size and the value of PRE obtained in our model comparison. However, it is possible to conduct a *sensitivity analysis.* A sensitivity analysis, as the name implies, estimates the sensitivity of a given model comparison by determining the smallest effect size for which the existing test would have had adequate power. In our lottery ticket example, if we define adequate power as .80, we can use Figure 5.3 to find the smallest value of η^2 for which power exceeds .80. In this example, we know PC = 0, PA = 1, and $n = 20$. Looking at the row for which $n = 20$, we find that the rightmost column ($\eta^2 = .3$) shows almost-adequate power of .79. With a more precise table (or with software like R, G*Power, or SAS), a sensitivity analysis would show that this test had adequate power only for $\eta^2 >= .32$, which is a fairly large value.

Improving Power

In general, we want to increase the power of statistical inference. There are four basic strategies for improving power: increase the size of the effect, reduce error, increase α, and/or increase the number of observations. We consider each in turn.

Increasing the Size of the Effect

Any discussion of power requires that the effect of interest does, in fact, exist. Careful development of a research hypothesis based on existing theory, evidence, and logic improve the chances that the effect exists, and they therefore go a long way toward addressing concerns about power. Further, within the realm of effects that exist, it is easier to detect effects that are greater in magnitude. Imagine that researchers want to determine whether a vaccine induces a fever. They might inject a very small dose of the vaccine, which would have a tiny effect. Alternatively, they might inject a moderate dose, which would have a larger effect. Holding other factors equal, their power to detect the (real) effect of the vaccine will be greater if they use a larger dose (though there may be ethical concerns with increasing the dose too much).

Reducing Error

One way to reduce error is to control as many of the possible random perturbations as possible. In our lottery example, one might reduce error and obtain more power by providing clear instructions to participants, making sure participants were well rested, eliminating distractions in the bidding environment, and using a more reliable bidding procedure. In other words, error is reduced by obtaining data of better quality. In the equation

DATA = MODEL + ERROR

the model will account for a higher proportion of the data if the data are of higher quality and hence have less error. Less error allows us to obtain a more powerful look at our data. Although reducing error by such means may be the most effective method for improving power, the techniques for doing so are usually domain-specific and outside the scope of this book.

Another way to reduce error is to improve the quality of the model. Again, in the equation

DATA = MODEL + ERROR

error will be smaller for data of fixed quality if the model can be improved to account for more of the data. How to use models more complex than the simple models we have been considering in these beginning chapters is the subject of the remainder of the book, so we cannot give too many details here. The general idea is to build models that make predictions conditional on additional information we have about the observations. In the lottery example we might know, say, which bidders had participated in lotteries before and which ones, if any, had ever won a lottery. If having participated in a lottery before makes a difference in the amount an individual is willing to pay for a lottery ticket, then we can make different predictions conditional on whether (or perhaps how often) the bidder had participated in lotteries previously. By doing so we will have, in essence, removed what was formerly a random perturbation—previous participation in lotteries—from the error and included it in the model. Again, the reduced error will give us a more powerful look at our data. In later chapters we explicitly consider the addition of parameters to the model for the purpose of improving power.

Increasing α

A different way to improve power is to increase α, the probability of a Type I error. The probabilities of Type I and II errors are linked in that if we choose a critical value that increases (decreases) α then we simultaneously and unavoidably decrease (increase) the probability of a Type II error. For our lottery data with $n = 20$, Figure 5.4 shows power as a function of η^2 and α. As α increases from .001 to .25 the critical values for F and PRE decrease. It becomes easier for the values of F and PRE calculated from the data to beat these critical values, so the power increases as α increases. For example, if we do a "what if" power analysis with $\eta^2 = .2$, then the power at $\alpha = .05$ is .56, but if we increase α to .1, then power increases to .70.

Editors of scientific journals are wary of Type I errors and will seldom accept the use of $\alpha > .05$ for statistical inference. However, there are many practical data analysis problems when a higher α is justified to increase power. Characteristics of such data analyses are (a) that increasing power in any other way is infeasible, (b) that rejection of Model C if it is indeed false would have important practical consequences, and (c) that the costs associated with a Type I error are not great. Consider, for example, the statistical decision problem faced by a researcher testing the effectiveness of an innovative new curriculum. It may be difficult for the researcher to control any other sources of error to increase power. It would certainly be important to identify a curriculum that could improve student learning. The consequence of a Type I error would probably be further trial use of the curriculum in several classrooms the following year, which may not involve a significant cost. Thus, the researcher might well adopt a higher α to choose the critical values for PRE and F.

			Prob(PRE > critical value)							
	Critical values		*True PRE, η2*							
α	*F*	*PRE*	*0*	*.01*	*.03*	*.05*	*.075*	*.1*	*.2*	*.3*
.001	15.08	.443	.00	.00	.00	.01	.01	.02	.09	.22
.005	10.07	.346	.01	.01	.02	.03	.05	.07	.21	.43
.01	8.18	.301	.01	.02	.03	.05	.08	.11	.30	.54
.025	5.92	.238	.03	.04	.07	.10	.15	.20	.44	.69
.05	4.38	.187	.05	.07	.12	.16	.23	.29	.56	.79
.1	2.99	.136	.10	.13	.20	.26	.34	.42	.70	.88
.25	1.41	.069	.25	.29	.38	.46	.55	.63	.85	.96

Figure 5.4 Power for PA = 1, PC = 0, and $n = 20$ for various levels of α

Conversely, reducing α also reduces power. In the lottery example, lowering α to .01 would increase our protection against the possibility of a Type I error but would reduce power to less than .30, a level for which it would not likely be worth conducting the study. If the costs of a Type I error are very high—for example, if reliance on the research findings involved extensive teacher retraining in the case of an innovative new curriculum—then the use of restrictive values of α may be appropriate. However, in some cases, such as when the sample size is very small, as it is for our lottery study, reducing α would reduce the power so much that it would no longer be worthwhile to do the study—the chances of seeing anything are too low with such a low-powered "microscope."

Increasing n

Probably the most common technique for increasing the power of statistical tests is to increase the number of observations. Figure 5.5 is a graphical display of the 95% ($\alpha = .05$) and 99% ($\alpha = .01$) critical values of Figure 4.9 as a function of the number of observations. The value of PRE required to reject Model C drops dramatically as the number of observations increases. For example, had there been 51 participants instead of 20 in our lottery study, a PRE of only .075 would have been required ($\alpha = .05$) to reject the Model C that assumed $B_0 = 50$ instead of the PRE of .187 required in our example. The drop in the critical values for F and PRE needed to reject Model C corresponds to an increase in power, as we have noted several times. For example, for $\eta^2 = .2$ as a "what if" value of the true proportional reduction in error, the power for 20 observations is .56 (see Figure 5.3), but with 51 observations the power increases to .93.

Though higher power is generally advantageous, in some cases, power is so high that statistically significant results can be misleading. By turning up the power on our metaphorical statistical microscope to extraordinary levels, we might detect flaws in Model C that are statistically reliable but are trivial substantively. For example, with 1000 observations any PRE greater than .004 would lead us to reject Model C in favor of Model A. However, the 0.4% reduction in error means that Model A may be a trivial improvement over Model C. For this reason, one should always report not just the statistical inference about whether Model C is or is not rejected but also

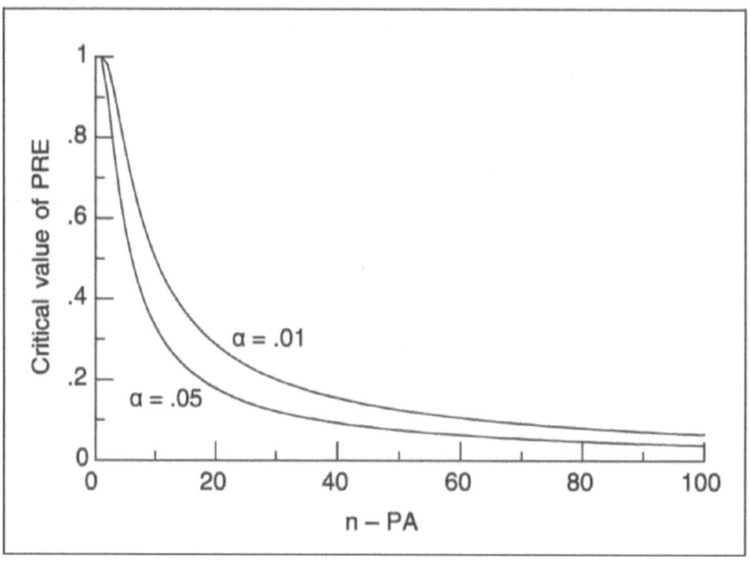

Figure 5.5 Critical values of PRE as a function of the number of observations

the obtained values for F and PRE (which is called an effect size) as well as a confidence interval so that the reader can evaluate the magnitude by which Model A improves on Model C. All else being equal, more statistical power is always better; we just need to be careful in interpreting the results as substantively or practically important when power is very high. At the same time, it is important not to dismiss a small but reliable PRE *just* because it is small. Rejecting Model C in favor of Model A may be important theoretically even if Model A provides only a slight improvement. Whether a given PRE is substantively interesting depends on theory and prior research experience in the particular domain.

Confidence Intervals

Confidence intervals provide an alternative way to consider statistical inference. Although, as we shall see later, confidence intervals are exactly equivalent to statistical inference as previously described, they reorganize the information in a way that can give useful insights about the data and our model.

A *confidence interval* simply consists of all possible values of a parameter that, when used as a hypothesis for Model C, would not cause us to reject Model C. For example, the mean for the 20 lottery bids is 42.5 and estimates the parameter β_0 in Model A. We have already determined that 50 is not in the confidence interval for β_0 because, when we used 50 as the prediction for Model C, we obtained an unlikely PRE value that would occur less than 5% of the time. Any value greater than 50 would produce an even larger PRE, so none of those values is included in the confidence interval. Conceptually, we could find the boundary of the confidence interval by trying increasingly higher values (i.e., higher than b_0) for B_0 in Model C until we found a value for B_0 that produced a PRE that would not cause us to reject Model C. For example, if we tried $B_0 = 45$, the respective sums of squared errors for Models C and A would be 4392 and 4267, yielding PRE = .028, which is below the critical value for $\alpha = .05$; hence 45 is in the confidence interval for β_0, and the boundary must be somewhere between 45 and 50. We would also need to search for the lower boundary below the estimated value of 42.5. We can avoid this iterative search because it can be shown that the boundaries are given by

$$b_0 \pm \sqrt{\frac{F_{crit;1,n-1;\alpha}\,\text{MSE}}{n}}$$

For the simple model of the lottery bids, MSE = s^2 = SSE/$(n - 1)$ = 4267/19 = 224.58. $F_{\text{crit};1;n-1;\alpha}$ is the critical value at level α for F with 1 degree of freedom for the numerator (i.e., the difference between the number of parameters in Models C and A) and $n - 1$ degrees of freedom for the denominator (i.e., the number of observations minus the number of parameters in Model A). For α = .05, the critical value of $F_{1,19;.05} = 4.38$. For these data, b_0 = the mean = 42.5, so the boundaries of the confidence interval are given by

$$42.5 \pm \sqrt{\frac{(4.38)224.58}{20}} \quad \text{or} \quad 42.5 \pm 7.01$$

Thus, the lower boundary is 42.5–7.01 = 35.49 and the upper boundary is 42.5 + 7.01 = 49.51. We are therefore 95% $(1 - \alpha)$ confident that the true value for β_0 is in the interval [35.49, 49.51]. Any B_0 not in this interval that we might try for Model C would, with the present data, produce values of PRE and F that would lead us to reject Model C. Any B_0 in this interval would produce values of PRE and F below the critical values, so we would not reject Model C.

When estimating statistical power, we described how to do an a priori power analysis. Those "what if" power analyses are necessarily inexact. Confidence intervals are useful for describing post hoc the actual statistical power achieved in terms of the precision of the parameter estimates. Wide confidence intervals represent low statistical precision, and narrow intervals represent high statistical precision. In this case, they suggest we have a more precise estimate of the true mean.

Effect Size

PRE is a standardized measure of the magnitude of the difference between Model A and Model C. It is a standardized effect size. We use PRE throughout this book because it can be applied to a wide variety of model comparisons, from the very simple to the very complex. We also use it because it has a fairly intuitive meaning: It is the proportional reduction in error. We view PRE as a good all-purpose effect size. PRE can be computed from DATA and used to perform an inferential test, using the machinery described in Chapter 4. Adjusted PRE also serves as an estimate of the *true* proportional reduction in error η^2, allowing us to perform power analysis using the procedures described in this chapter. Finally, PRE can be reported in a journal article to communicate the magnitude of an effect.

However, it is important to acknowledge other measures of standardized effect size that are used for power analysis and reported in journal articles. Certain measures of effect size are conventionally associated with particular tests. For example, a *t*-test is conventionally associated with a measure called Cohen's *d*. Analysis of variance is conventionally associated with f^2. In Chapter 4, we introduced the simplest model comparison and showed its equivalence to a one-sample *t*-test. We can therefore discuss the interpretation of Cohen's *d* in the context of that test, for which Cohen's *d* is computed as

$$d = \frac{\beta_0 - B_0}{s}$$

It thus reflects the magnitude of the difference between the mean and the value predicted by Model C in standard deviation units. A value of 0.5 would suggest that the difference between the mean and B_0 is half of one standard deviation. As we introduce other more complex model comparisons, we will discuss other applicable effect size measures. These measures simply reflect different ways of conceptualizing the size of the effect. Accordingly, given a value of PRE, one can compute Cohen's *d* (and vice versa). For reference, we provide translations between PRE and other measures in Figure 5.6.

We present these alternative effect size measures because we want readers to be familiar with and understand effect sizes that are commonly reported in social science journal articles. We prefer η^2 because, as the proportional reduction in error, it is a relatively intuitive measure, and it applies easily to any statistical test that falls into the general linear model. Other effect sizes (e.g., Cohen's *d*) are generally used only for particular analyses. For example, Cohen's *d* is not easy to conceptualize in models with continuous predictors. We thus view η^2 (or PRE) as a general-purpose measure that complements the unifying approach of this book.

An Example

In this section we illustrate the techniques of this chapter using the internet data that were presented in Figure 1.1. Recall that in Chapter 4 we analyzed the data to test marketing researchers'

	Cohen's effect size to η^2	*η^2 to Cohen's effect size*
t-test	$\eta^2 = \frac{d^2}{d^2+4}$	$\eta^2 = \frac{2\eta^2}{\sqrt{1-\eta^2}}$
Correlation/regression	$\eta^2 = r^2$	$r = \sqrt{\eta^2}$
ANOVA/ANCOVA	$\eta^2 = \frac{f^2}{1+f^2}$	$f = \sqrt{\frac{\eta^2}{1-\eta^2}}$
Multiple regression	$\eta^2 = \frac{f^2}{1+f^2}$	$f^2 = \frac{\eta^2}{1-\eta^2}$

Figure 5.6 Conversions to/from η^2 and Cohen's effect size measures

projection that 90% of individuals across the 45 countries would use the internet in the year 2021. To do so, we compared the following two models:

MODEL A: $Y_i = \beta_0 + \varepsilon_i$
MODEL C: $Y_i = 90 + \varepsilon_i$

We calculated $\hat{Y} = 83.982$ (see Chapter 2) so the estimated model comparison and the sums of squared errors were

MODEL A: $\hat{Y}_i = 83.982$ SSE(A) = 5392.86
MODEL C: $\hat{Y}_i = 90.00$ SSE(C) = 7022.595

From the aforementioned information, we calculated MSE, which in this case is equal to s^2, as 122.565 and determined our critical value, $F_{1,\,44,\,\alpha=.05} = 4.06$.

We can use this information to calculate the 95% confidence interval for β_0, the true average percentage of individuals who used the internet across the 45 countries. Substituting the appropriate values into

$$b_0 \pm \sqrt{\frac{F_{crit;1,n-1;\alpha}\ \text{MSE}}{n}}$$

yields

$$83.982 \pm \sqrt{\frac{4.06(122.565)}{45}} = 83.982 \pm 3.325$$

which gives an interval of [80.657, 87.307]. Using this interval, we can easily ask other questions. For example, had the marketing researchers projected that $B_0 = 85$, we would not conclude that the actual percentage of individuals who used the internet was significantly less than the projection of 85%, because 85 is included in the 95% confidence interval.

We can also calculate an unbiased estimate of η^2 from the formula provided earlier in this chapter:

$$\text{Unbiased estimate of } \eta^2 = 1 - (1 - \text{PRE})\left[\frac{n - \text{PC}}{n - \text{PA}}\right]$$

Recall that PRE for this example was calculated as

$$\text{PRE} = \frac{\text{SSE}(\text{C}) - \text{SSE}(\text{A})}{\text{SSE}(\text{C})} = \frac{7022.595 - 5392.86}{7022.595} = \frac{1629.735}{7022.595} = .232$$

The number of parameters estimated in Model C (PC) = 0, the number of parameters estimated in Model A (PA) =1, and $n = 45$. Substituting these values in our formula yields

$$\text{Unbiased estimate of } \eta^2 = 1 - (1 - .232)\left[\frac{45-0}{45-1}\right] = 1 - .768(1.023) = .214$$

If a researcher wanted to replicate this finding in a new study, she should use this unbiased estimate of η^2, that is, .214. We used a statistical program to calculate power, and our estimate of power for this test is greater than .999, assuming $\alpha = .05$, $\eta^2 = .214$, PA-PC = 1, and n – PA of 44. In other words, we would have a nearly 100% chance of rejecting Model C in favor of Model A, assuming that Model C is indeed incorrect in the population.

Summary

In this chapter, we have provided a broad overview of how to estimate power, including the pieces of information that inform power analyses (i.e., the study design, α, power, effect size, and sample size) and ways that researchers can increase power. Researchers generally select *α*, the probability of a Type I error, and they generally try to reduce the chances of a Type II error or equivalently increase power. Power can be increased by collecting better data, building better models, and increasing the number of observations. We also paid considerable attention to how one might select an appropriate effect size for a study, and how researchers might do a "what if" analysis to estimate power under various conditions. All of our power analyses were conducted in the context of a simple Model A/C comparison in which we compare a mean to an a priori value. However, the procedures that we outlined here generalize to more complex analyses, which we consider in subsequent chapters.

Section A Recap

Statistical Machinery

In Chapters 1 through 5, we introduced the basic statistical processes that are the foundation of this book. This process involves specifying two models, each of the form DATA = MODEL + ERROR. The more complex model, called Model A, contains every parameter in the more compact model, called Model C, along with at least one additional parameter. For each model, we determine the aggregate ERROR, measured as the sum of squared errors. The amount of ERROR reduced in Model A relative to Model C provides an index of the improvement conferred by the more complex model. If the magnitude of that improvement is not surprising, we retain Model C as an expression of our understanding about the nature of the DATA. However, if Model A reduces ERROR by a surprising degree, we reject Model C and adopt Model A as our new understanding of DATA.

The model comparison approach offers a clear conceptualization of ordinary least squares data analysis as it is typically performed in the behavioral sciences, and researchers will benefit from a clear understanding of this method. However, we want to be clear that ordinary least squares estimation is only one of many approaches. Scholars argue about which approaches are better for different situations, and researchers should conscientiously choose the analysis that makes most sense given the research question and data at hand. Next, we briefly describe a few alternatives researchers might consider. In their own way, each of these alternatives involves a different estimate of something like aggregate ERROR. Accordingly, each can be naturally conceptualized in terms of a model comparison.

Minimizing the Sum of Absolute Error

As discussed in Chapter 3, researchers sometimes use the sum of absolute error (SAE), rather than SSE, to aggregate errors (e_i) across observations in the dataset. This approach yields statistics related to the median, rather than the mean. As we demonstrated, using the sum of squared error as the measure of aggregate error is advantageous when individual errors are normally, independently, and identically distributed. When those assumptions about the errors do not hold, when the data cannot be transformed, or when a researcher—for some other reason—wants to adopt an approach that penalizes large individual errors less severely, it is possible to use SAE in place of SSE. (Recall that, in SSE, each error is squared, so large errors make an especially large contribution to the sum.) Model comparison proceeds in much the same way: We compute SAE for Models A and C, compute PRE, and reach our statistical decision.

Weighted Sum of Squared Errors (WSSE)

Measures of aggregate error, like SSE and SAE, give equal weight to each observation. For many reasons, we may want to give more weight to some observations and less weight to others when

DOI: 10.4324/9781003438335-8

calculating aggregate error. For example, we may have reason to believe that certain observations are questionable because they were collected with less precision or less reliability than the other observations. Or we may not want to count an error of 10 as being the same when $\hat{Y}_i = 1000$ as when $\hat{Y}_i = 5$. In the former instance, the error of 10 amounts to only a 1% error, while in the latter instance it is an error of 200%. Or we might be suspicious of a couple of extreme observations just because they are "outliers" with respect to other observations. Whatever our reasons, it is easy to incorporate a weight, w_i, for each observation into the formal definition:

$$\text{ERROR} = \sum_{i=1}^{n} w_i e_i^2 = \sum_{i=1}^{n} w_i \left(Y_i - \hat{Y}_i\right)^2 = \sum_{i=1}^{n} w_i \left(Y_i - b_0\right)^2$$

The weights w_i might be assigned a priori based on judgments of data quality or some formal index of each observation's reliability. One possibility is to weight all observations equally, in which case $w_i = 1$ for all i. This weighted sum of squared errors becomes simply the sum of squared errors.

Statisticians have created some very clever ways of defining weights to solve a variety of complicated problems. We will encounter examples of those weights later in the context of specific problems. Generally, the use of weights gives us a great deal of flexibility in defining the aggregate measure of error. Again, model comparison can easily incorporate this alternative measure of ERROR, allowing us to employ the same basic machinery to compare Models A and C.

Maximum Likelihood Estimation

Maximum likelihood estimation offers another extremely flexible strategy for estimating how well a model describes the data. Like SSE, maximum likelihood provides a simple measure of model fit, allowing us to compare Models A and C. However, as we will see later in the book, maximum likelihood estimates can be used in a variety of situations where SSE will not suffice.

Earlier we saw that, in a simple one-parameter model, if we minimize the sum of squared errors in estimating β_0, then our best estimate equals the sample mean. Thus, the mean is the best *least-squares* estimate in the sense that it minimizes the sum of squares. It turns out that, for this simple model, the mean is also what is called the *maximum likelihood* estimate. That is, the mean maximizes the joint likelihood of all the data in the sample, given an underlying normal distribution with some constant variance of errors, σ^2.

To unpack this phrase, imagine that we assume that the data in a sample come from an underlying normal distribution. Given some true value of μ and σ for that underlying distribution, we can calculate the likelihood of each data point in that distribution. This likelihood is given by the density function for the normal distribution given earlier:

$$f(x) = \frac{1}{\sigma\sqrt{2\pi}} e^{-(x-\mu)^2/2\sigma^2}$$

Assuming all the data points in a sample come from the same distribution and are independent of each other, one can calculate the *joint* likelihood of all the data points by taking the product of all of the likelihoods of each of the individual data points. A *maximum likelihood* estimate is that estimate of β_0 or μ that maximizes that joint likelihood, given some value of σ. And it turns out that, if the data are normally distributed, that best estimate is $\bar{Y}$, the sample mean.

To illustrate this, consider the sample of 20 data points that Mother Nature sampled where the true value of β_0 was 50 and where each data point had an error ticket tacked onto it. Those 20 data points are reproduced here.

41	78	33	72	85
38	27	51	60	61
56	26	31	24	19
46	58	70	36	48

Let us assume that the standard deviation of the error tickets in Mother Nature's bag is 15. Then it becomes possible to calculate the likelihood of each individual data point under the normal distribution with different estimates of β_0 or μ.

For instance, if we started by estimating μ as 55, then the likelihood of each of the individual data points (given the normal distribution, μ estimated as 55, and $\sigma = 15$) is as follows:

.0172	.0082	.0091	.0140	.0036
.0140	.0047	.0257	.0252	.0246
.0265	.0041	.0074	.0031	.0015
.0222	.0261	.0161	.0119	.0239

Each of these likelihoods is calculated using the normal distribution function to solve for $f(x)$. And the joint likelihood of all 20 data values is 6.695×10^{-40} (the product of all 20 individual likelihoods), an incredibly small number!

We could then try other estimates of μ and recalculate the likelihoods of each of the individual data points and their joint likelihood. What we would like to do is find the estimate of β_0 or μ that maximizes the joint likelihood. And it turns out that the best estimate is the sample mean, which for this sample of 20 data points is 48. The joint likelihood of the 20 data points given the estimate of 48 is 5.909×10^{-39}, which is still a very small number. But the point is that this joint likelihood is larger than what we got when we used the estimate of 55, and in fact it is larger than any other estimate we might use. For the purposes of statistical inference, we can use the joint likelihood to derive a new measure akin to ERROR and use it to compare Models A and C.

In this simple model, the maximum likelihood estimate, since it is $\bar{Y}$, is the same as the least-squares estimate. And in most of the models we consider in this book, this will continue to be the case. That is, in general, the parameter estimate that minimizes SSE in a sample of data will also be the maximum likelihood estimate. However, at later points in the book, when we consider models for data that are dichotomous (Chapter 15) or those for data with multiple sources of error (Chapter 13), then we will need to rely on maximum likelihood estimates rather than least-squares estimates.

Bayesian Analysis

The process of inference we have described requires the researcher to make an inferential leap: If the data are surprising under the assumptions of Model C, then reject Model C in favor of Model A. This leap is a bit dubious because of Type I error. We may reject Model C either because Model C is wrong, or because Model C is true *and a surprising event occurred.* So which is it? This procedure does not answer that question. Further, we are presumably interested in the likelihood that Model A is correct, but we never estimate that likelihood. These concerns are relevant to the distinction between traditional null hypothesis statistical analysis (sometimes called *frequentist* analysis) and Bayesian analysis.

Frequentist analysis takes Model C as an assumption about the truth, and then, based on Model C, it estimates the plausibility of DATA (and rejects Model C if the data are surprising). Bayesian analysis approaches the question differently. It does not rely on rejecting a null hypothesis at all. Based on DATA, it estimates the plausibility of a MODEL. Bayesian analysis has other interesting features and constraints. For example, it requires the analyst to formally specify a preliminary representation of the phenomenon in question, called a *prior*. This representation might be specified based on prior research, but it can also be specified in a manner that reflects an almost complete lack of prior knowledge. The analysis then determines how this prior should be updated in light of the data. In practice, a researcher might specify Model A and Model C and examine the degree to which each model seems plausible in light of DATA. A detailed discussion of Bayesian analysis is beyond the scope of this book, but it is important to note that model comparison still works within this framework. Once we have estimated the plausibility of Model C and the plausibility of Model A, those two estimates can be compared using something called a Bayes factor. A Bayes factor thus indicates whether Model C is more or less plausible than Model A. Though a Bayes factor differs from PRE in important ways, it can thus guide decisions about which model is preferable.

Section B

Increasingly Complex Models

Introduction to Section B

Increasingly Complex Models

In Section A, we introduced a procedure to perform a statistical comparison between a compact Model C and an augmented Model A. The logic behind this model comparison process was developed in the context of two simple models that each make a constant prediction for all observations in the dataset, one off-the-rack estimate, and one tailored estimate. These simple models are useful for conveying the fundamental ideas of model comparison, but researchers often have questions that involve more complex differences and relationships. These questions cannot all be tested with simple models.

The chapters in Section B introduce a series of increasingly complex models that allow researchers to test various kinds of questions. In general, the topics covered in each chapter serve as the foundation for the subsequent chapter. Often, what was Model A in one chapter will become Model C in the next, as we build toward very complex and flexible models in Chapter 11. Each chapter is associated with a particular kind of standard statistical test (e.g., simple regression, *t*-test, ANOVA, etc.) that is typically used to answer a particular question (e.g., "Is there a relationship between two variables?," "Do the group means differ?," etc.). But we will see that the Model As in each chapter can be compared to a variety of Model Cs. Each of these comparisons answers a distinct type of question. For example, although the Model A associated with a simple regression might typically be used to test whether two variables are related, that model can also be used to perform a powerful test of the intercept. Understanding this flexibility is a primary advantage of the model comparison framework developed in Section A.

Chapter 6 introduces a Model A that allows a single continuous predictor variable. This model is involved in tests of simple regression and correlation, which test relationships between two variables. We will see that this Model A can be used to provide a powerful test of the intercept (a potential improvement on the approach described in Chapter 4 and 5).

Chapter 7 introduces a Model A that allows two or more continuous predictor variables. This model involves multiple regression. This chapter introduces the important concept of redundancy among predictors.

Chapter 8 introduces a Model A that allows for interactions between continuous predictors, such that the effect of one predictor variable can vary as a function of another predictor variable. This chapter also addresses models in which the relationship between a predictor and the outcome is nonlinear.

Chapters 9 and 10 recapitulate the concepts presented in Chapters 6, 7, and 8, adapting them to models that use categorical rather than continuous predictors. The tests in these chapters are often referred to as one-way and factorial analysis of variance (ANOVA).

Chapter 11 introduces a Model A that can combine all of the tests described previously mentioned. These models allow for both continuous and categorical predictors, as well as interactions, including nonlinear relationships and interactions between categorical and continuous predictors.

DOI: 10.4324/9781003438335-10

Throughout this section, we will assume that the model's errors satisfy the assumptions introduced in Chapter 3. That is, we will assume that the data are independent, unbiased, and normally distributed with constant variance. In Section C, we will discuss how to analyze data when the errors do not conform to these assumptions.

6 Simple Regression

Estimating Models With a Single Continuous Predictor

We have used the simple single-parameter model to illustrate the use of models, the notion of error, and inference procedures to be used in comparing augmented and compact models. We have focused on this single-parameter model in so much detail because the estimation and inference procedures that we developed within this very simple context generalize to much more complicated models. That is, regardless of the complexity of a model, estimation from here on will be done by minimizing the sum of squared errors, just as we did in the single-parameter case, and inference will be done by comparing augmented and compact models using PRE and F.

However, as we noted, single-parameter models are only infrequently of substantive or theoretical interest. The example from the last chapter, which tested the null hypothesis that people were willing to pay the expected value of $50, is unusual in the behavioral sciences. More frequently such a priori values do not exist, and instead we might ask whether the mean in one group of respondents (e.g., those who were trained in the meaning of expected values) differs from the mean in another group of respondents (e.g., those who received no such training). Or, returning to the data on internet use, although it is possible that we would be interested in testing whether some a priori percentage (e.g., 90%) is a good estimate of mean internet use, it is much more likely that we would be interested in examining the determinants or correlates of internet use rates. In other words, our interest is more likely to focus on attempts to explain the internet use data than on tests of alternative values for the mean rate.

To examine these types of substantive issues, we need to consider models that have more than a single parameter. Initially, we will consider two-parameter models, which take the following form:

$$Y_i = \beta_0 + \beta_1 X_i + \varepsilon_i$$

The exact definition of the terms in this two-parameter model will be detailed further. For the present we simply note that we are making predictions of Y_i conditional upon some other variable X_i, because the model's predictions from this two-parameter model change as X_i changes, assuming that β_1 takes on some value other than zero.

There are two variations on this two-parameter model, each of which is illustrated by one of the two examples we have just discussed. Let us first consider our lottery ticket example. We might wish to test whether people who receive training in the meaning of expected values influences are willing to pay more or less than those who do not. This amounts to asking whether we need different predictions from the model for the two groups of respondents, or whether a single prediction suffices regardless of whether training was received or not. In other words, we want to compare a Model A in which predictions are conditional on whether a given respondent received training with a Model C that makes the same prediction for all respondents. For this comparison, the augmented

DOI: 10.4324/9781003438335-11

model is a two-parameter model, defining X_i *in such a way that* X_i *identifies whether the respondent received training. We might, for instance, define* X_i as follows:

$X_i = -1,$	if a respondent did not receive training
$X_i = +1,$	if a respondent did receive training

If the estimated value of β_1 in the aforementioned two-parameter model is something other than zero, the model makes different predictions for participants in the two groups. For example, if b_0 (the estimated value of β_0) equals 46 and b_1 (the estimated value of β_1) equals 3, then the prediction for the respondents without training is

$$\hat{Y} = b_0 + b_1X_i = 46 + 3(-1) = 46 - 3 = 43$$

and the prediction for the respondents who receive training is

$$\hat{Y} = b_0 + b_1X_i = 46 + 3(1) = 46 + 3 = 49$$

Notice that there are only two possible values for X_i in this example, and hence only two predicted values. Respondents either receive training or not, and our decision about the numerical values used to represent this training is arbitrary. For instance, had we defined X_i differently, giving respondents with training a value of 2 on X_i and those without training a value of 4, the two-parameter model would still generate different predictions for the two groups of students, assuming the estimated value of β_1 does not equal zero.

Now consider the second example. Suppose we wanted to explain variation in the internet use data and we suspected that the percentage of employees in service industries might be an explanatory factor. In other words, we thought that internet use rates might be higher in countries that had a higher percentage of employees in service industries. If we define X_i as the service employment rate, we can use the two-parameter model to make predictions for countries conditional on service employment. This variable has many possible values, and it would be unlikely that any two countries would have exactly the same percentage of employees in service industries and, hence, the exact same values on X_i. Therefore, instead of making two different predictions as in the lottery example, our two-parameter model now is likely to make a different prediction for each country because each country is likely to have a unique value of X_i. Another difference between this two-parameter model and the lottery example is that here values on X_i are less arbitrary than they were in the lottery example. Each country has an actual percentage of employees in service industries that can be compared with other countries, and the information about country-to-country differences needs to be represented in the values we assign to X_i for each country.

The difference between these two examples lies in the nature of the units of measurement of the predictor variable, the variable upon which the predictions are conditional. In the lottery example, no training versus training is a categorical variable, in the sense that all participants are in one group or the other. Although X_i needs to code the distinction between the two training conditions, it does not matter which group of respondents is given the higher value on X_i nor does it matter which two values are used. In contrast, the percentage of employees in service industries is what we call a continuous predictor variable, in the sense that different countries have different values and the differences among these values are meaningful.

All procedures for building models, estimating parameters, and comparing augmented and compact models can be used regardless of whether the predictor variable (or variables) is categorical or continuous. However, it is conceptually useful to treat models with categorical predictor variables separately from models whose predictors are assumed to be continuously measured.

In the current chapter, we consider two-parameter models that have a single, continuously measured predictor variable. Then, in Chapters 7 and 8, we consider models with multiple continuously measured predictors. In traditional statistical jargon these three chapters deal with simple and multiple regression, including polynomial and moderated regression models. Then in Chapters 9 and 10 we turn our attention to models having categorical predictors. Again, in traditional statistical jargon, these chapters deal with analysis of variance. Finally, in Chapter 11 we consider models in which some predictors are categorical variables and some are continuous variables. Such models are traditionally referred to as analysis of covariance models. Across chapters, our approach to each type of model will be uniform. Rather than describing seemingly different statistical techniques for multiple regression, analysis of variance, and analysis of covariance, we will estimate parameters just as we have done in the simple single-parameter case, and we will test hypotheses by comparing augmented and compact models. So although our treatment of categorial predictors is located in different chapters from our treatment of continuous predictors, the same procedures will be used throughout.

Defining a Linear Two-Parameter Model

We now confine our attention to two-parameter models with a single continuous predictor variable. As an example, we will use the data contained in Figure 6.1 to ask whether differences between countries in internet use are predictable from differences in their service employment rates. As we speculated earlier, it seems reasonable that internet use may be higher in countries that have stronger service economies.

Figure 6.2 is a scatterplot of the tabular data from Figure 6.1. The vertical axis represents the percentage of the population who used the internet, and the horizontal axis represents the percentage of employees in service industries. Each point in this plot represents one of the 45 countries. The question that we would like to ask is whether we can use service employment rates to predict internet use. Or, expressed differently, do our predictions of internet use improve by making those predictions conditional on knowledge of service employment?

We will use a simple linear model to generate such conditional predictions. As already discussed, this model is

$$Y_i = \beta_0 + \beta_1 X_i + \varepsilon_i$$

where Y_i is a country's internet use and X_i is its service employment rate. Returning to our generic equation

Data = Model + Error

we see that the model in this two-parameter equation is represented by

$$\beta_0 + \beta_1 X_i$$

Country	*Internet use (% of population)*	*Service employment rate (% of employees in service industries)*
ARG	87.15	72.40
AUS	96.24	78.81
BLR	86.89	59.36
BEL	92.79	79.69
BWA	73.50	62.21
BRA	80.69	69.68
BGR	75.27	62.86
CAN	92.83	79.40
CHN	73.05	47.35
COL	73.03	63.92
HRV	81.25	64.41
DNK	98.87	78.61
RCU	76.20	50.99
EGY	72.06	51.09
FIN	92.81	74.47
FRA	86.10	77.99
DEU	91.43	71.12
GRC	78.49	73.36
GTM	50.84	48.76
HUN	88.64	64.16
ISL	99.69	77.13
IDN	62.10	49.25
IRL	95.17	76.70
ISR	90.30	82.12
ITA	74.86	69.31
JPN	82.91	73.12
JOR	82.76	79.12
MYS	96.75	62.09
MEX	75.63	62.02
NLD	92.05	83.79
NIC	57.15	52.90
NOR	99.00	78.50
PER	71.11	55.12
POL	85.37	60.67
PRT	82.31	70.90
ROU	83.59	51.00
RUS	88.21	67.32
ESP	93.90	75.75
SWE	88.31	79.88
CHE	95.57	77.94
THA	85.27	45.90
TUR	81.41	56.33
UKR	79.22	60.87
GBR	96.68	81.00
USA	91.75	79.15

Figure 6.1 Internet use and service employment rates

Source: World Bank World Development Indicators from the International Telecommunication Union (ITU) World Telecommunication/ICT Indicators Database. Service employment rates are International Labour Organization (ILO) estimates from 2023. https://databank.worldbank.org/reports.aspx?source=world-development-indicators#

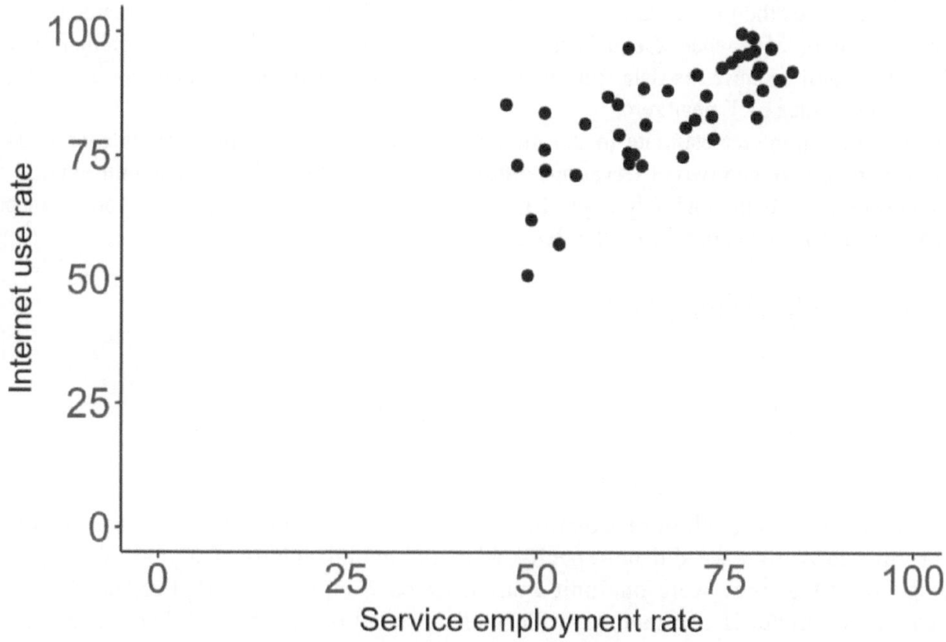

Figure 6.2 Scatterplot of internet use and service employment rates for the 45 countries

In terms of estimated parameter values, the predictions made by this model for each country's service employment rate are given by

$$\hat{Y}_i = b_0 + b_1 X_i$$

Because ERROR equals DATA minus MODEL, the residuals in this two-parameter model can be expressed as follows:

$$e_i = Y_i - \hat{Y}_i$$

$$e_i = Y_i - (b_0 + b_1 X_i)$$

Let us examine each of the parameter estimates in this model and see what each is telling us. First, consider b_0. In the single-parameter model, we saw that b_0 equaled the mean value of Y_i, assuming that we define error as the sum of squared errors. Another way of saying the same thing is that in the single-parameter model b_0 is our predicted value for each country. However, in this two-parameter model we wish to take further information into account in making each country's prediction. We are making each country's prediction conditional on its percentage of employees in service industries. Therefore, b_0 is not the predicted value for each country because the predictions vary as a function of service employment rates. There is one case, however, when this model predicts an internet use rate equal to b_0. This is when X_i equals zero because

$$\hat{Y}_i = b_0 + b_1(0) = b_0$$

This, then, provides the interpretation for the parameter estimate b_0 in this two-parameter model: b_0 is our prediction of Y_i when X_i equals zero. As we will see for our example, this prediction may not be very useful because the data from which we estimate this parameter may not include data points having values of X_i near zero.

The second parameter estimate in the model, b_1, tells us how our predictions change as X_i changes. Suppose we had two observations differing in their values on X_i by one unit, with X_i for the first observation being one unit larger than X_i for the second. According to the model, our predictions for the two data points would differ by b_1 since

$$
\begin{aligned}
\hat{Y}_1 - \hat{Y}_2 &= (b_0 + b_1X_1) - (b_0 + b_1X_2) \\
&= b_1X_1 - b_1X_2 \\
&= b_1(X_1 - X_2) \\
&= \mathrm{b}_1
\end{aligned}
$$

So b_1 tells us by how much our predictions of Y_i change as X_i increases by one unit. Notice that in this derivation we did not specify what the actual values of X_1 and X_2 were. We only specified that they were one unit apart from each other. This implies that b_1 in this two-parameter model is constant, regardless of the level of X_i. This is what was meant by the definition of this sort of two-parameter model as a linear model. As X_i changes by some set amount, our predictions of Y_i change by a constant amount, regardless of the starting value of X_i.

To review, b_0 and b_1 tell us very different things: b_0 is a predicted value (a $\hat{Y}_i$) at a particular value of X_i, namely when X_i equals zero; b_1 is not a predicted value, rather it is the difference between two predicted values as we move from a smaller X_i to one that is one unit larger.

Let us look at this two-parameter model graphically for the example in which internet use is predicted from service employment rates. Figure 6.3 presents the graph of the model set against the data. All the predictions $\hat{Y}_i$ lie on the line defined by the model. Errors of prediction, e_i, as in the single-parameter model, are defined as vertical distances between the line and actual observations. That is, an error or residual is the difference between Y_i and $\hat{Y}_i$. b_0 is the value of $\hat{Y}_i$ when X_i equals zero; it is frequently called the intercept because it is the value on the vertical axis of the graph where the prediction function crosses or "intercepts" it. b_1 is the difference in $\hat{Y}_i$ for each unit increase in X_i. We can think of it as the slope of the line since algebraically it is the difference in predicted values between any two points on the line per their difference in X_i values: *rise over run*:

$$
b_1 = \frac{(\hat{Y}_2 - \hat{Y}_1)}{(X_2 - X_1)}
$$

where the subscripts designate any two points on the line. Notice that the slope can take on any positive or negative value. If the slope is positive, it means that the model predicts higher values of Y_i as X_i increases. If the slope is negative, the model predicts lower values of Y_i as X_i increases.

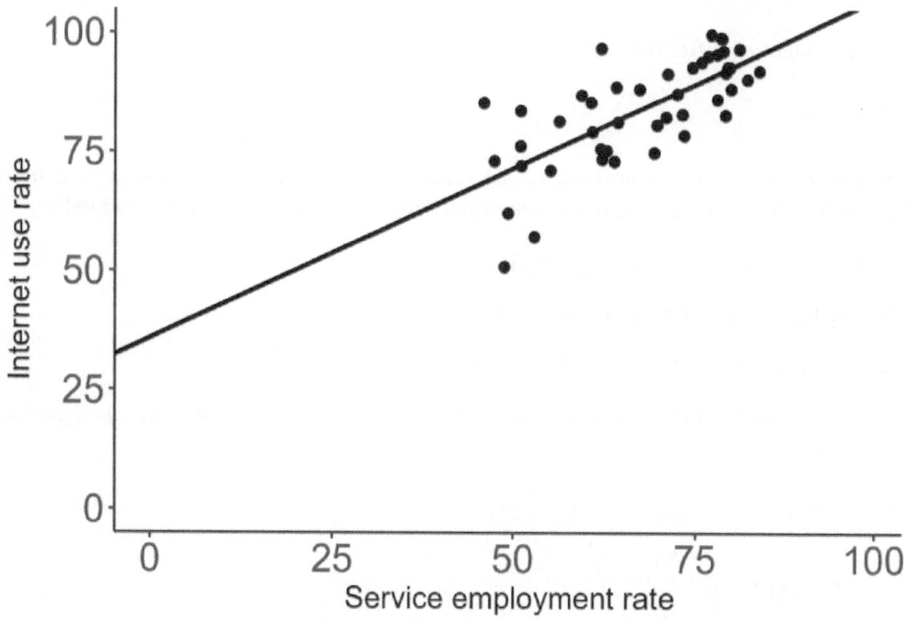

Figure 6.3 Scatterplot with two-parameter model predicting internet use from service employment rates

Estimating a Linear Two-Parameter Model

Given some sample of data, how do we estimate the parameters of this sort of model? We want to use our sample of data to generate values of b_0 and b_1 in the equation

$$\hat{Y}_i = b_0 + b_1 X_i$$

that are good estimates of the true (but unknown) parameters β_0 and β_1. To do this, we decided in Chapter 3 that for the single-parameter model we would derive estimates that minimize the sum of squared errors. We made this decision because if the errors are normally distributed, least-squares parameter estimates are unbiased, consistent, and relatively efficient. This continues to be the case in the context of the present two-parameter model and will continue to be the case in more complicated models with many parameters, which we consider in subsequent chapters. For now, we want to derive estimated values of β_0 and β_1 that minimize $\Sigma(Y_i - \hat{Y}_i)^2$. The resulting *least-squares* parameter estimates are given as

$$b_1 = \frac{\Sigma\left(X_i - \bar{X}\right)\left(Y_i - \bar{Y}\right)}{\Sigma\left(X_i - \bar{X}\right)^2}$$

$$b_0 = \bar{Y} - b_1 \bar{X}$$

The derivation of these estimates is given in Box 6.1.

Box 6.1 Algebraic Derivation of Least-Squares Estimates of β_0 and β_1

$$\text{SSE} = \Sigma(Y_i - \hat{Y}_i)^2 = \Sigma(Y_i - b_0 - b_1X_i)^2$$

given that $\hat{Y}_i = b_0 + b_1X_i$. We now add $(\bar{Y} - \bar{Y})$ and $b_1(\bar{X} - \bar{X})$ inside the parentheses to this expression for SSE. Since both expressions equal zero, we have not changed the equality. Thus

$$\text{SSE} = \Sigma(Y_i - \bar{Y} + \bar{Y} - b_0 - b_1X_i + b_1\bar{X} - b_1\bar{X})^2$$

Grouping terms yields the equivalent expression:

$$\text{SSE} = \Sigma[(Y_i - \bar{Y}) + (\bar{Y} - b_0 - b_1\bar{X}) - b_1(X_i - \bar{X})]^2$$

If we square the term in brackets and distribute the summation sign, this gives the equivalent expression:

$$\begin{aligned}\text{SSE} &= \Sigma(Y_i - \bar{Y})^2 + 2(\bar{Y} - b_0 - b_1\bar{X})\,\Sigma(Y_i - \bar{Y})\\ &= -2b_1\,\Sigma(Y_i - \bar{Y})(X_i - \bar{X}) + n(\bar{Y} - b_0 - b_1\bar{X})^2\\ &= -2b_1(\bar{Y} - b_0 - b_1\bar{X})\,\Sigma(X_i - \bar{X}) + b_1^2\,\Sigma(X_i - \bar{X})^2\end{aligned}$$

Since both $\Sigma(Y_i - \bar{Y})$ and $\Sigma(X_i - \bar{X})$ equal zero, this expression reduces to

$$\text{SSE} = \Sigma(Y_i - \bar{Y})^2 - 2b_1\,\Sigma(Y_i - \bar{Y})(X_i - \bar{X}) + n(\bar{Y} - b_0 - b_1\bar{X})^2 + b_1^2\,\Sigma(X_i - \bar{X})^2$$

Since the third term in this expression, $n(\bar{Y} - b_0 - b_1\bar{X})^2$, is necessarily positive, to minimize SSE we would like to set it equal to zero. Therefore, we want values of b_0 and b_1 such that

$$n(\bar{Y} - b_0 - b_1\bar{X}) = 0$$

Dividing both sides of this equality by n gives us

$$\bar{Y} - b_0 - b_1\bar{X} = 0$$

or, equivalently

$$b_0 = \bar{Y} - b_1\bar{X}$$

We have now reduced our expression for SSE, assuming the desire to minimize it, to:

$$\begin{aligned}\text{SSE} &= \sum(Y_i - \bar{Y})^2 - 2b_1\sum(Y_i - \bar{Y})(X_i - \bar{X}) + b_1^2\sum(X_i - \bar{X})^2\\ &= \sum(Y_i - \bar{Y})^2 + \sum(X_i - \bar{X})^2\left[b_1^2 - 2b_1\frac{\Sigma(Y_i - \bar{Y})(X_i - \bar{X})}{\Sigma(X_i - \bar{X})^2}\right]\end{aligned}$$

Let us now add to and subtract from this expression the quantity

$$\sum\left(X_i-\bar{X}\right)^2\left(\frac{\Sigma\left(X_i-\bar{X}\right)\left(Y_i-\bar{Y}\right)}{\Sigma\left(X_i-\bar{X}\right)^2}\right)^2$$

Thus

$$\begin{aligned}\text{SSE}=&\sum\left(Y_i-\bar{Y}\right)^2\\&+\sum\left(X_i-\bar{X}\right)^2\left[b_1^2-2b_1\frac{\Sigma\left(X_i-\bar{X}\right)\left(Y_i-\bar{Y}\right)}{\Sigma\left(X_i-\bar{X}\right)^2}+\left(\frac{\Sigma\left(X_i-\bar{X}\right)\left(Y_i-\bar{Y}\right)}{\Sigma\left(X_i-\bar{X}\right)^2}\right)^2\right]\\&-\sum\left(X_i-\bar{X}\right)^2\left(\frac{\Sigma\left(X_i-\bar{X}\right)\left(Y_i-\bar{Y}\right)}{\Sigma\left(X_i-\bar{X}\right)^2}\right)^2\end{aligned}$$

Rearranging terms and taking the square root of the term in brackets gives us

$$\begin{aligned}\text{SSE}=&\sum\left(Y_i-\bar{Y}\right)^2-\sum\left(X_i-\bar{X}\right)^2\left(\frac{\Sigma\left(X_i-\bar{X}\right)\left(Y_i-\bar{Y}\right)}{\Sigma\left(X_i-\bar{X}\right)^2}\right)^2\\&+\sum\left(X_i-\bar{X}\right)^2\left[b_1-\frac{\Sigma\left(X_i-\bar{X}\right)\left(Y_i-\bar{Y}\right)}{\Sigma\left(X_i-\bar{X}\right)^2}\right]^2\end{aligned}$$

The last term in this expression for SSE is necessarily positive. Therefore, to minimize SSE we want this last term to equal zero. This occurs if

$$b_1=\frac{\Sigma\left(X_i-\bar{X}\right)\left(Y_i-\bar{Y}\right)}{\Sigma\left(X_i-\bar{X}\right)^2}$$

We can think about the formula for the estimated slope in a couple of different ways. One way is to divide both the numerator and denominator of the formula by $n-1$:

$$\begin{aligned}b_1&=\frac{\Sigma\left(X_i-\bar{X}\right)\left(Y_i-\bar{Y}\right)/(n-1)}{\Sigma\left(X_i-\bar{X}\right)^2/(n-1)}\\&=\frac{s_{XY}}{s_X^2}\end{aligned}$$

In this last expression, the numerator of the slope, s_{XY}, is known as the covariance of X_i and Y_i, and the denominator is the variance of X_i.

Examining the crossproduct of $(X_i-\bar{X})(Y_i-\bar{Y})$ for any given observation helps to conceptualize the meaning of the covariance between two variables. To aid with this conceptualization, in

Figure 6.4 we have added a horizontal line at the mean of Y_i and a vertical line at the mean of X_i to the scatterplot of the data. Any given observation will have a positive value for its crossproduct if it lies either in the upper right quadrant of the scatterplot or the lower left quadrant of the scatterplot, where the quadrants are defined by the intersecting lines at the two means. Positive values of the crossproduct thus come from observations that are either above the mean on both variables or below the mean on both. On the other hand, observations with negative values for their crossproducts will lie in the other two quadrants, having values that are below the mean on one variable but above the mean on the other. The covariance is (roughly) the average of all these individual crossproducts. Given that the denominator of the slope is always positive in value, the sign of the slope is determined by the sign of the covariance. This means that the slope will be positive if most of the observations in a scatterplot are either above the means on both variables or below the means on both variables. A negative slope happens when most of the observations fall into the other two quadrants: above the mean on one variable but below it on the other. A slope near zero would occur when the observations are randomly distributed throughout all four quadrants.

The other way to think conceptually about the meaning of the slope involves applying a bit of algebra to the original formula for the slope given above, yielding

$$b_1 = \sum w_i \left[\frac{Y_i - \bar{Y}}{X_i - \bar{X}} \right], \text{where } w_i = \frac{\left(X_i - \bar{X}\right)^2}{\Sigma\left(X_i - \bar{X}\right)^2}$$

The term in the brackets in this equation can be thought of as the slope suggested by each individual observation—it is the rise over run of a line that goes between the joint mean of the

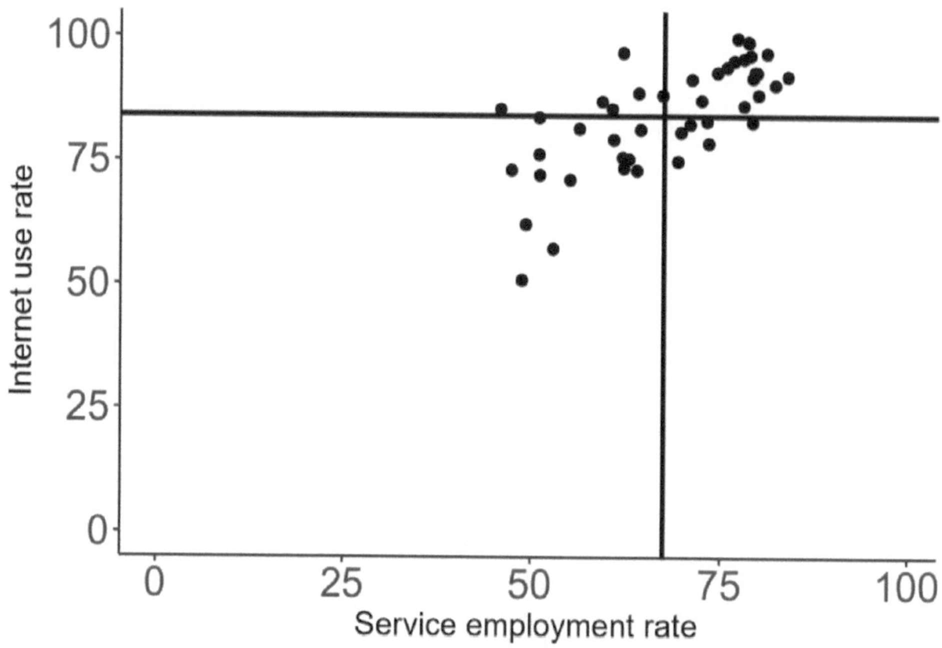

Figure 6.4 Scatterplot with horizontal line at $\bar{Y}$ and a vertical line at $\bar{X}$

two variables and a particular observation. In Figure 6.5 we have added a few of these individual "slopes" for a few observations. The w_i can be thought of as a weight assigned to each observation, where the weight represents the proportion of the total sum of squares of X_i that is attributable to the particular observation. In essence, we can think of each observation as having a slope that it "prefers" (between the joint mean and itself), that gets a certain weight or vote in determining the value of the slope for all the observations. The more extreme the observation is on X_i, the greater the vote. Using the aforementioned formulas to calculate both parameter estimates, b_0 and b_1, or more efficiently using a regression routine in a statistical software package, we can calculate the estimated intercept and slope for the model where we regress internet use on service employment rate. (Notice here that in regression terminology one regresses Y_i *on* X_i, not the other way around.) The resulting estimated two-parameter model for these data is

$$\hat{Y}_i = 35.903 + 0.712X_i$$

This prediction function is graphed in Figure 6.6 as a straight line on the scatterplot we saw before.

(This figure is identical to Figure 6.3 where we plotted the slope and intercept prior to indicating how they were estimated.)

Let us interpret each of the regression coefficients, b_0 and b_1, in this equation. The value of the intercept, 35.903, is the value predicted by the model for the country's internet use rate if the country's service employment rate were 0%. Although this intercept is the best unbiased estimate of this prediction based on a linear model of these data, it is a relatively meaningless value because no country in the data had a service employment rate of zero.

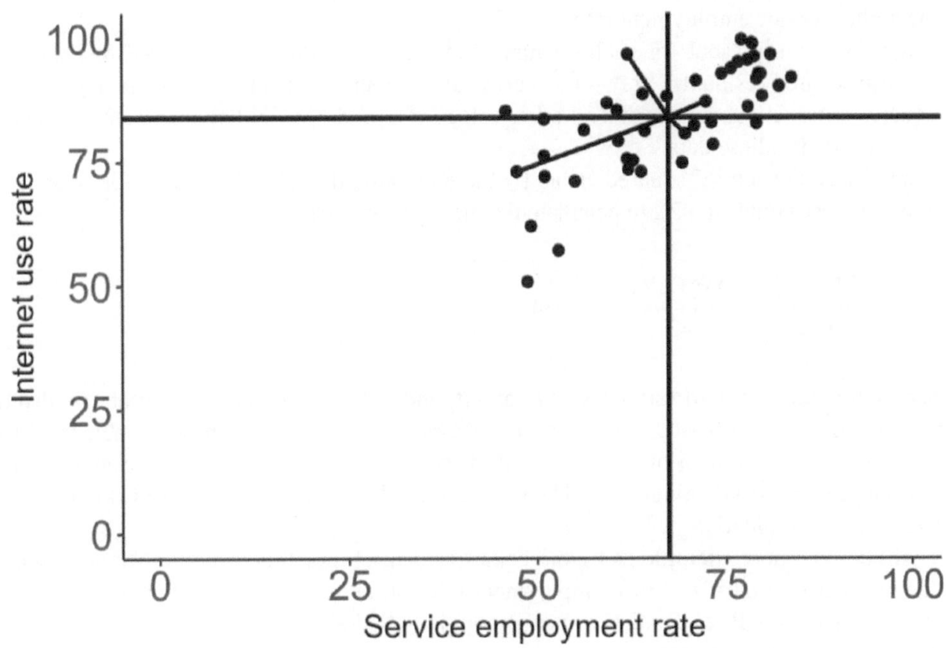

Figure 6.5 Scatterplot with individual slopes

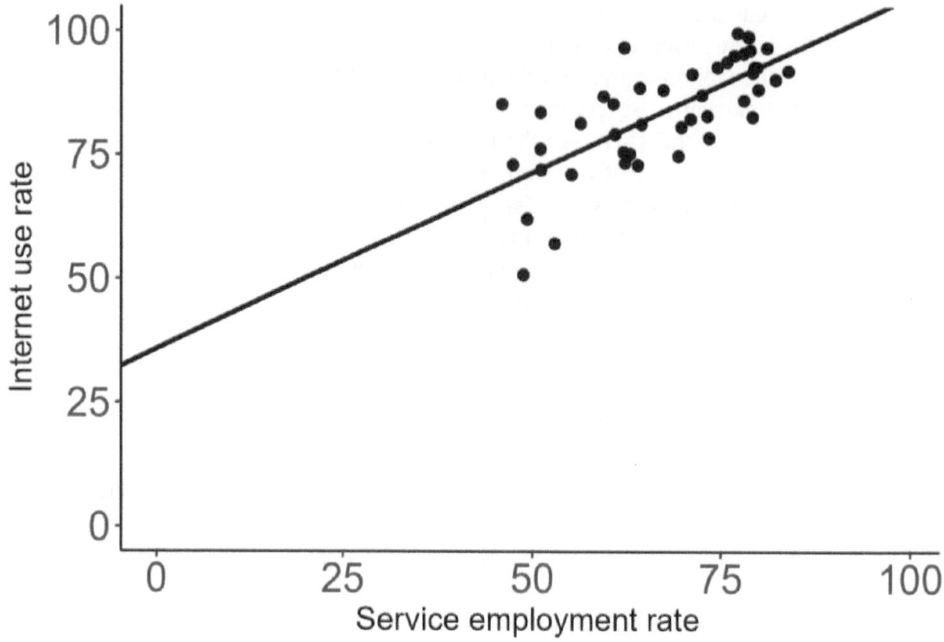

Figure 6.6 Scatterplot with two-parameter model: $\hat{Y}_i = 35.903 + 0.712X_i$

The value of the slope, 0.712, tells us that if we found two countries differing in service employment rates by 1%, our model predicts that the internet use would be .712% higher in the country with the higher service employment rate.

In Figure 6.7, we present for each country its Y_i, X_i, $\hat{Y}_i$ residual, and squared residual. The sum of these squared residuals, $\sum(Y_i - \hat{Y}_i)^2$, across all 45 countries is also given. Having used the least-squares criterion guarantees that no other values of b_0 and b_1 would give us a smaller sum of squared residuals for these data.

We can divide the sum of squared errors by the remaining degrees of freedom for error, $n - p$ (which in this case equals $n - 2$), to calculate the mean square error:

$$\text{MSE} = \frac{\Sigma\left(Y_i - \hat{Y}_i\right)^2}{n-2} = \frac{2581.474}{43} = 60.034$$

Just as b_0 and b_1 are unbiased estimates of β_0 and β_1 under the least-squares criterion, so also the mean square error is an unbiased estimate of the variance of ε_i. It estimates how variable the errors of prediction are at each level of X_i. As we will discuss, it is assumed that the variance of these errors is constant across all values of X_i. The square root of this mean square error is known as the *standard error of prediction*.

We will do one more example using the countries dataset. For this example, we are going to examine whether a country's service employment rate can be used to predict the country's gross domestic product or GDP per capita (i.e., per person) in U.S. dollars. More economically advanced countries typically have stronger service economies; one would therefore expect countries with higher service employment rates to produce more and thus have higher GDPs.

Country	Y_i	X_i	$\hat{Y}_i$	e_i	e_i^2
ARG	87.15	72.40	87.452	-0.302	0.091
AUS	96.24	78.81	92.016	4.224	17.845
BLR	86.89	59.36	78.167	8.723	76.085
BEL	92.79	79.69	92.642	0.148	0.022
BWA	73.50	62.21	80.197	-6.697	44.843
BRA	80.69	69.68	85.515	-4.825	23.282
BGR	75.27	62.86	80.659	-5.389	29.045
CAN	92.83	79.40	92.436	0.394	0.155
CHN	73.05	47.35	69.616	3.434	11.791
COL	73.03	63.92	81.414	-8.384	70.292
HRV	81.25	64.41	81.763	-0.513	0.263
DNK	98.87	78.61	91.873	6.997	48.954
RCU	76.20	50.99	72.208	3.992	15.937
EGY	72.06	51.09	72.279	-0.219	0.048
FIN	92.81	74.47	88.926	3.884	15.088
FRA	86.10	77.99	91.432	-5.332	28.429
DEU	91.43	71.12	86.540	4.890	23.908
GRC	78.49	73.36	88.135	-9.645	93.032
GTM	50.84	48.76	70.620	-19.780	391.253
HUN	88.64	64.16	81.585	7.055	49.774
ISL	99.69	77.13	90.820	8.870	78.685
IDN	62.10	49.25	70.969	-8.869	78.659
IRL	95.17	76.70	90.513	4.657	21.684
ISR	90.30	82.12	94.372	-4.072	16.585
ITA	74.86	69.31	85.252	-10.392	107.988
JPN	82.91	73.12	87.964	-5.054	25.547
JOR	82.76	79.12	92.236	-9.476	89.803
MYS	96.75	62.09	80.111	16.639	276.854
MEX	75.63	62.02	80.061	-4.431	19.636
NLD	92.05	83.79	95.561	-3.511	12.330
NIC	57.15	52.90	73.568	-16.418	269.544
NOR	99.00	78.50	91.795	7.205	51.912
PER	71.11	55.12	75.148	-4.038	16.309
POL	85.37	60.67	79.100	6.270	39.312
PRT	82.31	70.90	86.384	-4.074	16.596
ROU	83.59	51.00	72.215	11.375	129.391
RUS	88.21	67.32	83.835	4.375	19.142
ESP	93.90	75.75	89.837	4.063	16.508
SWE	88.31	79.88	92.778	-4.468	19.959
CHE	95.57	77.94	91.396	4.174	17.420
THA	85.27	45.90	68.584	16.686	278.429
TUR	81.41	56.33	76.010	5.400	29.160
UKR	79.22	60.87	79.242	-0.022	0.001
GBR	96.68	81.00	93.575	3.105	9.641
USA	91.75	79.15	92.258	-0.508	0.258
					SSE = 2581.490

Figure 6.7 Predicted values and residuals for internet use data by country

The parameter estimates from the regression model make clear how these variables are related:

$$\hat{Y}_i = -93252.0 + 1838.8X_i$$

Let us interpret both parameter estimates in this model. Doing so will make clear that their interpretation depends on the metric in which the two variables are measured. First, the intercept, –93,252.0, represents the GDP per capita that would be predicted if a country's service employment rate were zero. Of course, this number is not very informative because no country has a service employment rate that is zero. Yet it is the best linear prediction from these data, albeit well outside of the range of actual values found in the data. The slope, 1,838.8, is positive, meaning that in countries with stronger service economies, GDP is generally higher. The exact interpretation is that for every 1% increase in service employment rate (the measurement metric of X_i) we predict an increase in the GDP per capita of 1,838.8 U.S. dollars (the measurement metric of Y_i).

An Alternative Specification

It will prove useful at later points to be able to specify regression models in which the predictor variables have been put into "mean-deviation" form or "centered." What this means is that for every observation we have taken the value of the variable and subtracted from it the variable's mean value. Thus, if the predictor variable is X_i, the mean-deviated or centered variable is $(X_i - \bar{X})$. This centered variable will necessarily have a mean of zero, i.e., $(\bar{X} - \bar{X}) = 0$. We will then regress Y_i on $(X_i - \bar{X})$ rather than on X_i:

$$Y_i = b'_0 + b'_1(X_i - \bar{X})$$

How do these new parameter estimates, using a centered predictor differ from the parameter estimates using an uncentered predictor?

$$\hat{Y}_i = b_0 + b_1X_i$$

To answer this question, we can examine the formulas for the parameter estimates that we gave earlier, but this time with a centered X_i:

$$b'_1 = \frac{\Sigma\left(\left((X_i - \bar{X}) - (\bar{X} - \bar{X})\right)(Y_i - \bar{Y})\right)}{\Sigma\left((X_i - \bar{X}) - (\bar{X} - \bar{X})\right)^2}$$

$$b'_0 = \bar{Y} - b_1(\bar{X} - \bar{X})$$

Since $(\bar{X} - \bar{X}) = 0$, it follows that

$$b'_1 = \frac{\Sigma(X_i - \bar{X})(Y_i - \bar{Y})}{\Sigma(X_i - \bar{X})^2} = b_1$$

$$b'_0 = \bar{Y}$$

In other words, centering the predictor has no effect upon the slope, that is, $b_1' = b_1$, but it does change the intercept. The intercept with the predictor in mean-deviated or centered form will be the mean of Y_i.

Conceptually, if we think graphically about the scatterplot of data and the superimposed regression line, by centering all we are really doing is changing the scale of the X_i axis in the scatterplot, redefining the value of X_i so that its mean equals zero. Such a scatterplot with the X_i axis centered is given in Figure 6.8. As this makes clear we have not changed the observations in the scatterplot at all; we have simply shifted the origin point in the plot. In a fundamental sense, our prediction function has not changed at all; the same line, having the same slope, minimizes the squared errors of prediction. The only change derives from the change in the location of the zero point on the horizontal axis. That zero point is now at the mean of X_i and, accordingly, the value of the intercept (i.e., the value of Y_i where the prediction function crosses the vertical axis) changes. It is now $\bar{Y}$. This means that the regression line inevitably goes through the point defined by the joint means of the two variables.

In the case of the estimated model predicting internet use from service employment rate, when service employment rate is centered, the resulting least-squares parameter estimates are

$$\hat{Y}_i = 83.982 + 0.712(X_i - \bar{X})$$

The slope is unchanged by the centering of X_i, but the intercept has changed. It no longer equals 0.712, rather it equals $\bar{Y}$, which is 83.982. One can still use the conventional interpretation for the intercept: It remains the predicted value when the predictor equals zero, that is, when $(X_i - \bar{X})$ equals zero. And $(X_i - \bar{X})$ equals zero when $X_i = \bar{X}$.

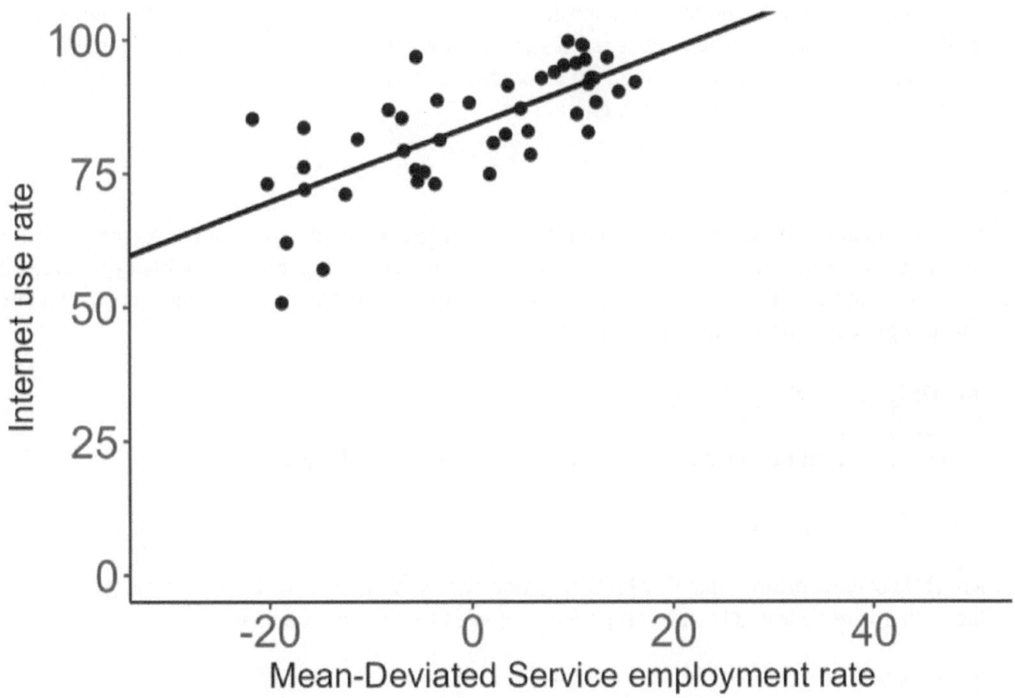

Figure 6.8 Scatterplot and prediction function with centered X

Because all that has changed with centering the predictor is the zero point on the horizontal access of the scatterplot, we are still dealing fundamentally with the same data and the same regression line, making the same predictions for each observation. Thus, centering the predictor leaves the mean squared error of the model unchanged. In a deep sense the regression results are unchanged by this transformation.

Statistical Inference in Two-Parameter Models

Now that the basics of estimation and interpretation in simple regression models are clear, we turn to the issue of statistical inference, asking questions about parameter values in such models. Our approach to statistical inference in the case of two-parameter models will be identical to the approach we adopted in the single-parameter case. That is, we will compare an augmented model (in which both parameters are estimated from the data) to a compact one (in which one or more parameters are fixed at a priori values). We will calculate the sum of squared errors associated with both the augmented and compact models; from these we will then compute PRE, the proportional reduction in error, as we go from the compact to the augmented model. This PRE, and its associated F statistic, can then be compared to their critical values, making the assumptions of normality, constant variance, and independence of residuals. Such a comparison permits a test of the null hypothesis that the values of the parameters fixed in the compact model are the true unknown parameter values. Put the other way around, we are testing whether the estimated parameter values in the augmented model depart significantly from their a priori values specified in the compact model.

Given that the augmented model for such comparisons is now the two-parameter simple regression model, there are alternative compact models with which it may be compared. On the one hand, one may be interested in asking questions about the slope, comparing the augmented model to a compact one in which the slope parameter has been set to some a priori value. On the other hand, there may arise occasions when one is interested in testing a null hypothesis about the intercept in this two-parameter model, comparing it to a compact one in which the intercept has been fixed at some a priori value. We consider each one in turn.

Inferences About β_1

To ask statistical inference questions about the slope is to ask about associations: As the predictor variable (X_i) increases, what is our conclusion about whether the dependent variable (Y_i) increases or decreases, and at what rate? Our augmented two-parameter model for such questions is the one we have been using throughout this chapter:

MODEL A: $Y_i = \beta_0 + \beta_1 X_i + \varepsilon_i$

The compact one, in its generic form, with which we will be making comparisons, is

MODEL C: $Y_i = \beta_0 + \mathrm{B}_1 X_i + \varepsilon_i$

where B_1 is some a priori value to which the slope parameter has been set. As before, the null hypothesis that is to be tested by the comparison between these two models is

$\mathrm{H}_0\colon \beta_1 = \mathrm{B}_1$

In words, the inferential test is whether the slope differs significantly from the a priori value.

By far the most common form this comparison takes is when the a priori value of the slope in the compact model equals zero. In this case the compact model is

$$\text{MODEL C: } Y_i = \beta_0 + 0X_i + \varepsilon_i$$
$$= \beta_0 + \varepsilon_i$$

and the null hypothesis is

$$H_0: \beta_1 = 0$$

The question being asked is whether X_i is a useful predictor of Y_i. If it is a useful predictor, then the predicted values of Y_i should change, either increasing or decreasing, as X_i changes. If we do no better with the augmented model (in which the slope is estimated from the data) than with the compact one (where the slope is constrained to equal zero) then it implies that the two variables are unrelated. We do just as well making a constant prediction of all Y_i values regardless of an observation's X_i value as we do making conditional predictions.

For this comparison, the compact model is what was considered as the augmented model in the previous chapter. This will frequently be the case throughout the remainder of the book as we consider more complex models and additional inferential questions: What is for one question the augmented model becomes the compact model for a different question. So, in Chapter 4, we tested null hypotheses about a constant predicted value for every observation. Now we are testing whether we need to make conditional predictions, and the compact model is one in which we estimate from the data a constant predicted value for each observation. That constant predicted value, when we estimate b_0 in Model C from the data, will not be the same as the intercept in Model A, when we estimate that model in our data. The best least-squares estimate of β_0 in Model C will be the mean of Y_i, just as it was (when we treated it as the augmented model) in Chapter 4. In the estimated augmented model (with two parameters), however, the estimate of β_0 will in general not be the mean of Y_i (unless the predictor has been centered). This makes clear another important point that will remain true as we consider more complex models in later chapters: The best estimate for a given parameter in general depends on what other parameters are estimated in the model. In the compact single-parameter model, the best estimate of β_0 will not in general be the same as the best estimate of β_0 in the two-parameter augmented model.

To ask whether the slope equals zero, and thus whether X_i is a useful predictor of Y_i, is the most common inferential question that one might ask about the slope. But it is certainly not the only question one might ask. Other Model Cs, with other values of B_1, and thus other null hypotheses, might occasionally also be of interest. For instance, there are occasions when we might be interested in testing the null hypothesis that

$$H_0: \beta_1 = 1$$

In this case, the augmented model remains the same two-parameter simple regression model, but Model C becomes

$$\text{MODEL C: } Y_i = \beta_0 + 1X_i + \varepsilon_i$$

By casting all statistical inference questions in the form of Model A/Model C comparisons, testing such a null hypothesis becomes entirely feasible, even if standard statistical texts and software programs do not routinely provide such tests.

Testing the Null Hypothesis That $\beta_1 = 0$

In the context of our model that predicted countries' internet use from their service employment rates, let us test the first of the previous null hypotheses, asking whether predictions of internet use are improved by making them conditional on service employment rates, compared to a compact model that sets the slope at zero and thus makes a constant prediction for internet use for all countries.

In terms of parameters, the models to be compared are

MODEL A: $Y_i = \beta_0 + \beta_1 X_i + \varepsilon_i$
MODEL C: $Y_i = \beta_0 + \varepsilon_i$

These are estimated as

MODEL A: $\hat{Y}_i = 35.903 + 0.712X_i$
MODEL C: $\hat{Y}_i = 83.982$

In Figures 6.9 and 6.10, we present the results of these two models for each country. In the first two columns of values in Figure 6.9 both Y_i (internet use rate) and X_i (service employment rate) values for each country are given. Then the next three columns provide the results of Model C. The first of these provides the predicted values, $\hat{Y}_{iC}$. These are necessarily the same value for every country because Model C predicts simply the mean internet use rate for each. Next, for each country we give its error, e_i, and its squared error, e_i^2. Across all 45 countries, the sum of squared errors for Model C equals 5392.860. This model has a single estimated parameter, hence n – PC equals 44 (with 45 countries). Thus, the mean square error for Model C is

$$5392.860/44 = 122.565$$

Given that this is the simplest single-parameter model, as defined in the previous chapter, this mean square error is also called the variance of Y_i.

The next three columns in Figure 6.9 give the parallel results for Model A, first the predicted values ($\hat{Y}_{iA}$) and then errors of prediction and squared errors. Notice in this case that the predicted values are not constant; rather, they are a linear function of the value of X_i for each country. And, again, the sum of the squared errors across the 45 countries is SSE for Model A, which equals 2581.490. In Model A we have estimated two parameters, accordingly PA equals 2, n – PA equals 43, and the mean square error for this model is 60.034.

Because Model A uses an additional parameter to model the data, it necessarily does better than Model C in terms of sums of squared errors. That is not to say, however, that in every country the prediction made by Model A is better than that made by Model C. Examine Jordan (JOR) for instance. Its internet use rate is 82.76%. Model C predicts it to be the mean of 83.982, which is a difference of −1.222. Model A, on the other hand, making a prediction conditional on Jordan's service employment rate, predicts it to be 92.236, missing by 9.476. For this particular country, the Model A prediction is not as accurate as the Model C prediction. Yet on average across countries

Country	*Internet use*	*Service employment*	$\hat{Y}_{iC}$	e_{iC}	e^2_{iC}	$\hat{Y}_{iA}$	e_{iA}	e^2_{iA}	$(\hat{Y}_{iA}-\hat{Y}_{iC})^2$
ARG	87.15	72.40	83.982	3.168	10.036	87.452	-0.302	0.091	12.040
AUS	96.24	78.81	83.982	12.258	150.259	92.016	4.224	17.845	64.541
BLR	86.89	59.36	83.982	2.908	8.456	78.167	8.723	76.085	33.811
BEL	92.79	79.69	83.982	8.808	77.581	92.642	0.148	0.022	75.000
BWA	73.50	62.21	83.982	-10.482	109.872	80.197	-6.697	44.843	14.330
BRA	80.69	69.68	83.982	-3.292	10.837	85.515	-4.825	23.282	2.351
BGR	75.27	62.86	83.982	-8.712	75.899	80.659	-5.389	29.045	11.040
CAN	92.83	79.40	83.982	8.848	78.287	92.436	0.394	0.155	71.467
CHN	73.05	47.35	83.982	-10.932	119.509	69.616	3.434	11.791	206.376
COL	73.03	63.92	83.982	-10.952	119.946	81.414	-8.384	70.292	6.594
HRV	81.25	64.41	83.982	-2.732	7.464	81.763	-0.513	0.263	4.924
DNK	98.87	78.61	83.982	14.888	221.653	91.873	6.997	48.954	62.273
RCU	76.20	50.99	83.982	-7.782	60.560	72.208	3.992	15.937	138.630
EGY	72.06	51.09	83.982	-11.922	142.134	72.279	-0.219	0.048	136.958
FIN	92.81	74.47	83.982	8.828	77.934	88.926	3.884	15.088	24.440
FRA	86.10	77.99	83.982	2.118	4.486	91.432	-5.332	28.429	55.501
DEU	91.43	71.12	83.982	7.448	55.473	86.540	4.890	23.908	6.546
GRC	78.49	73.36	83.982	-5.492	30.162	88.135	-9.645	93.032	17.250
GTM	50.84	48.76	83.982	-33.142	1098.392	70.620	-19.780	391.253	178.540
HUN	88.64	64.16	83.982	4.658	21.697	81.585	7.055	49.774	5.746
ISL	99.69	77.13	83.982	15.708	246.741	90.820	8.870	78.685	46.752
IDN	62.10	49.25	83.982	-21.882	478.822	70.969	-8.869	78.659	169.338
IRL	95.17	76.70	83.982	11.188	125.171	90.513	4.657	21.684	42.659
ISR	90.30	82.12	83.982	6.318	39.917	94.372	-4.072	16.585	107.961
ITA	74.86	69.31	83.982	-9.122	83.211	85.252	-10.392	107.988	1.612
JPN	82.91	73.12	83.982	-1.072	1.149	87.964	-5.054	25.547	15.860
JOR	82.76	79.12	83.982	-1.222	1.493	92.236	-9.476	89.803	68.136
MYS	96.75	62.09	83.982	12.768	163.022	80.111	16.639	276.854	14.984
MEX	75.63	62.02	83.982	-8.352	69.756	80.061	-4.431	19.636	15.372
NLD	92.05	83.79	83.982	8.068	65.093	95.561	-3.511	12.330	134.084
NIC	57.15	52.90	83.982	-26.832	719.956	73.568	-16.418	269.544	108.456
NOR	99.00	78.50	83.982	15.018	225.540	91.795	7.205	51.912	61.043
PER	71.11	55.12	83.982	-12.872	165.688	75.148	-4.038	16.309	78.032
POL	85.37	60.67	83.982	1.388	1.927	79.100	6.270	39.312	23.834
PRT	82.31	70.90	83.982	-1.672	2.796	86.384	-4.074	16.596	5.769
ROU	83.59	51.00	83.982	-0.392	0.154	72.215	11.375	129.391	138.462
RUS	88.21	67.32	83.982	4.228	17.876	83.835	4.375	19.142	0.022
ESP	93.90	75.75	83.982	9.918	98.367	89.837	4.063	16.508	34.281
SWE	88.31	79.88	83.982	4.328	18.732	92.778	-4.468	19.959	77.362
CHE	95.57	77.94	83.982	11.588	134.282	91.396	4.174	17.420	54.972
THA	85.27	45.90	83.982	1.288	1.659	68.584	16.686	278.429	237.105
TUR	81.41	56.33	83.982	-2.572	6.615	76.010	5.400	29.160	63.553
UKR	79.22	60.87	83.982	-4.762	22.677	79.242	-0.022	0.001	22.463
GBR	96.68	81.00	83.982	12.698	161.239	93.575	3.105	9.641	92.026
USA	91.75	79.15	83.982	7.768	60.342	92.258	-0.508	0.258	68.489

Figure 6.9 Model comparison for internet use data

the squared errors of prediction must be at least as small from Model A as they are from Model C, simply because an additional parameter has been estimated.

Figure 6.10 presents the results graphically. The horizontal line superimposed on the data points is the prediction function of Model C. The other line is the prediction function of Model A. The inferential question that we now want to ask is whether the reduction in the SSEs when we replace Model C with Model A has been worth the reduction in the error degrees of freedom due to estimating an additional parameter. Graphically, the question is: Do we do sufficiently better with the sloped line to merit the added complexity over and above the horizontal prediction function? To answer this, we compute PRE, the proportional reduction in error as we move from Model C to Model A:

$$\text{PRE} = \frac{\text{SSE}(\text{C}) - \text{SSE}(\text{A})}{\text{SEE}(\text{C})} = \frac{5392.860 - 2581.490}{5392.860} = .5213$$

Thus, when we make conditional predictions in this example, the total errors of prediction are reduced by more than 52% compared to simply predicting the mean level of internet use for each country.

The numerator of PRE is the sum of squares reduced (SSR) and can be calculated directly by taking the squared difference between the predicted value for Model A and that for Model C, and summing these across the 45 countries:

$$\text{SSR} = \text{SSE(C)} - \text{SSE(A)} = \Sigma(\hat{Y}_{iA} - \hat{Y}_{iC})^2$$

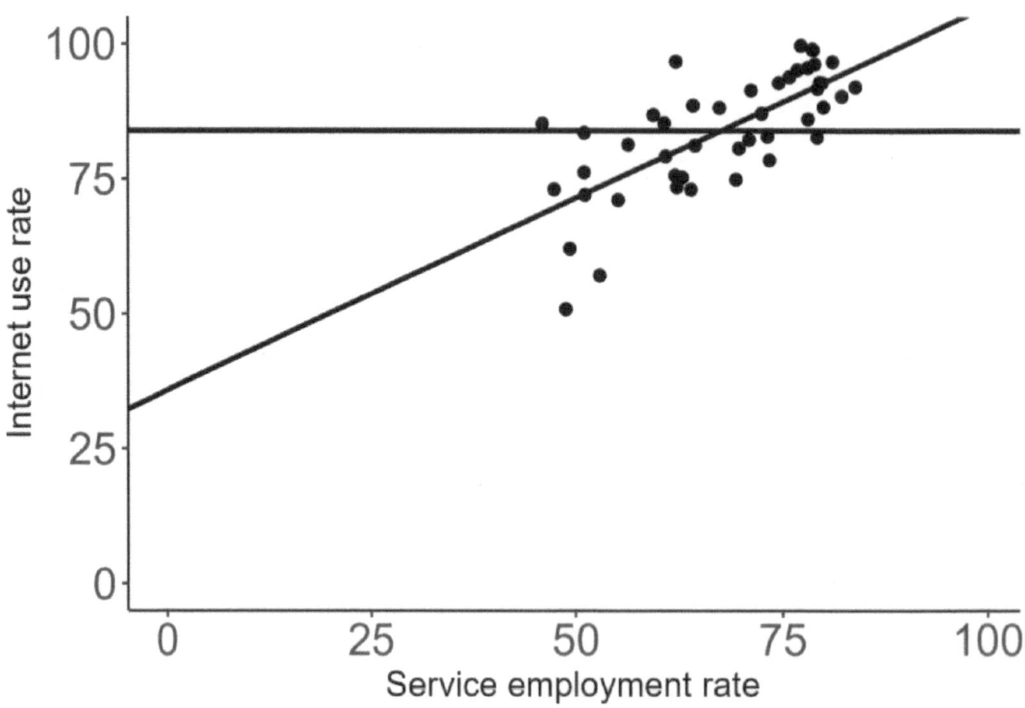

Figure 6.10 Model C and Model A

In the final column of Figure 6.9, we present these squared differences in predicted values for every state. The sum of the numbers in this final column necessarily equals (within rounding error) the difference in the sum of squared errors between the two models, i.e., 5392.86 –2581.49 or 2811.37.

We can also compute the F statistic associated with the comparison between these two models. Here we do this with the equivalent expressions for F, either in terms of PRE or in terms of the sums of squares of the two models:

$$F = \frac{\text{PRE}/(\text{PA}-\text{PC})}{(1-\text{PRE})/(n-\text{PA})} = \frac{.5213/1}{(1-.5213)/43} = 46.83$$

$$F = \frac{\text{SSR}/(\text{PA}-\text{PC})}{\text{SSE}(\text{A})/(n-\text{PA})} = \frac{2811.370/1}{2581.490(43)} = 46.83$$

Using the appropriate tables in the Appendix, we compare the computed values of either PRE or F with their critical values, given that the assumptions of normality, constant variance, and independence of errors are met. The critical values at $\alpha = .05$ (with 1 and 43 degrees of freedom) are approximately .088 for PRE and 4.07 for F. Clearly, our computed values exceed these, and hence we can reject Model C in favor of the two-parameter Model A that makes conditional predictions. Our conclusion is that service employment rates are a useful predictor of internet use rates in these data. Further, once we have rejected the null hypothesis of no relationship between the two variables, we can make a conclusion about the direction of the relationship between them, based on the sign of the estimated slope: In countries where the service employment is higher, there are higher rates of internet use.

Figure 6.11 summarizes the results of the statistical analysis in an ANOVA table of the same type as we developed in Chapter 4. The first row provides information about the reduction in error achieved by including X_i as a predictor in the model (i.e., by using Model A instead of Model C). The entry in the SS column for that row is the SSR computed earlier. Associated with this SSR is a single degree of freedom, PA – PC = 1, for this comparison. The next row provides information about the error remaining in Model A and its associated degrees of freedom (n – PA). The final row provides similar information for Model C. Calculating MS = SS/df provides the basis for calculating F according to the sum of squares formula presented earlier. In the p column is indicated the fact that the computed PRE and F exceed the critical value with $\alpha = .05$. Finally, the value of PRE is given, computed as the SSR divided by SSE for Model C.

One final comment is appropriate about the statistics we have just computed. Throughout this book, we will use PRE to refer to the proportional reduction in error when replacing any compact model with an augmented one. In this sense, PRE is a very general statistic that is informative regardless of the specifics of the models that are being compared. However, in the history of inferential statistics, a variety of specific names for PRE have been developed for particular model comparisons. For the case at hand, testing whether the slope in a single predictor model is different from zero, the square root of PRE is known as the Pearson correlation coefficient, commonly referred to simply as the correlation or r. Thus, the inferential test we have just conducted is equivalent to a test of whether the true correlation between two variables differs from zero. If PRE and F do not surpass their critical values, then it means that the true value of PRE (η^2) may equal zero. In fact, since the F statistic we have computed has only a single degree of freedom in its numerator, its square root is a t statistic with n – PA (n – 2 in this case) degrees of freedom. It is simple algebra to show that the square root of the F formula we have given is the same as the t that is typically given

Source	*SS*	*df*	*MS*	*F*	*p*	*PRE*
Reduction (using $b_1 = 0.712$)	2811.370	1	2811.37	46.83	< .001	.521
Error (using $b_1 = 0.712$)	2581.490	43	60.034			
Total error (using $b_1 = 0$)	5392.860	44	122.565			

Figure 6.11 ANOVA source table test of $H_0: \beta_1 = 0$ in simple regression context

in traditional statistics texts for testing whether a correlation between two variables is significantly different from zero:

$$t_{n-2} = \frac{r}{\sqrt{\frac{1-r^2}{n-2}}}$$

As a result, our conclusion in this case that Model A does significantly better than Model C is equivalent to concluding that the two variables are related, that β_1 differs from zero, that the true correlation differs from zero, and that the true PRE (η^2) differs from zero.

Testing Null Hypotheses for Other Values of β_1

There may be occasions when theory makes strong predictions about particular values of slopes other than zero. For instance, if two variables are measured in the same metric, then we might be interested in whether a one-unit difference on one variable is associated with a one-unit difference on the other. To illustrate, we continue to use the internet use/service employment example. Since they are both measured in the same metric (percentage of the country who used the internet or the percentage of service employment), we will ask the question of whether a 1% increase in service employment is associated with a 1% increase in internet use on average across the countries. Obviously, we do not have any strong theory that would make this prediction; we use it for illustrative purposes only. The model comparison for this question is

MODEL A: $Y_i = \beta_0 + \beta_1 X_i + \varepsilon_i$
MODEL C: $Y_i = \beta_0 + 1X_i + \varepsilon_i$

And the null hypothesis is that $\beta_1 = 1$.

Unfortunately, few statistical programs readily permit the estimation of parameters in a model in which some parameters are fixed at values other than zero. In this case, however, we can easily estimate β_0 in Model C by noting that if we subtract X_i from both sides of Model C we obtain

$$Y_i - X_i = \beta_0 + \varepsilon_i$$

so the mean of the constructed variable $Y_i - X_i = 16.46$ estimates β_0 in Model C. The estimated models are then

MODEL A: $Y_i = 35.903 + 0.712X_i + e_i$
MODEL C: $Y_i = 16.46 + 1X_i + e_i$

The SSE for Model C equals 3041.254, again with $n - \text{PC}$ equal to 44. Model A makes better predictions since as we saw earlier its sum of squared errors equals 2581.490. SSR equals 459.764, with PRE and F computed as

$$\text{PRE} = \frac{459.764}{3041.254} = .151$$

$$F_{1,43} = \frac{459.764/1}{2581.490/43} = 7.58$$

These values are larger than their respective critical values. Thus, we reject the null hypothesis—the true slope is significantly lower than 1.00.

Confidence Intervals of β_1

Recall from Chapter 5 that the confidence interval defines the range of values for the parameter for which we would fail to reject the null hypothesis. In other words, if we tested a null hypothesis that the parameter equals a value that lies within the confidence interval, we would fail to reject that null hypothesis. If the null hypothesis to be tested specifies that the parameter equals a value that lies outside of the confidence interval, we would reject that null hypothesis. In this sense, the confidence interval is entirely redundant with inferential tests about the value of a parameter.

In the case of β_1 in the two-parameter model we are considering, its confidence interval is given as

$$\text{CI} = b_1 \pm \sqrt{\frac{F_{crit;1,n-2;\alpha}\text{MSE}}{(n-1)s_x^2}}$$

where $F_{crit\ 1,\ n-2;\alpha}$ is the critical value of F at level α, with degrees of freedom of $\text{PA} - \text{PC} = 1$ and $n - \text{PA} = n - 2$. MSE is the mean square error from the augmented model, again based on $n - 2$ degrees of freedom. And s_x^2 is the variance of the predictor variable.

For the internet use/service employment example, the critical F with 1 and 43 degrees of freedom, using $\alpha = .05$, equals approximately 4.07, MSE from Model A equals 60.035, and the variance of the predictor variable (service employment rate) equals 126.022. Accordingly, the confidence interval for β_1 is

$$\text{CI}_{b_1} = b_1 \pm \sqrt{\frac{4.07(60.034)}{44(126.022)}} = 0.712 \pm 0.209$$

or

$$0.503 \leq \beta_1 \leq 0.921$$

Based on the present data, we can thus say we are confident that the true value for the slope in this two-parameter regression model, predicting internet use rates from service employment rates, lies somewhere between 0.503 and 0.921. Notice that both zero and 1 lie outside of this interval. Both of these are consistent with the results already reported for our two null hypotheses: Both the null hypothesis that the parameter equals zero and the null hypothesis that it equals 1.00 were rejected.

Although we might lament the fact that many statistical packages do not permit the estimation of models in which parameter values are fixed at values other than zero, the confidence interval permits a general approach for testing any null hypothesis about the slope. If the value that is specified by the null hypothesis lies within the interval, it would not be rejected. All other null hypotheses would be.

As we indicated in Chapter 5, the formula for the confidence interval provides some insights into the factors that influence statistical power—the probability of rejecting the null hypothesis when it is false—and how to improve it. In general, the narrower the confidence interval, the more precision we have in estimating a parameter, and statistical power means that we have greater precision, i.e., narrower confidence intervals. According to the formula, what are the factors that affect the width of the interval, and therefore power?

First, the critical value of F affects its width. If we use a smaller α level, thus reducing Type I errors, the critical value of F increases, thereby widening the confidence interval and resulting in less power and greater probability of Type II errors. This is the inherent tradeoff between Type I and Type II statistical errors. Second, the width of the confidence interval is affected by the mean square error from the augmented model. As the variability of errors of prediction (the MSE) is reduced, the confidence interval becomes narrower. Thus, whatever we can do to reduce error, such as improving the quality of measurement of Y_i, increases power. Third, as n increases, all else being equal, power increases. This is reflected by the fact that $n - 1$ appears in the denominator of the confidence interval.

What is new in simple regression is that we have a predictor variable X_i that also affects the width of the confidence interval; its variance appears in the denominator. As X_i becomes more variable, the interval narrows, and power improves. Given some predictor variable, we will examine its effects as a predictor with more precision (assuming a linear model) if we make sure that we sample widely across its values.

Power Analysis in Tests of Simple Regression

In Chapter 5 we performed "what if" power analyses for the simple model making a constant prediction for each observation. We can use exactly the same process to ask "what if" power analysis questions for simple regression models using one predictor variable. We perform "what if" power analyses for particular values of the true proportional reduction in error, η^2, which may be of interest, in exactly the same way as before. For example, for the internet use data, we might want to know the power of detecting a relationship between it and some variable—detecting that the slope for a predictor is different from zero—when in fact we think that $\eta^2 = .20$. To do this, we can use a software program, such as R, G*Power, or SAS, to calculate power as we mentioned in the last chapter. For this Model A/Model C comparison, $PA - PC = 1$ and $n - PA = 43$.

If $\eta^2 = .20$ then the power of our test is roughly .91. That is, if we were to do a study with an n of 45 and expect to find a relationship between a predictor and Y_i with a true PRE of .20, we would have roughly a 91% chance of correctly rejecting the null hypothesis.

Given that we now know the procedure for asking questions to determine the power with which we can assess whether one variable is significantly related to another in a simple regression model, we need to know what values of η^2 are appropriate and expected for such comparisons. As we noted in Chapter 5, prior experience in a research domain may provide values of PRE that can be used directly in a power analysis. For the simple regression case, we might have estimates of PRE available from previous research, typically reported as the correlation between two variables, rather than as PRE itself. In this case, we need to square the correlation coefficient to obtain the estimate of PRE, since PRE = r^2 for this question. As before, we would want to convert past empirical values of PRE to unbiased estimates of η^2, using the same adjustment formula as before:

$$\text{Unbiased estimate of } \eta^2 = 1 - \frac{(1-\text{PRE})(n-\text{PC})}{n-\text{PA}}$$

To illustrate, suppose prior research has reported a correlation between two variables of .33 based on a sample size of 30. We intend to do a study, examining the relationship between the same two variables, but we intend to use an n of 50. We would like to know the power of our planned study, if in fact the two variables are related as strongly as they were reported to be in the prior research. To do this, we first convert the previously reported correlation to PRE, by squaring it: $.33^2 = .11$. We then convert this PRE to an unbiased estimate of η^2 using the aforementioned formula and the sample size from the prior study that reported the .33 correlation:

$$\text{Unbiased estimate of } \eta^2 = 1 - \frac{(1-.11)(29)}{28} = .078$$

We then use the statistical software, R, to estimate our power in the new study we plan to conduct. With an n of 50, n – PA for our study will be 48, and with an anticipated η^2 of .078, our power is approximately .52. Given this result, we may want to think further about our anticipated study. It might be worthwhile, for instance, to recruit a larger sample to increase power.

Another approach for finding an appropriate value of η^2 for "what if" power analyses involves guesses about the parameter values and variances, just as in Chapter 5. Again, to have reasonable expectations about the parameter values and variances generally requires as much or more experience in a research domain as is necessary to know typical values of PRE. We present it, however, for completeness.

The formula that relates true values of the parameters to η^2, the true value of PRE, is

$$\eta^2 = \beta_1^2 \frac{\sigma_X^2}{\sigma_Y^2}$$

where β_1 is the true parameter value for the slope, σ_X^2 is the true variance of the predictor variable, and σ_Y^2 is the true variance of the dependent variable. In other words, given some alternative hypothesis that specifies what we think is the correct value for the slope, β_1, and given that we want to determine the power of our test of the null hypothesis that β_1 equals zero, we can calculate the corresponding η^2 using the previous expression, assuming we have estimates of the variances of both X_i and Y_i. We can then take that value, and the anticipated $n - 2$ for the study in the planning and estimate power.

Inferences About β_0

Our discussion so far has concentrated exclusively on inferences about the slope in the two-parameter simple regression model. But one may also compare this augmented two-parameter model with a compact one that fixes the intercept, rather than the slope, at some a priori value:

MODEL A: $Y_i = \beta_0 + \beta_1 X_i + \varepsilon_i$

MODEL C: $Y_i = \mathrm{B}_0 + \beta_1 X_i + \varepsilon_i$

where B_0 represents an a priori value for the intercept. The null hypothesis tested by such a comparison would then be

$\mathrm{H}_0: \beta_0 = \mathrm{B}_0$

Although perhaps not of frequent theoretical interest in the behavioral sciences, one particular form of this compact model involves what is known as "regression through the origin" in which the intercept is fixed at zero:

MODEL C: $Y_i = 0 + \beta_1 X_i + \varepsilon_i$

Recall that the intercept is defined as the value predicted by the model when X_i equals zero. Thus, regression through the origin means that the line goes through the (0, 0) point on the scatterplot.

Importantly, null hypothesis tests about the intercept are of a fundamentally different nature than those about the slope. When we test whether the slope is zero, we are asking whether there is any change in the predicted values when the predictor varies. In other words, we are testing whether X_i and Y_i are related. In contrast, inferences about the intercept are inferences about specific predicted values, not inferences about changes in predicted values. That is, we are asking whether the predicted value of Y_i *when* X_i *equals zero* differs significantly from some a priori value, be it zero (in regression through the origin) or some other value.

In many cases, there is no intrinsic interest in the predicted value of Y_i when X_i equals zero. This may be because the value of zero lies outside of the range of the X_i values represented in the dataset, as in the data that we have been using where no country has a service employment rate that is near zero. Even if zero is a value within the range of X_i in the dataset, it may not be a point that has much theoretical interest. However, with transformations of X_i, such as centering it or deviating it from its mean (discussed earlier), the zero value becomes potentially more interesting. Earlier in this chapter we saw that when X_i is centered the intercept in the estimated model equals the mean of Y_i (i.e., $\bar{Y}$). Accordingly, with a centered predictor, the following Model A/Model C comparison is equivalent to asking questions about the mean of Y_i:

MODEL A: $Y_i = \beta_0 + \beta_1(X_i - \bar{X}) + \varepsilon_i$

MODEL C: $Y_i = \mathrm{B}_0 + \beta_1(X_i - \bar{X}) + \varepsilon_i$

with the following null hypothesis:

$\mathrm{H}_0: \beta_0 = \mathrm{B}_0$ or $\mu_Y = \mathrm{B}_0$

where μ_Y is the true mean of Y_i.

Let us illustrate such a test with the internet use data that we have used throughout this chapter, asking the same question that we did at the end of Chapter 4—whether the projection of a mean rate of 90% was overly optimistic for the year 2021. But this time, we will ask the question in the context of a simple regression model that makes conditional predictions of internet use based on countries' service employment rates. We will then compare our test in this model with our results from Chapter 4 to examine how the present test differs from that used there.

Estimating a model in which internet use rates are regressed on service employment, with the latter variable in its centered or mean-deviated form, gives the following estimates:

$$\text{MODEL A: } \hat{Y}_i = 83.982 + 0.712(X_i - \bar{X})$$

with a sum of squared errors of 2581.490. As explained earlier, the intercept in this model is now the mean value for the internet use variable, while the slope has not changed compared to the model in which X_i was not centered. Again, in a deep sense, this model is identical to the one with X_i uncentered, in that it makes the same predictions and therefore necessarily has the same sum of squared errors.

To test the null hypothesis that the true mean of Y_i equals 90, we want to compare this model to a compact one in which the intercept has been fixed at 90:

$$\text{MODEL C: } \hat{Y}_i = 90 + b_1(X_i - \bar{X})$$

This comparison permits a test of the null hypothesis:

$$H_0\text{: } \beta_0 = 90 \text{ or } \mu_Y = 90$$

As we said previously, many computer packages for data analysis do not readily permit the estimation of models in which parameters have been fixed at values other than zero. In this case, with a centered predictor, it can be shown that the best least-squares estimate for β_1 does not change even if we fix the intercept at some a priori value.[1] Accordingly, Model C is estimated as

$$\text{MODEL C: } \hat{Y}_i = 90 + 0.712(X_i - \bar{X})$$

The sum of squared errors from this model can be directly computed across the observations. More simply, it can be computed by first calculating the sum of squares reduced (SSR) as we move from Model C to Model A.

Recall that $\text{SSR} = \Sigma(Y_{iA} - Y_{iC})^2$, accordingly,

$$\text{SSR} = \Sigma((83.982 + 0.712(X_i - \bar{X})) - (90 + 0.712(X_i - \bar{X})))^2$$

Because both predicted values have the same slope for the centered predictor, this reduces to

$$\text{SSR} = \Sigma((83.982) - (90))^2 = 45(-6.018)^2 = 1629.735$$

Accordingly, the sum of squared errors for the compact model is

$$\text{SSE(C)} = \text{SSE(A)} + \text{SSR} = 2581.490 + 1629.735 = 4211.225$$

Now that we have the sums of squared errors, we can calculate PRE and F for the comparison and the test of the null hypothesis:

$$\text{PRE} = \frac{\text{SSR}}{\text{SSE}(\text{C})} = \frac{1629.735}{4211.225} = .387$$

$$F = \frac{.387/1}{(1-.387)/43} = 27.15$$

All of this is summarized in the ANOVA source table of Figure 6.12.

Let us now compare these results with the test of the same null hypothesis reported at the end of Chapter 4, in the context of the simplest single-parameter model. There, estimated Models A and C were:

MODEL A: $\hat{Y}_i = 83.982$
MODEL C: $\hat{Y}_i = 90$

The sum of squared errors associated with Model A was 5392.860, that is, the total sum of squares of Y_i around its mean. While the SSR was found to be

$$\text{SSR} = \Sigma((83.982) - (90))^2 = 45(-6.018)^2 = 1629.735$$

Thus, the results of this test of the same null hypothesis yielded the ANOVA source table in Figure 6.13.

Although both tests resulted in the rejection of the null hypothesis, clearly the two approaches differ substantially in terms of the obtained PRE and F. Those values, in the context of the two-parameter simple regression model, are about twice what they were for the same test in the context of the single-parameter model of Chapter 4. And this difference is attributable entirely to the difference in the sum of squared errors for Model A (and the concomitant difference in n – PA). The sum of squared errors for Model A in the context of the simple regression model equals 2581.490, with n – PA equal to 43. The same values in the context of the single-parameter model

Source	*SS*	*df*	*MS*	*F*	*PRE*	*p*
Reduction	1629.735	1	1629.735	27.15	.387	< .001
Error	2581.490	43	60.034			
Total error	4211.225	44	95.710			

Figure 6.12 ANOVA source table test of H_0: $\beta_0 = \mu_Y = 90$ in simple regression context

Source	*SS*	*df*	*MS*	*F*	*PRE*	*p*
Reduction	1629.735	1	1629.735	13.29	.232	< .001
Error	5392.860	44	122.565			
Total error	7022.595	45				

Figure 6.13 ANOVA source table test of H_0: $\beta_0 = \mu_Y = 90$ in single-parameter model (from Chapter 4)

of Chapter 4 are SSE = 5392.860 and n – PA = 44. As a result, the MS error values (denominators of F) are markedly different: 60.034 (for the simple regression model) versus 122.565 (for the Chapter 4 single-parameter model). Importantly, the numerator of F is the same in both tests, with SSR equal to 1629.735 and PA – PC = 1.

The difference in the sums of squared errors for Model A in the two source tables is, not surprisingly, the sum of squares that is attributable to the predictor variable (service employment rates), on the basis of which conditional predictions are made in the simple regression model. That is, earlier, we reported that the SSR attributable to a Model A that made conditional predictions of internet use, using service employment rates as a predictor, compared to a Model C that predicted the mean value of Y_i was 2811.37. This is the difference between the sums of squared errors for the two Model As used in these tests. Model A for the Chapter 4 single-parameter version of the test makes unconditional predictions of Y_i; Model A for the simple regression version of the test makes predictions of Y_i conditional on (centered) X_i. Making these conditional predictions means that Model A in the simple regression context has one fewer degrees of freedom for error (n – PA = 43) than Model A in the Chapter 4 version of the test (where n – PA = 44). But the loss in degrees of freedom has been more than compensated for by the substantial difference between the sums of squared errors of the two Model As. As a result, the test in the context of the conditional simple regression model has substantially more statistical power than the same test conducted in the context of the single-parameter model of Chapter 4.

In Chapter 4 we mentioned that the test we reported is known as the one-sample *t-test*. The advantage of our model comparison approach is that we have been able to generalize this test to cases where conditional predictions are made by the models that are compared. In the jargon traditionally used in the statistical inference literature, we have just conducted a one-sample *t-test* while controlling for a "covariate."

A further advantage of our approach is that it permits us to conduct inferential tests about predicted values other than the mean. Suppose, for instance, that we had some reason to want to ask about internet use in countries where the service employment rate was 70%. Rather than centering the predictor around its mean value, one could deviate the predictor from the value of 70. Then one could estimate a Model A in which the predictor variable was this deviated variable:

$$\text{MODEL A: } Y_i = \beta_0 + \beta_1(X_i - 70) + \varepsilon_i$$

In a deep sense we would be dealing with the same conditional model; we have simply moved the zero point on the x-axis to what was the value of 70. Accordingly, the slope remains the same, while the estimated intercept in the model would be the predicted value of Y_i when X_i equals 70. This model might then be compared to a Model C that uses the same deviated predictor but fixes the intercept at some a priori value of interest.[2]

Confidence Interval for the Intercept

The confidence interval for the intercept represents the range of values for the intercept that would not be rejected by an inferential statistical test. In simple regression models, with a single predictor variable, the confidence interval for the intercept is

$$b_0 \pm \sqrt{\frac{F_{crit;1,n-2;\alpha}\text{MSE}\left(\Sigma X_i^2\right)}{n\Sigma\left(X_i - \bar{X}\right)^2}}$$

This is the confidence interval for the intercept regardless of whether the predictor has been deviated from some value or not. If it has been deviated, then the X_i terms in the confidence intervals are the new values following deviation.

When using a centered predictor, deviated from its mean value, then $\bar{X}$ for the centered predictor equals zero, and the aforementioned formula for the confidence interval reduces to

$$b_0 \pm \sqrt{\frac{F_{crit;1,n-2;\alpha}\text{MSE}}{n}}$$

which is the formula that we gave in Chapter 5 for the confidence interval for β_0 in the single-parameter model (except there the critical F value had 1 and $n - 1$ degrees of freedom). Of course, with a predictor variable in the model that is a useful predictor of Y_i, the MSE in the numerator of this confidence interval should be considerably smaller than the MSE in the numerator of the interval used in Chapter 5 without a predictor. This difference reflects the increase in power when conducting inferential tests about the mean in the context of a useful predictor compared to the same test in the single-parameter context of Chapter 5.

To illustrate, with the centered predictor from the simple regression model used in this chapter, the confidence interval for β_0 (equivalently for μ_y) equals

$$83.982 \pm \sqrt{\frac{4.07(60.034)}{45}} = 83.982 \pm 2.322$$

$$81.660 \leq \beta_0 \leq 86.304$$

In contrast, in the context of the single-parameter model of Chapter 5, the same confidence interval equals

$$83.982 \pm \sqrt{\frac{4.07(122.565)}{45}} = 83.982 \pm 3.329$$

$$80.653 \leq \beta_0 \leq 87.311$$

Clearly, the confidence interval for the mean of Y_i is smaller in the context of the simple regression model, reflecting the increase in power resulting from the inclusion of the predictor variable (with its associated reduction in errors of prediction).

Using the general formula for the confidence interval for β_0 in this two-parameter model given earlier:

$$b_0 \pm \sqrt{\frac{F_{crit;1,n-2;\alpha}\text{MSE}\left(\Sigma X_i^2\right)}{n\Sigma\left(X_i - \bar{X}\right)^2}}$$

We can generate confidence intervals for predicted values at all possible levels of the predictor variable (by deviating the predictor from each of those levels and then calculating the resulting confidence intervals for the varying intercepts). In Figure 6.14 we have graphed the 95%

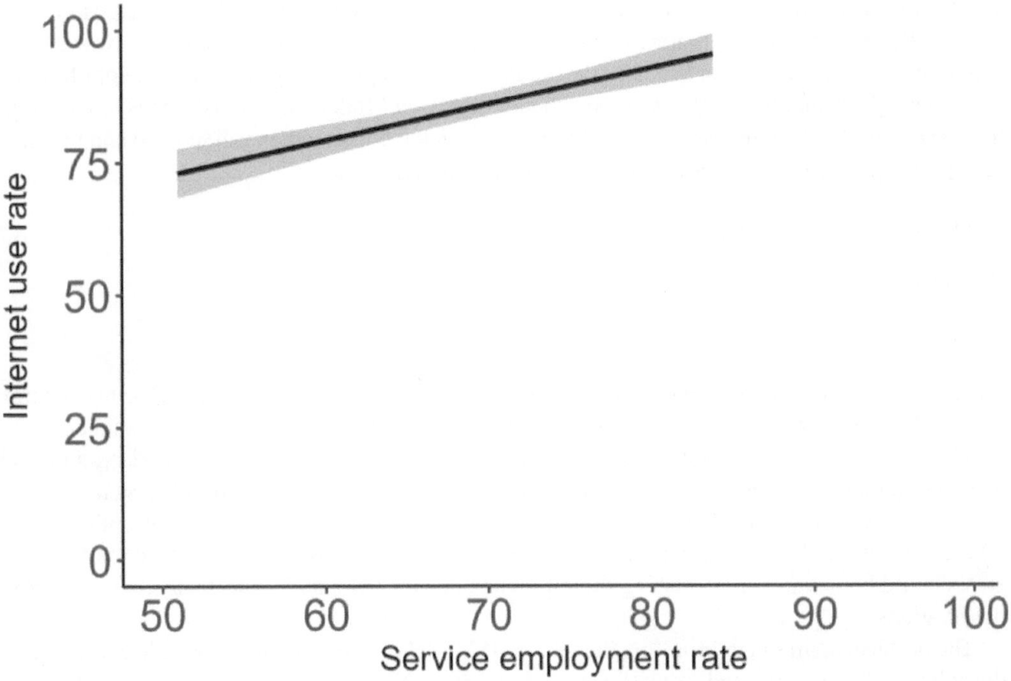

Figure 6.14 Confidence limits of predicted values

confidence limits for the predicted values of internet access at values of the predictor (service employment rate) ranging from 50% to about 85%. The middle straight line in the graph represents the predicted values and the shaded areas above and below it represent the upper and lower confidence limits. Notice that the confidence interval is narrowest near the mean value of the predictor, where the predicted value is the mean of the dependent variable, and it becomes wider as we depart in either direction from that mean. Thus, we have greater precision in inferring predicted values near the joint mean of the bivariate distribution than when we move further away along the horizontal axis.

Two-Parameter Model Comparisons

To conclude this chapter on two-parameter simple regression models, we should note that MODEL A/MODEL C comparisons are now also possible involving PA – PC > 1. For instance, suppose we wanted to simultaneously ask about fixed a priori values both for the intercept and for the slope:

MODEL A: $Y_i = \beta_0 + \beta_1 X_i + \varepsilon_i$

MODEL C: $Y_i = \mathrm{B}_0 + \mathrm{B}_1 X_i + \varepsilon_i$

where both B_0 and B_1 are fixed a priori values (zero or any other value). The resulting null hypothesis from this comparison has two different components to it:

H_0: $\beta_0 = \mathrm{B}_0$; $\beta_1 = \mathrm{B}_0$

To illustrate a situation where this sort of model comparison might be of interest, suppose we had data on the heights of each member of father–son pairs and we wanted to examine how these two height measures were related. We might estimate a Model A predicting each son's height from his father's height, estimating both the slope and intercept as in Model A previously. This model might then be meaningfully compared with a Model C in which the intercept was fixed at 0 and the slope at 1, yielding the following model comparison and null hypothesis:

MODEL A: $Y_i = \beta_0 + \beta_1 X_i + \varepsilon_i$

MODEL C: $Y_i = X_i + \varepsilon_i$

$H_0: \beta_0 = 0; \beta_1 = 1$

Such a comparison asks whether sons and fathers are perfect resemblances of each other in terms of height. Except for error, Model C assumes that sons' heights equal their fathers' heights.

There is nothing statistically wrong with such a two-parameter model comparison and such a compound null hypothesis. The resulting PRE and F would be computed in the same ways as always, albeit with PA – PC = 2 degrees of freedom for the numerator. Since there is nothing to estimate in Model C, PC = 0. And, as the model predicts X_i for each observation, the sum of squared errors could be obtained by squaring the difference between X_i and Y_i and then summing across observations.

The problem comes in interpreting the results. If Model C is rejected in favor of Model A (i.e., if the null hypothesis is rejected) we will not be able to be confident about *why* it was rejected. Maybe it is the case that the a priori value in Model C for the intercept is wrong. Maybe it is the case that the a priori value in Model C for the slope is wrong. Maybe both a priori values are wrong. All we can say is that Model A is preferred over Model C, but we will not know why.

It is for this reason that we prefer model comparisons where PA – PC = 1, where the numerator of the F statistic has only a single degree of freedom. There is nothing wrong with statistical tests involving more than one degree of freedom in the numerator of F, and we will occasionally discuss them and even recommend them. In general, however, we will refrain from such unfocused model comparisons. We simply note that in the context of the present two-parameter simple regression model they become possible for the first time.

Summary

In this chapter we considered models with a single predictor, measured more or less continuously. We started by considering the definitions of both the intercept and slope in such models, with the former being a particular predicted value when the predictor equals zero and the latter being the per unit difference between predicted values. We then provided formulas for estimating these parameters and used these to illustrate alternative ways of thinking about what a slope estimate means. Finally, we considered models in which the predictor is centered, that is, put into mean-deviation form. In such cases, the slope of the predictor is unchanged while the intercept equals the mean of Y_i.

The second half of the chapter was devoted to model comparisons, treating the two-parameter, single predictor model as Model A and comparing it to an alternative Model C, testing inferences about the slope and the intercept and then both simultaneously. Inferential tests of the slope most frequently make comparisons with a Model C that fixes the slope at zero, thus testing the null hypothesis that the predictor is not a useful one, or equivalently that the predictor and Y are unrelated to each other. There are occasions, however, when other null hypotheses about the slope are of interest, and we illustrated these. Inferences about the intercept are most frequently of interest when the predictor has been centered, thus permitting inferences about the mean of Y_i. Testing null hypotheses about the mean of Y_i will be more powerful in the presence of a predictor when that predictor is a useful predictor; that is, when it explains a significant amount of variation in Y_i. This was illustrated by making comparisons with the simplest model comparisons of Chapter 4.

Notes

1 It is only in the case of a centered predictor in simple regression models that the estimate of the slope will remain constant regardless of whether the intercept is estimated or fixed at various a priori values. Unless predictors are centered around their mean, this will generally not be the case.

2 Importantly, the estimated slope in such a Model C will differ from the estimated slope in Model A. As we mentioned previously, only when the predictor variable is centered around its mean will the estimated slope remain the same regardless of whether the intercept is estimated or fixed. For all other cases, the slope estimate will vary.

7 Multiple Regression

Models With Multiple Continuous Predictors

In the previous chapters we considered the simple model that makes the same prediction for all the observations in the data and the simple regression model. We now increase complexity by considering models that make predictions conditional on multiple predictor variables. For example, in our discussion of the rates of internet use, we suggested several plausible hypotheses about variables that might be predictive of each country's internet use—service employment, authoritarian versus democratic government, and GNP. The simple regression model limited us to testing one predictor variable at a time. To test our different hypotheses using the simple regression model, we would have to first test the model with one predictor, say, service employment, and then test another one-variable model with, say, GNP as the predictor. It is obviously more efficient to test more than one predictor simultaneously. More importantly, understanding differences in the countries' internet use will likely require more than one predictor. For example, we might need to base our predictions on *both* the service employment rate and GNP. The solution we consider in this chapter is *multiple regression.*

Multiple regression is very similar to simple regression. There are, however, a few important differences. We begin by briefly considering these similarities and differences with respect to the model, the new problem of redundancy among the predictors, and statistical inference. Then we will consider an extended example designed to illustrate the interpretation of the parameters in the multiple regression model.

Multiple Regression Model

We want to consider models of the form

$$Y_i = \beta_0 + \beta_1 X_{1i} + \beta_2 X_{2i} + \ldots + \beta_{p-1} X_{p-1i} + \varepsilon_i \tag{7.1}$$

where, as before, Y_i represents an observed data value and ε_i represents the error disturbance. X_{ji} represents the value of the *i*th observation on the *j*th predictor variable and β_j is the *partial regression coefficient* representing the weight we should give to X_{ji} in making our predictions conditional on X_{ji}. In other words, the partial regression coefficients are the degree to which we adjust our prediction $\hat{Y}_i$ by each observation's X_{ji} value. In the multiple regression model the coefficients are called *partial* regression coefficients because, as we shall see later, the value of, say, β_1 may well depend on whether, say, the predictor X_{2i} and its parameter β_2 are included in the model. To remind us that the meaning of β_j is conditional on the other predictor variables included in the equation, we sometimes use the notation $\beta_{j.123\ldots p-1}$. The letter before the dot in the subscript represents the variable with which the regression coefficient β is associated and the numbers after the dot represent the other variables that are simultaneously included in the model equation. But this notation is

DOI: 10.4324/9781003438335-12

cumbersome, so we, and certainly most computer programs, more often just use β_j with the understanding that its meaning depends on all the other parameters and predictors included in the model.

We often refer to the model of Equation 7.1 as the multiple regression model or sometimes as the general linear model. Linear means that the separate components, after first being weighted by their respective β_j, are simply added together. The general linear model is indeed very general and constitutes an extremely important tool for data analysis. As we shall see in subsequent chapters, even the "linear" part is not much of a restriction because many apparently nonlinear models can be represented in terms of Equation 7.1 by a suitable choice of the predictor variables. It is also clear that the models of the previous chapters are simply special cases of the multiple regression model. In fact, a great many statistical procedures can be represented in terms of Equation 7.1 with clever construction of the X_{ij} predictor variables. The following chapters are devoted to consideration of several interesting and very important special cases, such as analysis of variance, and a number of special problems in the context of this model, such as violation of the assumption that the ε_i error terms are independent. In this chapter, we present the basics of the general model—basics that we will use again and again.

Redundancy

The power and generality of the multiple regression model do not come without a cost. That cost is a definite increase in complexity and the introduction of some special problems that do not arise in the case of the simple models that we considered previously. One such problem is redundancy among the predictors. Suppose, for example, that we were attempting to predict the weights of a sample of elementary school children with two predictor variables: height measured in inches and height measured in centimeters. Our two predictor variables are obviously completely redundant—either would do as well as the other for predicting weight, and if we had already used one as a predictor there would clearly be no benefit to adding information from the other predictor.

Redundancy among the predictor variables is seldom so extreme as measuring height in inches and centimeters. A more typical example would be predicting the weights of the elementary school children using height and age as predictors. Height and age are obviously related—knowing children's ages would likely allow us to make reasonable guesses about their heights—but they are not completely redundant because our guess of height from age would not be perfect. This means that height and age share in their ability to reduce error in the predictions of weight. If we first used height as a predictor, we would be improving our predictions (and reducing our error) in some of the same ways we would have if we had first used age as a predictor. Sorting out the effects of predictor variables and their relative utility is clearly going to be difficult when there is redundancy among our predictors. In general, there is redundancy when it is possible to predict, at least somewhat, one or more of the X_{ji} with some combination of the other predictor variables. In our analysis of multiple regression models, we will have to be alert for redundancy and be very careful about the model interpretations we make in such cases. We will consider special techniques for doing so.

Statistical Inference

Although we will have more choices of models to compare and test with multiple predictor variables, the multiple regression model poses no new problems for statistical inference. To make statistical inferences we use what is now our standard operating procedure for any statistical model. We first fit our model to the data by estimating parameters so as to minimize error. Next, we calculate PRE for our augmented model (Model A) relative to some suitably chosen compact model (Model C) that is defined by the null hypothesis that we wish to test. Then we calculate F, an index of how much PRE we get per extra parameter. We then compare these calculated statistics (PRE and F) to their critical values—that is, the values from their sampling distributions that would surprise us if Model C (and the null hypotheses) were true. As we discussed in Chapter 4, what constitutes surprising in this case depends on our alpha level, that is, what we define as an acceptable Type I error rate. If the obtained values of PRE and F are surprising, then we reject Model C in favor of Model A. If the obtained values of PRE and F are not surprising, then we do not reject Model C. We consider, in turn, each of these steps for the multiple regression model, but only the details of estimating parameters are new. The statistical inference process itself is *exactly* the same as before.

Estimating Parameters in Multiple Regression

At the conceptual level, the problem of estimating parameters for the multiple regression model is the same as for previous models that we have considered. We want to find the least-squares estimators of $\beta_0, \beta_1, \ldots, \beta_{p-1}$ that will minimize SSE, the sum of squared errors. That is, we want to find the estimates $b_0, b_1, \ldots, b_{p-1}$ in the estimated model:

$$\hat{Y}_i = b_0 + b_1 X_{1i} + \ldots + b_{p-1} X_{p-1i}$$

so that

$$\text{SSE} = \Sigma(Y_i - \hat{Y}_i)^2$$

is as small as possible. In that way we will ensure that the model fits the data as closely as possible to make the error (as measured by SSE) as small as possible.

If there is no redundancy among the predictors, then the same procedures we used in Chapter 6 to estimate slopes in the simple regression model will produce the appropriate estimates for the parameters in the multiple regression model. That is, the best estimate for each slope is

$$b_j = \frac{\Sigma\left(X_{ij} - \bar{X}_j\right)\left(Y_i - \bar{Y}\right)}{\Sigma(X_{ij} - \bar{X}_j)^2}$$

and the best estimate of the intercept is

$$b_0 = \bar{Y} - b_1 \bar{X}_j - b_2 \bar{X}_j - \ldots - b_{p-1} \bar{X}_{p-1}$$

These formulas apply whenever there is no redundancy among the predictors. To say there is no redundancy among the predictors is equivalent to saying that the correlation between every pair of predictors equals zero. Unless the predictor variables are specially constructed to meet this condition (as in some experimental designs), at least some redundancy among the predictors will always be present. In the presence of redundancy, deriving the least-squares parameter estimates is considerably more complicated. We can no longer use the procedures outlined in Chapter 6 and given by the aforementioned formula for b_j to estimate those parameters. There are well-defined algorithms for finding the unique least-squares estimates when there is redundancy among the predictors. However, these algorithms are extremely tedious on hand calculators for even just a few predictors and arc, on the whole, neither very conceptual nor helpful in aiding our understanding of multiple regression. We therefore leave the calculation of parameter estimates to the computer.

All standard computer statistical packages have multiple regression procedures that can estimate the parameters of any of the models we consider in this book.

Before turning over all of our calculations to the computer, however, it is important that we have a firm understanding of the underlying concepts. For the simple single-parameter model we found the single value b_0 (a point, geometrically) that minimized SSE. For simple regression we found two values b_0 and b_1 that defined a line: b_0 is the intercept of that line, and b_1 is its slope. For a multiple regression model with two predictors the three estimates b_0, b_1, and b_2 define a plane, as depicted in Figure 7.1. The two predictor variables define two of the axes in the three-dimensional space (the two horizontal ones), and the data variable provides the vertical axis.

In simple regression we conceptually moved the line around, changing its slope and its intercept, until we found the position of the line that minimized total error. Similarly, for the multiple regression model we can move the plane around until we find the location that minimizes total error. Actually, we want to minimize squared error, but we have not complicated the picture in Figure 7.1 by adding the appropriate squares; conceptually the idea is the same. The intersection of the plane with the y-axis defines b_0, the slope of the plane with respect to X_1 defines b_1, and the slope with respect to X_2 defines b_2. Conceptually the complicated computer algorithms simply find the location of the plane that minimizes the sum of squared errors, where an error is defined as the vertical distance from each observation to the plane that represents the model. For more than two predictors, the model is equivalent to a hyperplane in spaces with four or more dimensions. Although it is impossible to draw a picture of such hyperplanes, the concept is the same. The best

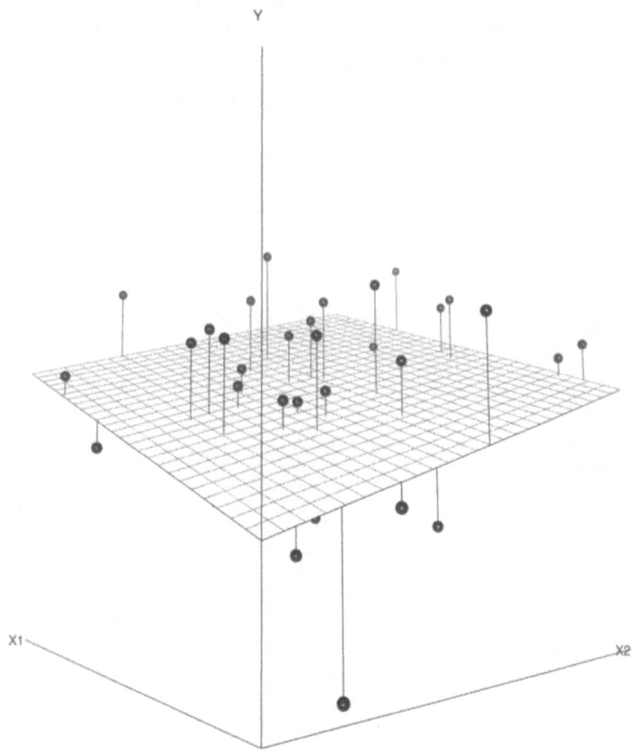

Figure 7.1 Plane defined by multiple regression model with two predictors

location for the hyperplane minimizes the sum of squared errors, and the hyperplane's intercept and slopes define the parameter estimates.

Interpreting Parameters in Multiple Regression

We will use a detailed examination of the data in Figure 7.2 to develop our understanding of the meaning of the partial regression coefficients in a multiple regression model. This dataset, which appears in many SAS examples (SAS Institute, 2015) and in other examples found on the web, contains the weight, height, age, and gender of 19 middle school or junior high students. We use this simplistic dataset because we have good intuitions about the relationships among these variables. Our goal will be to develop a model of the variable weight. We generally expect taller students to weigh more than shorter students. We expect older students to be heavier, except if the younger students happen to be exceptionally tall. We will see that such intuitions underlie the interpretation of multiple regression models.

We will begin by considering age and height as predictors of weight. That is, our model is

$$Wt_i = \beta_0 + \beta_1 Age_i + \beta_2 Ht_i + \varepsilon_i$$

where β_1 is the amount we adjust each person's weight estimate up or down depending on his or her age and β_2 is the amount we adjust each person's weight estimate up or down depending on his or her height. Figure 7.3 depicts data for the weights, heights, and ages for these 19 students as a three-dimensional scatterplot. As landmarks, the graph denotes the points for Philip (72 inches tall, age 16, and 150 pounds) and Joyce (51.3 inches tall, age 11, and 50.5 pounds). It is easy to see that the points in Figure 7.3 are not randomly scattered but instead tend to fall along a plane. More importantly, it is easy to see that older, taller students tend to be heavier. Another common way to display data for multiple predictors, especially when having more than two predictors precludes

Name	*Age*	*Height (in.)*	*Weight (lb)*
Alfred	14	69.0	112.5
Antonia	13	56.5	84.0
Barbara	13	65.3	98.0
Camella	14	62.8	102.5
Henry	14	63.5	102.5
Jamal	12	57.3	83.0
Jane	12	59.8	84.5
Janet	15	62.5	112.5
Jayden	13	62.5	84.0
John	12	59.0	99.5
Joyce	11	51.3	50.5
Judy	14	64.3	90.0
Louise	12	56.3	77.0
Mary	15	66.5	112.0
Philip	16	72.0	150.0
Robert	12	64.8	128.0
Sequan	15	67.0	133.0
Thomas	11	57.5	85.0
William	15	66.5	112.0

Figure 7.2 Age, height, and weight of 19 middle school students

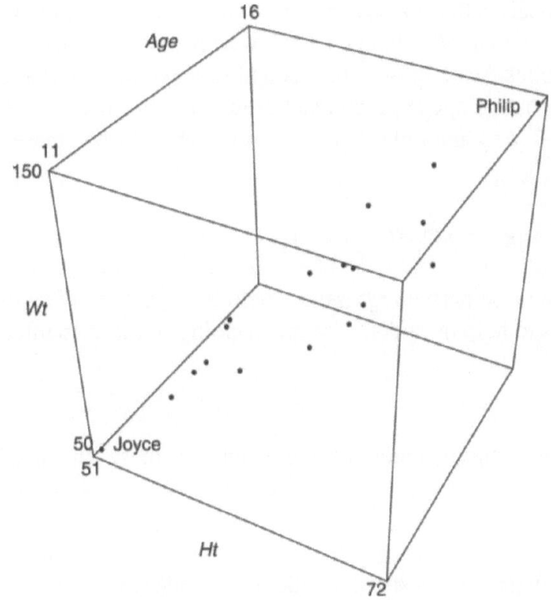

Figure 7.3 Three-dimensional scatterplot of the relationship of age and height to weight

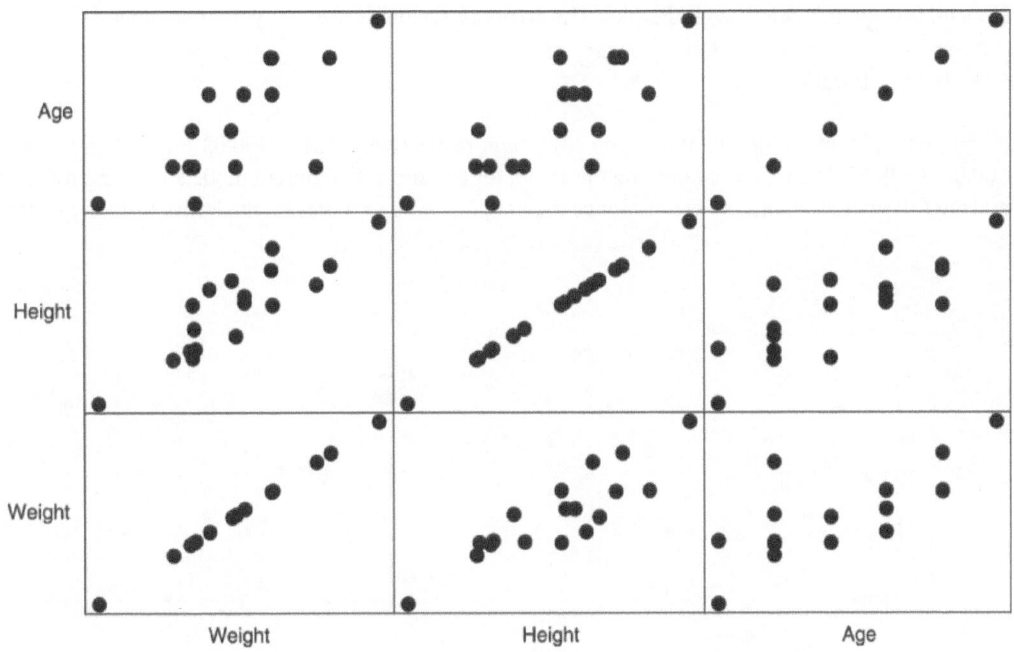

Figure 7.4 Matrix of two-dimensional scatterplot of the relationships among age, height, and weight

viewing the *n*-dimensional scatterplot, is a matrix of two-way scatterplots as depicted in Figure 7.4. In the last row we can see that taller students tend to be heavier and also that older students tend to be heavier. In the rightmost scatterplot in the second row, we can see that older students also tend to be taller; that is, height and age are redundant predictors. Because of the redundancy we cannot use the simple formulas. Any multiple regression program provides these estimates of the model parameters:

$$\hat{Wt}_i = -141.22 + 1.28Age_i + 3.60Ht_i$$

But what exactly do these partial regression coefficients mean? We will use a series of simple models and regressions to help us understand the meaning of these coefficients.

A Model of Weight

We begin by considering a simple model for the weights of these students. That is

$$Wt_i = \beta_0 + \varepsilon_i$$

Of course, the best estimate for this simple model is the mean, so

$$\hat{Wt}_i = \overline{Wt} = 100.03$$

The weights of the 19 students, arrayed along the x-axis in the same order as they are listed in Figure 7.2, are displayed in the left panel of Figure 7.5. The horizontal line at 100.03 represents the simple model for these data. The points for Philip, the heaviest student, and Joyce, the lightest student, are again labeled as landmarks. The errors or residuals are easily computed as

$$e_i = Wt_i - 100.03$$

For example, the errors for Philip and Joyce are, respectively, 151 − 100.03 = 50.97 and 50.5 − 100.03 = −49.53. We will be examining the errors for a number of different models for weight in this and the following sections, so we introduce the notation *Wt.*0 to represent the errors from the simple

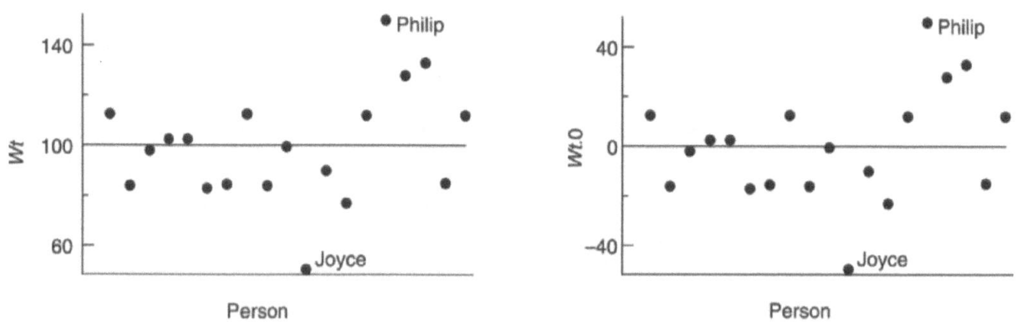

Figure 7.5 A simple model for weight

model for weight. You can read "*Wt.*0" as "the part of weight that remains after using a model with β_0." The right panel of Figure 7.5 displays these values of *Wt.*0. A positive value (i.e., those points above the horizontal line at zero) of *Wt.*0 indicates that the student is heavier than the average student, and a negative value (i.e., those points below the line) indicates that the student is lighter than the average student.

A Model of **Wt.0** *Using Age*

The values of *Wt.*0 in the right panel of Figure 7.5 represent when a student's weight is unexpectedly or surprisingly heavy relative to our simple model of weight. If we are to improve on the simple model, then we need a variable that will help us predict when a student's weight is unusually heavy or light among this group of students. The variable Age is an obvious candidate. For reasons of consistency and subsequent interpretation, we will construct *Age.*0 in the same way we constructed *Wt.*0. That is, a simple model of age is

$$Age_i = \beta_0 + \varepsilon_i$$

and the best estimate of the model is the mean, or

$$\hat{Age_i} = \overline{Age} = 13.32$$

Then we can compute the errors or residuals as

$$e_i = Age.0_i = Age_i - 13.32$$

For example, *Age.*0 for Philip and Joyce, respectively, are 16–13.32 = 2.68 and 11–13.32 = –2.32. The value of *Age.*0 for Philip means that he is 2.68 years older than the typical student in this group, and the negative value for Joyce means that she is 2.32 years younger than the typical student.

It is now natural to ask whether a student who is older than average also tends to be heavier than average. That is equivalent to a model using *Age.*0 to predict *Wt.*0, or

$$Wt.0_i = \beta_1 Age.0_i + \varepsilon_i$$

Note that we have omitted the intercept β_0 in this model. We know that the means of both *Wt.*0 and *Age.*0 must be zero because the sum of the deviations around the mean will always equal zero. Given that the regression line must go through the point defined by the two means, the intercept is necessarily zero. Hence, there is no reason to include the intercept in the model. Figure 7.6 presents all the values of *Wt*, *Wt.*0, *Age*, and *Age.*0 for the 19 students. Our question is then whether a student being older than average predicts being heavier than average? In other words, does *Age.*0 predict *Wt.*0? Figure 7.7 provides the scatterplot for these two variables. Clearly, as *Age.*0 increases, so does *Wt.*0. We can use the formula for the slope from Chapter 6 or a simple regression program to estimate the model (we urge the reader to do the computations for this and subsequent simple regressions in this section using a hand calculator or using a computer program). The resulting model is

$$\hat{Wt}.0 = 11.31 Age.0$$

Name	*Wt*	*Wt.0*	*Age*	*Age.0*
Alfred	112.5	12.47	14	0.68
Antonia	84.0	−16.03	13	−0.32
Barbara	98.0	−2.03	13	−0.32
Camella	102.5	2.47	14	0.68
Henry	102.5	2.47	14	0.68
Jamal	83.0	−17.03	12	−1.32
Jane	84.5	−15.53	12	−1.32
Janet	112.5	12.47	15	1.68
Jayden	84.0	−16.03	13	−0.32
John	99.5	−0.53	12	−1.32
Joyce	50.5	−49.53	11	−2.32
Judy	90.0	−10.03	14	0.68
Louise	77.0	−23.03	12	−1.32
Mary	112.0	11.97	15	1.68
Philip	150.0	49.97	16	2.68
Robert	128.0	27.97	12	−1.32
Sequan	133.0	32.97	15	1.68
Thomas	85.0	−15.03	11	−2.32
William	112.0	11.97	15	1.68
Mean	100.03	0	13.32	0

Figure 7.6 Values of *Wt*, *Wt*.0, *Age*, and *Age*.0

In other words, for each year older (or younger) than average, we expect the student to weigh 11.31 pounds more (or less).

For someone like Philip who is 16, or 2.68 years older than the average student, we expect him to weigh, according to the model, 11.31(2.68) = 30.31 pounds more than the average student. He in fact weighs 49.97 pounds more than the average student. Hence, the error after using both an intercept in the model and age as a predictor is *Wt.*0,*Age* = 49.97–30.31 = 19.66 pounds. Another way to say this is that Philip, "adjusting for" or "controlling for" his age, is 19.66 pounds heavier than expected. Similarly, Joyce, who is 2.32 years younger than the average student, is expected to weigh 11.31(–2.32) = –26.24 pounds. Put differently, she is expected to weigh 26.24 pounds less than the average student. However, she is in fact 49.53 pounds lighter so she weighs *Wt.*0,*Age* = –49.53 – (–26.24) = –23.29, or 23.39 pounds less than expected for her age. The values of *Wt.*0,*Age* for all 19 students are displayed in Figure 7.8 (due to rounding errors in the hand calculations, the values are slightly different from those computed previously). Note that although most of the errors become smaller (i.e., *Wt.*0,*Age* < *Wt.*0), the errors for a few students become larger. For example, although John's weight is close to the mean weight (i.e., his *Wt.*0 = –0.53), he is 14.35 pounds heavier than we would expect based on his relatively young age of 12 (i.e., his *Wt.*0,*Age* = 14.35). On the whole, however, the errors are smaller so the sum of squared errors decreases by making predictions conditional on age. In Figure 7.8, the sum of squares for *Wt.*0 (i.e., the SSE for a simple model making unconditional predictions) is 9335.74, but decreases to 4211.25 for *Wt.*0,*Age* (i.e., the SSE for a model making predictions conditional on age); the proportional reduction in error is 0.55. In other words, PRE = .55 for this model comparison

MODEL A: $Wt_i = \beta_0 + \beta_1 Age + \varepsilon_i$

MODEL C: $Wt_i = \beta_0 + \varepsilon_i$

Verifying this value for PRE using a simple regression program is a useful exercise for the reader.

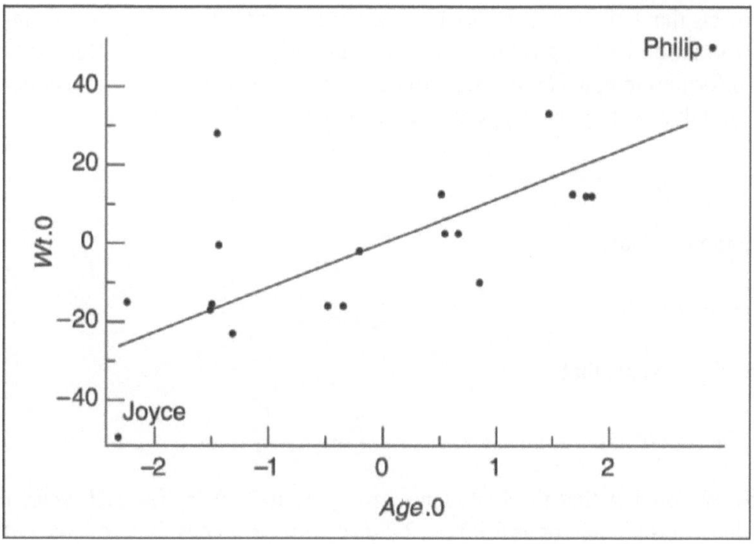

Figure 7.7 Relationship between *Age*.0 and *Wt*.0

Name	*Wt*	*Wt.0*	*Wt.0,Age*	*Ht.0,Age*
Alfred	112.5	12.47	4.74	4.76
Antonia	84.0	−16.03	−12.46	−4.96
Barbara	98.0	−2.03	1.54	3.84
Camella	102.5	2.47	−5.26	−1.44
Henry	102.5	2.47	−5.26	−0.74
Jamal	83.0	−17.03	−2.15	−1.37
Jane	84.5	−15.53	−0.65	−1.13
Janet	112.5	12.47	−6.56	−4.53
Jayden	84.0	−16.03	−12.46	1.04
John	99.5	−0.53	14.35	0.33
Joyce	50.5	−49.53	−23.35	−4.58
Judy	90.0	−10.03	−17.76	0.06
Louise	77.0	−23.03	−8.15	−2.37
Mary	112.0	11.97	−7.06	−0.53
Philip	150.0	49.97	19.63	2.18
Robert	128.0	27.97	42.85	6.13
Sequan	133.0	32.97	13.94	−0.03
Thomas	85.0	−15.03	11.15	1.62
William	112.0	11.97	−7.06	−0.53
Mean	100.03	0	0	0
SS		9335.74	4211.25	
PRE			0.55	

Figure 7.8 SSE for simple and conditional models

***A Model of* Wt.0, Age *Using* Ht.0, Age**

Why is Philip heavier than expected and Joyce is lighter than expected for their ages? Perhaps Philip is unusually tall and Joyce is unusually short for their ages. We might be tempted simply to add Height to the model. However, our intuitions and visual inspection of the height by age scatterplot suggests that height and age are partially redundant. That is, height and age share some of the same information that might be used to predict weight. The only useful part of the height information—information than can further reduce error—is the part that is not redundant with age. We can easily find that part by using age to predict height in this model:

$$Ht = \beta_0 + \beta_1 Age + \varepsilon_i$$

and computing the residuals:

$$e_i = Ht.0,Age_i = Ht_i - \hat{Ht}$$

A simple regression reveals that

$$\hat{Ht} = 25.22 + 2.79Age_i$$

In other words, we predict that these students grow about 2.79 inches each year. In particular, the model predicts Philip's height as 25.22 + 2.79(16) = 69.9 inches, but he is actually 72 inches tall. Hence, he is indeed unusually tall for his age by *Ht.*0,*Age* = 72–69.9 = 2.1 inches. Because he is taller than expected for his age, we might want to consider a model in which we adjust his predicted weight upward. Joyce, based on her age of 12, is expected to be 25.22 + 2.79(11) = 55.9 inches tall. However, she is only 51.3 inches tall, or *Ht.*0,*Age* = 51.3–55.9 = –4.6 inches shorter than expected for her age. And so perhaps her weight prediction should be adjusted downward. In general, we want to ask whether students who are taller than expected for their age (or when "adjusting for" or "controlling for" age) tend to be heavier than expected for their age. In other words, we are asking whether there is a relationship between *Ht.*0,*Age* (the part of height that is unrelated to age) and *Wt.*0,*Age* (the part of weight that is unrelated to age). The values for *Ht.*0,*Age* for all 19 students are listed in Figure 7.8. We want to consider the model,

$$Wt.0,Age_i = \beta_1 Ht.0,Age_i + \varepsilon_i$$

Figure 7.9 shows the data to answer this question in a scatterplot. Indeed, students who are taller than expected for their age (*Ht.*0,*Age* is high) tend to be heavier than expected for their age (*Wt.*0,*Age* is also high). The estimated model is

$$Wt.0,Age_i = 3.6Ht.0,Age_i$$

As with any slope, this means that as *Ht.*0,*Age* increases by one unit, the prediction of *Wt.*0,*Age* increases by 3.6 pounds. Let us be more precise, even though it is a bit wordy: For each inch students are taller than expected for their age, we expect them to weigh an additional 3.6 pounds more than expected for their age. Another way to say this is that after adjusting height and weight for age, each additional inch predicts an increase of 3.6 pounds. Note that for students whose heights were exactly what we expected based on their age (i.e., *Ht.*0,*Age* = 0), we would expect them to weigh exactly what we expected based on their age (i.e., $\hat{Wt}.0,Age = 0$). Philip is 2.1 inches taller than expected for his age, so we expect him to weigh 3.6(2.1) = 7.6 pounds more than expected for his age, but he actually weighs 19.6 pounds more than expected for his age. Or, Philip weighs *Wt.*0,*Age,Ht* = 19.6–7.6 = 12 pounds more than expected based on a model using both age and height. Similarly, Joyce, 4.6 inches shorter than expected for her age, weighs *Wt.*0,*Age,Ht* = –23–3.6(–4.6) = –23 – (–16.6) = –6.4 pounds less than expected based on her

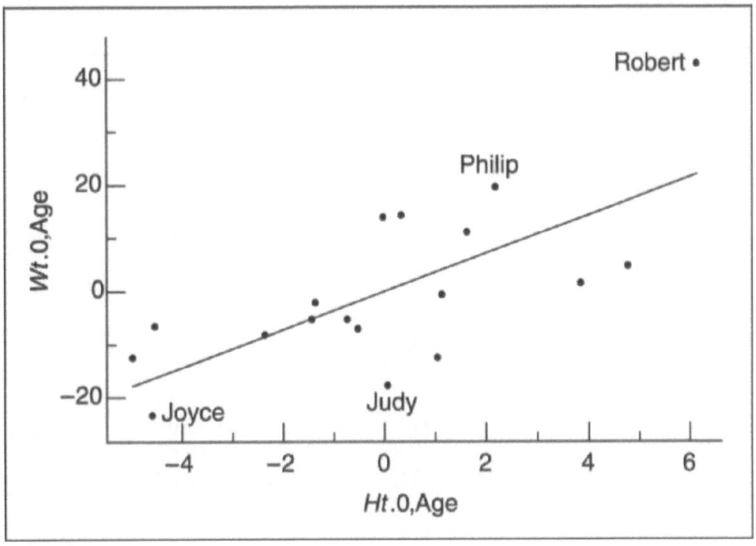

Figure 7.9 Relationship between *Ht*.0,*Age* and *Wt*.0,*Age*

age and height. When we just considered weight, Philip and Joyce were by far the heaviest and lightest students, respectively; naively we may have thought that they were overweight and underweight because they were so far from the average weight of these students. However, after accounting for (or adjusting for) their age and height, both Philip and particularly Joyce are close to the expected average weight. On the other hand, Robert and Judy, both marked in Figure 7.9, neither of whom appeared extreme when we only considered average weight, are the furthest from the weights that would be expected given their heights and ages. Note that Robert is unusually tall for his age of 12, but he is still heavier (by almost 21 pounds) than expected because of his extra height. Judy is almost exactly the height expected for her age of 14, but she is about 16 pounds lighter than expected for her combination of age and height. In other words, it is only after adjusting our weight predictions for age and height that we can see who is overweight (far above the regression line in Figure 7.9) or underweight (far below the regression line).

For each of the 19 students, Figure 7.10 lists their weights and their residuals from the various models we have considered so far. As a touchstone, let us consider the row of values for Philip. His weight of 150 pounds is about 50 (*Wt*.0) pounds heavier than average, about 19.6 (*Wt*.0,*Age*) pounds heavier than expected for students of his age, and only about 12 (*Wt*.0,*Age*,*Ht*) pounds heavier than expected for students of his age and height. Thus, the deviation or error between Philip's actual weight (the data) and our model of weight steadily decreases from 50 to 20 to 12. On the whole, although there are exceptions (such as Judy), the errors decrease as more predictors are included in the model (i.e., *Wt*.0,*Age* < *Wt*.0,*Age*,*Ht*). Most importantly, the sum of squared errors steadily decreases. As we saw before, 55% of the squared error for the simple model is reduced when we add age. Now, 50% of the squared error for the model with age is reduced when we add the part of height not redundant with age. Overall, relative to the simple model of weight, using both age and height reduces error by 77%.

$$\text{PRE} = \frac{9335.74 - 2121.04}{9337.74} = .77$$

Name	*Wt*	*Wt.0*	*Wt.0,Age*	*Wt.0,Age,Ht*
Alfred	112.5	12.47	4.74	–12.38
Antonia	84.0	–16.03	–12.46	5.38
Barbara	98.0	–2.03	1.54	–12.27
Camella	102.5	2.47	–5.26	–0.08
Henry	102.5	2.47	–5.26	–2.60
Jamal	83.0	–17.03	–2.15	2.78
Jane	84.5	–15.53	–0.65	–4.71
Janet	112.5	12.47	–6.56	9.73
Jayden	84.0	–16.03	–12.46	–16.20
John	99.5	–0.53	14.35	13.16
Joyce	50.5	–49.53	–23.35	–6.88
Judy	90.0	–10.03	–17.76	–17.98
Louise	77.0	–23.03	–8.15	0.37
Mary	112.0	11.97	–7.06	–5.15
Philip	150.0	49.97	19.63	11.79
Robert	128.0	27.97	42.85	20.80
Sequan	133.0	32.97	13.94	14.05
Thomas	85.0	–15.03	11.15	5.32
William	112.0	11.97	–7.06	–5.15
Mean	100.03	0	0	0
SS		9335.74	4211.25	2121.04
PRE			0.55	0.50

Figure 7.10 Students' weights and residuals for each of three models

of the original squared error. We will return to these values of PRE for different model comparisons when we consider statistical inference for multiple regression models later in this chapter.

Conclusion: Interpretation of Partial Regression Coefficients

The goal of this series of simple regressions using residuals is to better understand partial regression coefficients. Now that we have a good understanding of the data from this detailed examination, it is time to return to the original goal. Earlier we noted that the estimate of the model from multiple regression is

$$\hat{Wt}.0 = -141.22 + 1.28Age_i + 3.60Ht_i$$

The coefficient of 3.6 for *Ht* is *exactly* the same as the slope in Figure 7.9 where it is the coefficient predicting *Wt.0,Age* from *Ht.0,Age*. Therefore, the meaning of the 3.6 in the multiple regression equation is exactly the same as earlier: For each inch students are taller than expected for their age, we expect them to be 3.6 pounds heavier than expected for their age. Or we might say, after adjusting both height and weight for age, for each inch taller, a student is expected to weigh 3.6 pounds more. In other words, the coefficient of 3.6 in the multiple regression model describes the relationship between that part of height that is not redundant or shared with age and that part of weight that is not redundant or shared with age. It is common in reports of multiple regression to write something simple like: "When controlling for age, weight increases 3.6 pounds for each additional inch of height." However, it is important to remember the more precise meaning of the regression coefficient that we developed in the previous section and to remember that it is the slope in the scatterplot of two residuals in Figure 7.9. Statisticians and statistics programs often refer to plots like Figure 7.9 as *partial regression plots* because of their basis for defining the partial regression slope or coefficient. The wise data analyst will routinely examine partial regression plots, whose many benefits include (a) a visual representation of the regression coefficient and (b) visual identification of any unusual patterns or observations (such as Robert and Judy).

To interpret the coefficient of 1.28 for age in the multiple regression equation, we do not need to do the complete series of simple regressions parallel to those that we did previously. Instead, we can reverse the role of height and age in the aforementioned statements interpreting the coefficient for height, and we can examine the computer-generated partial regression plot in Figure 7.11 relating *Age*.0,*Ht* to *Wt*.0,*Ht*. Thus, for each year older students are than we would guess from their heights, we expect them to be 1.28 pounds heavier than expected for their height. Or we might say, after adjusting both age and weight for height, for each year older, a student is expected to weigh 1.28 pounds more. Still other language would be: After statistically equating two students on height or "holding height constant," we expect the student older by one year to weigh only 1.28 pounds more than the younger student. In that context, 1.28 pounds does not seem like much; indeed, the slope in Figure 7.11 appears as if it might not differ significantly from a slope of zero (we will check this later).

Note that, as must be the case, Robert and Judy have the largest residuals (i.e., greatest distance from the regression lines in both of the partial regression plots) because they are unexpectedly heavy and light, respectively, given their heights and ages. And again, Philip and Joyce are closer to their expected weights (i.e., closer to the regression lines) based on their heights and ages. In other words, *Wt*.0,*Age*,*Ht* = *Wt*.0,*Ht*,*Age*. Although overall predictions of weight have improved, considerable error remains. We might now ask whether there is a variable that could predict when students are unusually heavy (or light) for their heights and ages. Looking at the names above the regression line (i.e., unusually heavy for their height and ages) and those below (i.e., unusually light for their height and ages) suggests one variable to consider.

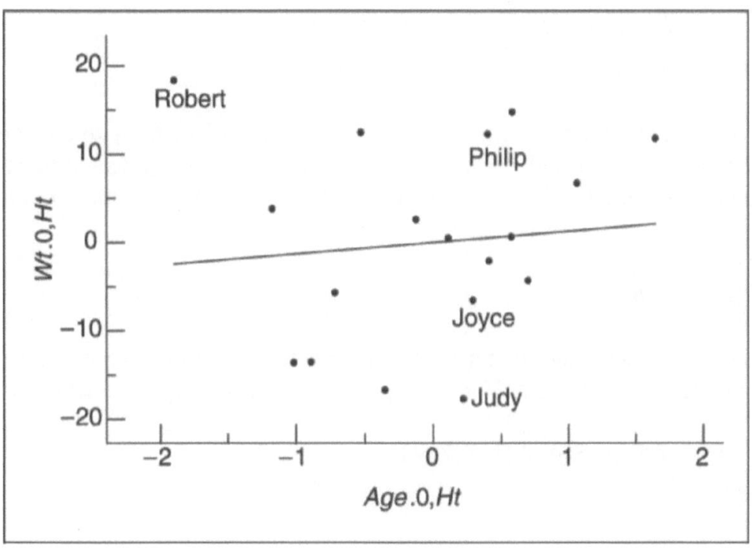

Figure 7.11 Partial regression plot between *Age*.0,*Ht* and *Wt*.0,*Ht*

Statistical Inference in Multiple Regression

Now that we understand the meaning of the partial regression coefficients, the remaining issue is how to determine whether those coefficients are significantly different from zero and to ask other model comparison questions involving models with multiple predictors. The general strategy for doing statistical inference or asking questions about data is exactly the same as before. In particular, the calculations of PRE and F pose no new problems as they are defined exactly as before. The only change is that the extra parameters in the multiple regression equation give us lots of freedom in defining Model A and Model C. As we shall see, different choices allow us to ask different questions and test different hypotheses. The only difficulty then is selecting the appropriate compact and augmented models to test a specific hypothesis. Once we have the appropriate models, estimation of the parameters (using a computer statistical package) and calculations of PRE and F are straightforward. Each question will have its own PRE and F.

For any set of data there will be many possible questions. There are several generic types of questions that occur frequently in multiple regression models. We consider each type separately.

Testing an Overall Model

Sometimes we want to ask whether our predictors *as a group* are any better than the simple model, which uses no predictors and predicts the mean for every observation. For example, U.S. college admissions offices have the practical and interesting question of whether student information such as high school percentile rank and Scholastic Aptitude Tests (SAT), both verbal and quantitative, predict performance in college. To answer that question, we could compare a model that made predictions of first-year cumulative grade point average conditional on high school rank, SAT verbal, and SAT quantitative to a simple model that made a single unconditional prediction of grade point average using the mean.

In general, we are comparing the following Models C and A:

$$\text{MODEL A: } Y_i = \beta_0 + \beta_1 X_{1i} + \beta_2 X_{2i} + \ldots + \beta_{p-1} X_{p-1i} + \varepsilon_i$$

$$\text{MODEL C: } Y_i = \beta_0 + \varepsilon_i$$

To be a test of the overall model, Model C must be the simple model with only one parameter. The null hypothesis being tested—that Model C is sufficient—is equivalent to saying that it is not useful to make the predictions conditional on all of the $p - 1$ predictor variables as a group. Formally,

$$H_0: \beta_1 = \beta_2 = \ldots = \beta_{p-1} = 0$$

Model C has one parameter and Model A has p parameters—one for each of the $p - 1$ predictors plus one parameter for the intercept (β_0). The appropriate sums of squared errors are given by

$$\text{SSE(C)} = \Sigma(Y_i - \bar{Y})^2 \text{ and SSE(A)} = \Sigma(Y_i - \hat{Y}_{Ai})^2$$

where

$$\hat{Y}_{Ai} = b_0 + b_1 X_{i1} + b_2 X_{i2} + \ldots + b_{p-1} X_{i,p-1}$$

And the difference between these two, SSR, is given by

$$\text{SSR} = \text{SSE(C)} - \text{SSE(A)} = \Sigma(\hat{Y}_{Ai} - \hat{Y}_{Ci})^2 = \Sigma(\hat{Y}_{Ai} - \bar{Y})^2$$

The PRE for testing the overall model relative to the simple mean model (the aforementioned Model A versus Model C) has the special name "coefficient of multiple determination" and is usually represented by the symbol R^2. All multiple regression computer programs, which estimate the parameters, will also provide R^2 directly so it is seldom necessary to calculate PRE by hand for the overall model. However, it is exceedingly important to remember that R^2 is nothing more than PRE for this special case of testing an overall model against the simple mean model. Also, for many other questions we will want to ask about our data, the computer programs will not provide the relevant PRE for the model comparison of interest; hence, it is important to know how to calculate PRE directly from SSE(A) and SSE(C).

The calculation of F for this case is straightforward once PRE = R^2 is known; that is

$$F = \frac{\text{PRE}/(\text{PA}-\text{PC})}{(1-\text{PRE})/(n-\text{PA})} = \frac{R^2/(p-1)}{(1-R^2)/(n-p)}$$

Alternatively, in terms of the sums of squares,

$$F = \frac{\text{SSR}/(\text{PA}-\text{PC})}{\text{SSE}(\text{A})/(n-\text{PA})}$$

If we make the normal assumptions that the errors ε_i have mean zero, are from one normal distribution, and are independent, then we can compare the F we calculated to the sampling distribution for the F statistic with $p - 1$ and $n - p$ degrees of freedom. Most computer programs that report R^2 will also report F and the "significance of F," which is the probability that a value of F this large or larger would be found if the null hypothesis (Model C) were true. If this probability is lower than our desired level for α (the Type I error probability) then we can reject the null hypothesis that none of the predictors are useful.

The probability tables for PRE and F that we developed in Chapter 4 are for the special case when the number of parameters in Models A and C differ by exactly one (i.e., PA – PC = 1). However, Model A might have many more parameters than the simple mean model, which serves as Model C. We therefore need new tables. The logic of developing these tables is exactly the same as for the tables developed in Chapter 4. In the Appendix, Figures A.2 and A.4 give the critical values for PRE and F, respectively, for $\alpha = .05$ for selected values of PA – PC.

An Example of an Overall Model Test

An example will help to make the statistical inference process for the overall model less abstract. Returning to our earlier question, college admissions offices want to know whether student information such as high school percentile rank and SAT, both verbal and quantitative, predict performance in college. In this example, we have the cumulative grade point average during the first year of college (GPA), high school rank (HSRank), SAT verbal (SATV), and SAT math (SATM) scores for 414 engineering students at a U.S. university. We want to know whether we can use those variables to make useful conditional predictions of each student's GPA or whether we would do just as well if we simply used the group mean as an unconditional prediction of everyone's GPA. In other words, does Model A, which uses all the predictor variables, do better than Model C, which uses none of the predictors? In terms of equations we are comparing

MODEL A: $GPA_i = \beta_0 + \beta_1 HSRank_i + \beta_2 SATV_i + \beta_3 SATM_i + \varepsilon_i$

MODEL C: $GPA_i = \beta_0 + \varepsilon_i$

For this comparison, PA = 4 and PC = 1. Model A and Model C would be the same if β_1, β_2, and β_3 all equal zero. Thus, comparing Models C and A is equivalent to testing the null hypothesis:

$H_0: \beta_1 = \beta_2 = \beta_3 = 0$

Computer programs provide these estimated models and associated SSEs:

MODEL A: $\widehat{GPA}_i = -1.74 + 0.027 HSRank_i + 0.011 SATV_i + 0.022 SATM_i$

SSE(A) = 172.99

MODEL C: $\widehat{GPA}_i = 2.70$

SSE(C) = 221.78

Although the program producing these estimates also provided PRE = R^2, F, and p, we will calculate them to reinforce the basic process that applies to all model comparisons involving multiple predictors:

$$\text{PRE} = \frac{\text{SSE(C)} - \text{SSE(A)}}{\text{SSE(C)}} = \frac{221.78 - 172.99}{221.78} = \frac{48.79}{221.78} = .22$$

The difference between the two sums of squared errors, 48.79, is equivalent to SSR. We can now compute

$$F_{PA-PC,n-PA} = F_{3,410} = \frac{\text{PRE}/(\text{PA}-\text{PC})}{(1-\text{PRE})/(n-\text{PA})} = \frac{.22/3}{.78/410} = \frac{.073}{.0019} = 38.54$$

$$F_{PA-PC,n-PA} = \frac{\text{SSR}/(\text{PA}-\text{PC})}{\text{SSE(A)}/(n-\text{PA})} = \frac{48.79/3}{172.99/410} = \frac{16.26}{0.422} = 38.54$$

In other words, the proportional reduction in error per additional parameter in the overall model of .073 exceeds by a factor of 38.5 the proportional reduction in error per parameter of .0019 that we would expect for a useless predictor. Consulting the tables in the Appendix for $\alpha = .05$, PA − PC = 3, and n − PA = 200 (the largest value for n − PA less than 410 available in the table), we find that the PRE and F for the data easily exceed the critical values of .038 and 2.65, respectively. Hence, we reject Model C in favor of Model A and conclude that the overall model is indeed significantly better than the simple mean model. Equivalently, we reject the null hypothesis that all the predictor coefficients equal zero.

Problems With Overall Model Tests

On the whole, overall model tests such as this one are not particularly useful for two reasons. First, it is easy when testing a large overall model to be in the position of losing a needle in a haystack. For example, suppose we were testing a large model in which only one predictor was actually useful; the other predictors are of no value in reducing error. We would get approximately the same value of PRE for this example whether we were testing the overall model or the simple regression model in which we used only the one good predictor. However, PRE per parameter (the numerator of F) would be much reduced for the overall model because we would be dividing approximately the same PRE by a larger number of extra, unnecessary parameters. Simultaneously, the remaining error (1 – PRE) per unused potential parameter (the denominator of F) would be larger for the overall model with the extra useless predictors because we would be dividing approximately the same residual error by a smaller number of potential unused parameters ($n - p$ instead of $n - 2$). Both of these effects will cause F to be considerably smaller than it otherwise would have been without the extra useless predictors. From the tables in the Appendix we see that the magnitude of the critical value of F decreases as the numerator degrees of freedom increase, so F may sometimes still be significant. However, in general, we risk losing the needle—the one good predictor—in a haystack of useless predictors. The essence of the problem is that by devoting parameters to useless predictors, we lose statistical power.

A second reason for avoiding overall tests of models is that the results are ambiguous. If the null hypothesis is rejected, then we only know that at least one of the partial regression coefficients is not equal to zero, but we do not know which predictor or predictors are useful. For example, for the GPA data we concluded that β_1, β_2, and β_3 are not all equal to zero, but we do not know precisely which ones are not equal to zero. There are seven different ways in which the overall null hypothesis could be false:

$$\beta_1 \neq 0, \beta_2 \neq 0, \beta_3 \neq 0$$

$$\beta_1 \neq 0, \beta_2 \neq 0$$

$$\beta_1 \neq 0, \beta_3 \neq 0$$

$$\beta_2 \neq 0, \beta_3 \neq 0$$

$$\beta_1 \neq 0$$

$$\beta_2 \neq 0$$

$$\beta_3 \neq 0$$

That is, HSRank, SATV, and SATM as a group are useful predictors of GPA, but we do not know whether all are needed or what combination is best. For instance, maybe SATV and SATM are highly redundant, so only one SAT score needs to be included in the regression equation.

The question we ask with the test of the overall model is so diffuse that we are generally unsure what the answer means. To increase power and reduce ambiguity, it is almost always better to ask more specific or, in the words of Rosenthal and Rosnow (1985), more *focused* questions of our data. Focused questions will generally ask about a single predictor. That does not mean that we will ignore all the other predictors, because, as we shall soon see, we can ask about the usefulness of a particular predictor in the context of all the other predictors.

So, for reasons of statistical power and specificity, we prefer null hypotheses that pertain to a single parameter, that is, where PA and PC differ by 1. The F for testing such null hypotheses will always have one degree of freedom in the numerator. If the compact model representing the null hypothesis is rejected, then the answer is unambiguous—the extra parameter in Model A is required. Next, we consider an important one-degree-of-freedom test.

One Predictor Over and Above the Other Predictors

We know that redundancy complicates questions about individual predictors. Regardless of whether a predictor is useful by itself in a simple regression, we often want to know whether it is useful over and above other predictors in the model. In other words, if we have already made conditional predictions based on a set of predictors, is there further improvement by adding one more predictor variable? In other words, we are comparing

$$\text{MODEL A: } Y_i = \beta_0 + \beta_1 X_{i1} + \beta_2 X_{i2} + \ldots + \beta_{p-1} X_{i,p-1} + \beta_p X_{ip} + \varepsilon_i$$
$$\text{MODEL C: } Y_i = \beta_0 + \beta_1 X_{i1} + \beta_2 X_{i2} + \ldots + \beta_{p-1} X_{i,p-1} + \varepsilon_i$$

That is, Model C and Model A are the same except that Model A is augmented by exactly one extra parameter (β_p) and predictor variable (X_p). Thus, Model C has p parameters, and Model A has $p + 1$ parameters. Most importantly, then, PA – PC = 1.

By comparing Model A with Model C, we are asking whether we need the predictor X_p given that we are already using predictors $X_1, X_2, \ldots, X_{p-1}$. The β values in this model comparison are partial regression coefficients in that their values depend on what other β values are included in the model. That is, the estimated value of, say, β_2 is likely to depend on whether β_1 is also included in the model. Thus, we should use the more complete notation that reminds us of this dependence. In particular, the extra coefficient added to Model A is represented by $\beta_{p.12\ldots p-1}$. The subscript before the dot indicates the particular coefficient, and the subscripts after the dot indicate the other coefficients also included in the model. Clearly, we do not need the extra predictor if its partial regression coefficient equals zero, so our null hypothesis is

$$H_0: \beta_{p.12\ldots p-1} = 0$$

Note that this is merely a generalization of the test that we did for the simple regression model with one predictor. There we asked whether adding a predictor was useful over and above the simple mean model. Here we ask whether adding a predictor is useful over and above the other predictors already in the model.

One strategy for model building is to start simple and add complexity only when necessary. Thus, a frequent question is whether the addition of another predictor would be worthwhile. As always, we answer the question about whether the extra parameter is worthwhile by estimating the model parameters to minimize the sum of squared errors and then using PRE and F to compare the two models. The PRE for the addition of exactly one parameter has the special name "coefficient of partial determination" and is represented as

$$r^2_{Yp.123\ldots p-1}$$

The square root of this PRE with the sign from the partial regression coefficient is often called the "partial correlation coefficient" because it is the simple correlation between Y_i and X_{pi} when controlling or holding constant the other $p - 1$ predictors.

Given F, we can compute a confidence interval for $\beta_{p.12\ldots p-1}$ using the same logic we used to construct a confidence interval for β_1 in the case of simple regression in Chapter 5. That is, for a specific set of data we can ask how extreme a null hypothesis could be so that F would still be less than the critical value of F for a given α under the usual assumptions. The resulting equation is

$$\text{CI} = b_{p.123\ldots p-1} \pm \sqrt{\frac{F_{crit;1,n-p-1;\alpha}\text{MSE}}{(n-1)s_{X_p}^2\left(1-R_p^2\right)}}$$

As before, MSE is the mean squared error for Model A, $F_{crit1;n-p-1;\alpha}$ is the appropriate critical value of F from the tables in the Appendix, and s_{Xp}^2 is the variance of the predictor variable. For whatever level of α is chosen, the aforementioned equation provides the $(1 - \alpha)\%$ confidence interval for the partial regression slope. The formula for the confidence interval is the same as the one for a simple regression slope from Chapter 6 except for the addition of the term $1 - R_p^2$ in the denominator. R_p^2 is simply the R^2 or PRE when all the other $p - 1$ predictors are used to predict X_p. Thus, R_p^2 is a measure of the redundancy between X_p and the other predictors already included in the model.

Conversely, the term $1 - R_p^2$ which has the special name of *tolerance* in some computer outputs, is a measure of X_p's uniqueness from the other predictors. Only the unique part of X_p might be useful for reducing the error further. If the tolerance is low, then it will be difficult for X_p to be useful. This is reflected in the formula for the confidence interval by including it in the denominator. A low tolerance in the denominator will cause the confidence interval to be large, and a large confidence interval will be more likely to include zero. Because a low tolerance makes the confidence interval wider, some programs report the variance inflation factor (VIF), which equals $1/(1 - R_p^2)$, the inverse of tolerance.

There is an exact equivalence between comparing F with its critical value for a certain α to test the addition of a parameter and whether the confidence interval constructed using that same critical value includes zero. The confidence interval excludes zero if and only if F exceeds the critical F.

It is seldom necessary to calculate by hand the F for the addition of a parameter because most regression programs routinely produce F (or sometimes the equivalent $t_{n-p-1} = \sqrt{F_{1,n-p-1}}$). Many programs also give the 95% confidence interval (corresponding to $\alpha = .05$) for $\beta_{p.12...p-1}$. Few programs give the corresponding PRE, which in this case is the proportional reduction in error due to adding the pth predictor. However, almost all programs will give the partial correlation coefficient, which for this model can be squared to yield PRE. The important things to remember therefore are not the computational formulas but that the coefficient of partial determination or, equivalently, the squared partial correlation, is simply PRE for the special case of comparing a compact model with an augmented model that has one extra predictor. Similarly, the associated F or t-tests whether the partial regression coefficient for X_p is zero. Thus, the following are equivalent null hypotheses:

$$H_0: \beta_{p.12...p-1} = 0 \text{ and } H_0: r_{p.12...p-1}^2 = 0$$

If one is false, then the other must be false also.

An Example of an Additional Predictor Test

We will again use the GPA data to illustrate the test for adding an additional predictor to the model. Suppose that we want to ask whether it is useful to add SATM to the model when HSRank and SATV are already in the model. This question not only has practical importance—it would be easier for the clerks in the admissions office if they only had to include HSRank and SATV in their calculations of predicted GPA—but it also is substantively interesting to know whether quantitative skills as measured by SATM make a contribution to the predicted GPA that is independent of HSRank and SATV. These particular data are for engineering students, so one would suppose that SATM as a measure of quantitative skills that engineers would need in their courses would be required even when HSRank and SATV as measures of overall ability were already included in a model. To test this question, we want to compare these two models:

$$\text{MODEL A: } GPA_i = \beta_0 + \beta_1 HSRank_i + \beta_2 SATV_i + \beta_3 SATM_i + \varepsilon_i$$

$$\text{MODEL C: } GPA_i = \beta_0 + \beta_1 HSRanki + \beta_2 SATV_i + \varepsilon_i$$

Note that PA = 4 and PC = 3, so PA – PC = 1; we therefore are asking a focused, one-degree-of-freedom question. Model A and Model C would be the same if β_3 equaled zero. (Note, we no longer use more complete $\beta_{3.12}$, with the implicit understanding that any parameter depends on the model in which it appears.) Thus, the Model A/Model C comparison is equivalent to testing the null hypothesis:

$$H_0: \beta_3 = 0$$

Model A here is the same as the Model A we used when testing the overall model. Thus, we already have the parameter estimates and SSE for Model A. A multiple regression program provides the parameter estimates and SSE for Model C. So

$$\text{MODEL A: } \widehat{GPA}_i = -1.74 + 0.027 HSRank_i + 0.011 SATV_i + 0.022 SATM_i$$

$$\text{SSE(A)} = 172.99$$

$$\text{MODEL C: } \widehat{GPA}_i = -0.71 + 0.028 HSRank_i + 0.018 SATV_i$$

$$\text{SSE(C)} = 181.48$$

To answer the question of whether SATM is useful over and above HSRank and SATV, we compute PRE and F in the usual way. That is,

$$\text{PRE} = \frac{\text{SSE(C)} - \text{SSE(A)}}{\text{SSE(C)}} = \frac{181.48 - 172.99}{181.48} = \frac{8.49}{181.48} = .047$$

and

$$F_{PA-PC,n-PA} = F_{1,410} = \frac{\text{PRE}/(\text{PA}-\text{PC})}{(1-\text{PRE})/(n-\text{PA})} = \frac{.047/1}{.953/410} = \frac{.047}{.0023} = 20.4$$

$$F_{PA-PC,n-PA} = \frac{\text{SSR}/(\text{PA}-\text{PC})}{\text{SSE(A)}/(n-\text{PA})} = \frac{8.49/1}{172.99/410} = \frac{8.49}{0.422} = 20.4$$

PRE = 0.047 and $F_{1,410} = 20.4$ easily exceed their respective critical values (for n – PA = 200, the largest value less than 410 in the Appendix tables) of 0.019 and 3.89. We therefore reject the null hypothesis that the partial regression coefficient for SATM is zero and conclude that it is worthwhile to add SATM to the model that already includes HSRank and SATV. In this case, the remaining error—the error left after using HSRank and SATV in the model—is reduced by 4.7%. The square root of PRE equals 0.22 and represents the partial correlation between GPA and SATM holding constant or "partialing out" HSRank and SATV. Knowing that the partial regression coefficient for SATM is not zero, we should therefore interpret it: When comparing students of equivalent HSRank and SATV scores, for each point higher on SATM (measured on a 20–80 scale as SAT scores are reported to U.S. universities), we predict that a student's GPA will be 0.022 points higher. In other words, controlling for HSRank and SATV, we would expect a student who scored 80 (800 as reported to the student) on SATM to have a GPA about two-thirds of a letter grade higher than a student who scored 50, i.e., (80–50) 0.022 = 0.66.

The ingredients for computing the confidence interval are available from various sources: MSE = SSE(A)/(n – PA) = 172.99/410 = 0.42, a descriptive statistics program gives the variance of SATM as 46.21, and a multiple regression program gives the PRE or $R^2_{3,12}$ for predicting SATM with HSRank and SATV as 0.11. Hence, the 95% confidence interval for β_3 is computed as

$$\begin{aligned}
\mathrm{CI}_{b_{SATM}} &= b_{SATM} \pm \sqrt{\frac{F_{crit;1,n-PA;\alpha}\,\mathrm{MSE}}{(n-1)s^2_{SATM}\left(1-R^2_{3.12}\right)}} \\
&= 0.022 \pm \sqrt{\frac{3.87(0.42)}{413(46.21)(1-0.11)}} \\
&= 0.022 \pm 0.0098 \\
&= [0.012,\ 0.032]
\end{aligned}$$

Hence, although our best estimate of the coefficient is 0.022, it might reasonably be as low as 0.012 or as high as 0.032.

The Overall Model Revisited

We noted earlier that a single test of the overall model was usually unsatisfactory because rejection of the null hypothesis does not indicate which of the many β_j significantly depart from zero. The test for adding another variable, which we have just considered, suggests a solution. The numbering of the predictor variables is arbitrary, so we can treat each predictor in turn as if it were the last variable to be added. In this way we will test separately for each predictor whether its partial regression coefficient (i.e., its coefficient with all the other variables in the model) is different from zero. Conceptually, this implies that we need to do separate regression analyses in which we first enter the other predictor variables and then test whether it is worthwhile to add the last. In practice, it is not necessary to do these separate analyses because the required information is available when the statistical program does the computations for the overall model. The results of these computations are often summarized in an analysis of variance table like Figure 7.12.

We will examine how this table summarizes the statistical tests we have done of (a) the overall model and (b) adding SATM over and above the other variables in the model. Then we will consider how it provides information for the tests of adding either SATV or HSRank last. In our test of the overall model we found

$$\text{MODEL A: } \widehat{GPA}_i = -1.74 + 0.027HSRank_i + 0.011SATV_i + 0.022SATM_i$$

$$\mathrm{SSE(A)} = 172.99$$

$$\text{MODEL C: } \widehat{GPA}_i = 2.70$$

$$\mathrm{SSE(C)} = 221.78$$

Source	*SS*	*df*	*MS*	*F*	*PRE*	*p*
Regression	48.789	3	16.263	38.5	.220	.0001
SATV	2.779	1	2.779	6.6	.016	.0106
SATM	8.483	1	8.483	20.1	.047	.0001
HSRank	29.209	1	29.209	69.2	.144	.0001
Error	172.994	410	0.422			
Total	221.783	413				

Figure 7.12 Analysis of variance summary table for the GPA data

The two sums of squared errors, SSE(A) = 172.99 and SSE(C) = 221.78, correspond, respectively, to the values in the *SS* column for Error and Total. Their difference is SSR = 221.78–172.99 = 48.79, which we used in the computation of PRE:

$$\text{PRE} = \frac{\text{SSE}(\text{C}) - \text{SSE}(\text{A})}{\text{SSE}(\text{C})} = \frac{221.78 - 172.99}{221.78} = \frac{48.79}{221.78} = .22$$

and which is listed in the *SS* column for Regression in Figure 7.12. The values in the df column for Regression, Error, and Total are, respectively, PA – PC, n – PA, and n – PC. The other values in the Regression row summarize the test of the overall model against the simple model. Note that the rows for Regression, Error, and Total in Figure 7.12 have the same role as they did in Figure 6.11 for simple regression. The only difference is that the Regression row now describes the reduction in error due to allowing all of the partial regression slopes to differ from zero and the Error row is the remaining error in that case. The Total row describes the error when all the partial regression slopes are set to zero.

Unlike the ANOVA table for simple regression (Figure 6.11), the one for multiple regression (Figure 7.12) has a row for each variable in the model. To understand these rows, we return to our earlier test of adding SATM to a Model C that already makes conditional predictions of GPA based on HSRank and SATV. Previously, we found

MODEL A: $\widehat{GPA}_i = -1.74 + 0.027HSRank_i + 0.011SATV_i + 0.022SATM_i$

SSE(A) = 172.99

MODEL C: $\widehat{GPA}_i = -0.71 + 0.028HSRank_i + 0.018SATV_i$

SSE(C) = 181.48

The difference between these two sums of squared errors, 181.48–172.99 = 8.48, appears in the *SS* column for SATM. The other columns in the row for SATM provide the statistics for comparing the aforementioned Models A and C. We computed PRE directly from these numbers before, but it is convenient when working with summary tables like Figure 7.12 to use this equivalent alternate formula:

$$\text{PRE} = \frac{\text{SSR}}{\text{SSR} + \text{SSE}(\text{A})} = \frac{8.483}{8.483 + 172.994} = \frac{8.483}{181.477} = .047$$

One way to think about this computation is that if SATM were not in the model, then the error it reduces would need to be put back into the model error—the computation in the denominator. In other words, the denominator corresponds to the total error for a model that included all the other variables except SATM.

Once we have this understanding of the analysis of variance summary table, it is easy to do other model comparisons. For example, we might want to know, for these engineering students, whether knowing their SATV scores improves predictions of GPA over and above a model based only on HSRank and SATM. In other words, our model comparison is

$$\text{MODEL A: } GPA_i = \beta_0 + \beta_1 HSRank_i + \beta_2 SATV_i + \beta_3 SATM_i + \varepsilon_i$$

$$\text{MODEL C: } GPA_i = \beta_0 + \beta_1 HSRank_i + \beta_3 SATM_i + \varepsilon_i$$

Instead of fitting these two models and computing their sum of squared errors we can obtain these values from Figure 7.12. Model A is as before, so we know SSE(A) = 172.994 from the Error row in the table. The value of SS = 2.779 for SATV corresponds to SSR for the earlier model comparison. Hence, SSE(C) = 2.779 + 172.994 = 175.773. We then obtain the PRE for this model comparison as

$$\text{PRE} = \frac{\text{SSR}}{\text{SSR} + \text{SSE}(\text{A})} = \frac{2.779}{2.779 + 172.994} = \frac{2.779}{175.773} = .016$$

And we can compute F for this model comparison either using the formula based on PRE or the sums of square formula (from the source table):

$$F_{PA-PC,n-PA} = F_{1,410} = \frac{\text{SSR}/(\text{PA}-\text{PC})}{\text{SSE}(\text{A})/(n-\text{PA})} = \frac{\text{MSR}}{\text{MSE}} = \frac{2.779}{0.422} = 6.6$$

The key values for MSR and MSE are available in the *MS* column of Figure 7.12. Computer programs for regression analysis often provide the probability values listed in the last column. In this case, the probability of obtaining a PRE or F this large or larger if Model C were correct is only 0.016. Using the usual cutoff of 0.05, we would reject Model C in favor of Model A; in other words, we would conclude that higher SATV is associated with higher first-year GPA (its coefficient is positive) over and above HSRank and SATM.

Now that we understand the information provided in the summary table, we can readily make the model comparison asking whether information about HSRank is useful over and above the predictive information provided by SATV and SATM. When comparing

$$\text{MODEL A: } GPA_i = \beta_0 + \beta_1 HSRank_i + \beta_2 SATV_i + \beta_3 SATM_i + \varepsilon_i$$

$$\text{MODEL C: } GPA_i = \beta_0 + \beta_2 SATV_i + \beta_3 SATM_i + \varepsilon_i$$

we see that SSR = 22.209, PRE = .144, and $F_{1,410} = 69.2$ with $p < .0001$. Clearly, HSRank significantly reduces error in predicting GPA over and above the error reduction provided by using SATV and SATM. We might say that performance information (HSRank) provided useful information beyond that provided by the aptitude information (SAT scores).

It is important to note in the ANOVA table of Figure 7.12 that when we add the SSRs for the individual variables, the sum (2.779 + 8.483 + 29.209 = 40.471) is less than the total SSR we obtain when using all the predictor variables together (48.789). The reason for this is the redundancy among the predictors. The SSRs when adding each predictor last captures only the *unique* contribution to the error reduction made by that predictor *over and above* the contribution from the other predictors. Hence, a contribution that is shared with other predictors appears in the total SSR but not in the unique SSR added by that variable. Had there been no redundancy among the predictors, then the sum of their SSRs would have equaled the SSR for the overall model. With certain patterns of positive and negative correlations among predictors, it is also possible for the sum of the individual SSRs to exceed the total SSR, a condition sometimes confusingly labeled as "suppression." Whether the sum of the SSRs is smaller or larger than the overall SSR is not particularly important. Simply interpret the individual SSRs as the unique contribution to the error reduction of each variable and the overall SSR as the total reduction in error when using all the variables together in a model.

Testing Other Model Questions

So far, we have asked questions about the overall model and about each variable if it were added last to the model. And we saw how these tests are summarized in the ANOVA table of Figure 7.12. However, this does not begin to exhaust the interesting questions we might ask. Another common question is to ask whether it is useful to add a set of variables to the model. For example, we previously observed that performance information (HSRank) added useful prediction over and above aptitude information (test scores). If we want to ask the corresponding question of whether, as a set, aptitude information provides useful prediction over and above performance information, then we would compare these two models:

$$\text{MODEL A: } GPA_i = \beta_0 + \beta_1 HSRank_i + \beta_2 SATV_i + \beta_3 SATM_i + \varepsilon_i$$

$$\text{MODEL C: } GPA_i = \beta_0 + \beta_1 HSRank_i + \varepsilon_i$$

In this case, the null hypothesis is that the parameters for *both* SATV and SATM are zero. That is,

$$H_0\text{: } \beta_2 = \beta_3 = 0$$

To answer this question, we need only find the appropriate SSEs. Using the most basic multiple regression program, we can simply estimate both models, obtain their SSEs, and do the computations of PRE and F manually. Many programs have options for testing sets of parameters directly. We will use the basic approach:

$$\text{MODEL A: } \widehat{GPA}_i = -1.74 + 0.027 HSRank_i + 0.011 SATV_i + 0.022 SATM_i$$

$$\text{SSE(A)} = 172.99$$

$$\text{MODEL C: } \widehat{GPA}_i = 0.18 + 0.028 HSRank_i$$

$$\text{SSE(C)} = 189.11$$

$$\text{PRE} = \frac{189.11 - 172.99}{189.11} = \frac{16.121}{189.11} = .085 \qquad F_{2,410} = \frac{.085/2}{(1 - .085)/410} = 19.04$$

The values for PRE and F easily exceed their critical values so we conclude that, as a set, the test scores SATV and SATM provide significant error reduction over and above that for HSRank alone.

As another example of the kind of questions we might ask beyond the obvious ones provided by most multiple regression programs, we might ask whether SATV and SATM need to be weighted differently. Or, in other words, we are asking whether the observed weights of 0.011 for SATV and 0.022 for SATM in the overall model are significantly different from each other. For predicting first-year GPA for engineering students, we might suspect that more weight should be given to SATM. Answering this question implies this model comparison:

$$\text{MODEL A: } GPA_i = \beta_0 + \beta_1 HSRank_i + \beta_2 SATV_i + \beta_3 SATM_i + \varepsilon_i$$

$$\text{MODEL C: } GPA_i = \beta_0 + \beta_1 HSRank_i + \beta_2 SATV_i + \beta_2 SATM_i + \varepsilon_i$$

Look closely to see that the only difference in the two models is that the same coefficient (β_2) is used for *both* SATV and SATM in Model C but separate coefficients (β_2 and β_3) are used in Model A. Note that this meets our rule for Model A/Model C comparisons because the parameters used in Model C are a subset of those used in Model A, even though both models have the same variables. In this case, the null hypothesis is

$$H_0: \beta_2 = \beta_3$$

Unlike the prior test where we asked whether both coefficients were zero, here we ask only if they are equal, but not necessarily zero. Another way to express this null hypothesis is that their difference is zero; that is

$$H_0: \beta_2 - \beta_3 = 0$$

Some multiple regression programs allow specification of complex null hypotheses like this. If not, then some ingenuity is required to obtain the necessary SSEs in order to compare the models. In this case we observe with an algebraic rearrangement that Model C is equivalent to

$$GPA_i = \beta_0 + \beta_1 HSRank_i + \beta_2(SATV_i + SATM_i) + \varepsilon_i$$

$$GPA_i = \beta_0 + \beta_1 HSRank_i + \beta_2(SATTOT_i) + \varepsilon_i$$

where SATTOT = SATV + SATM. Hence, we can get the SSE(C) by regressing GPA on HSRank and SATTOT:

$$\text{MODEL A: } \widehat{GPA}_i = -1.74 + 0.027 HSRank_i + 0.011 SATV_i + 0.022 SATM_i$$

$$\text{SSE(A)} = 172.99$$

$$\text{MODEL C: } \widehat{GPA}_i = -1.63 + 0.027 HSRank_i + 0.016 SATV_i + 0.016 SATM_i$$

$$\text{SSE(C)} = 173.86$$

$$\text{PRE} = \frac{173.86 - 172.99}{173.86} = \frac{0.87}{173.86} = .005 \quad F_{1,410} = \frac{.005}{(1 - .005)/410} = 2.06$$

Even for $n - PA = 500$, these values do not exceed the respective critical values of .008 and 3.86 for $\alpha = .05$. Hence, there is not sufficient evidence in these data to conclude that SATV and SATM ought to be weighted differently when predicting first-year GPA for engineering students.

These are but two of the many specialized, focused questions we might want to ask about our data that are beyond the obvious questions answered by the usual output from multiple regression programs. Other questions that might be of interest would be asking whether the new model is significantly better than the old model that might have been used in the decision to admit these students. Or we might want to ask questions about the mean GPA of these students, in which case we would convert our variables to mean-deviation form and do model comparisons parallel to those in Chapter 6 for the intercept. The important point is that you can answer a wide range of questions about your data by framing them in terms of a Model A/Model C comparison, obtaining the SSEs for those models, and then computing PRE and *F.*

Cautions About Using Multiple Regression

Multiple regression is likely the most widely used statistical procedure in the social sciences. Consequently, it is also the most frequently abused procedure. In this section we comment on some important cautions about using and especially interpreting multiple regression.

Causal Conclusions From Nonexperimental Data

We interpret the partial regression coefficient b_j as the amount the predicted value $\hat{Y}_i$ changes for each unit increase in the predictor variable, controlling for all other variables in the model. It is important to remember that this is only a prediction based on the existing data and may not represent a causal relationship between X_{ji} and Y_i. Users of multiple regression sometimes forget this fact and incorrectly presume that manipulating X_{ji} will *cause* a change in Y_i.

The following example highlights the causal ambiguity inherent in observational data. In almost any city with seasonal temperature variation it is easy to demonstrate that the number of ice cream cones sold in a month is a good predictor of the number of suicides in that month: The more cones sold the more suicides there will be. However, it would be ludicrous to suggest that banning ice cream sales would reduce the number of suicides. The problem with such a conclusion is that ice cream is not causing suicides; rather, it is probably just a proxy variable for temperature, which might be the cause of both increased suicides and increased ice cream sales. And even temperature is probably just a proxy variable for discomfort or unpleasantness that might prompt a person to attempt suicide.

The causal error is obvious in the ice cream example. But many researchers do not realize that the dangers of presuming that manipulation of the predictor variable will cause a change in Y_i are no less serious whenever the regression analysis is based on observational data.

On the other hand, it is also incorrect to assume that causal conclusions can never be based on regression analyses. If the researcher determines the values of the predictor variables and randomly assigns those values to the cases, then it is quite proper to presume that changing the predictor variables will cause a change in Y_i. Such a design, involving random assignment to levels of a predictor, is commonly known as an experimental design. For example, to test the sensitivity of credit card use to the interest rate charged, a bank might randomly assign, say, one of five different interest rates to different customers. If a regression analysis revealed that the interest rate was a significant predictor of the amount that customers charged to their credit cards, then one would reasonably presume that the manipulation of the interest rate by the bank causally affected the amount that people charged to their cards.

There are some other research designs, originally called *quasi-experiments* by Campbell and Stanley (1963), in which limited causal conclusions from regression analyses of observational data are warranted. For further treatment of these issues, see Shadish et al. (2002) and Judd and Kenny

(1981a). In general, however, one should be extremely cautious about making causal conclusions from regression analyses of nonexperimental data.

Relative Importance of Predictor Variables

Many researchers are tempted to make inferences about the relative importance of predictor variables in multiple regression models. It is not reasonable to compare directly the relative magnitudes of b_j because they clearly depend on the unit of measurement. By making the unit of measurement for a given predictor variable smaller (e.g., using centimeters instead of inches to measure height), we can make b_j arbitrarily large without changing the fundamental impact of that variable in the model.

Sometimes predictor variables will be measured on the same scale, enabling the same units of measurement for the variables. For example, consider trying to model attitudes toward a particular political party as a function of economic status. It might be reasonable to use both income and savings as predictor variables. We can easily measure both variables in dollars, so that we could be assured of using the same units. However, even here comparing b_j values as a measure of importance would probably not be appropriate because the ranges of the variables would be so different. One variable might have a smaller b_j—that is, a smaller predicted change in Y_i for a unit increase in X_{ji}—but still have a larger overall impact because it had a larger range over which it varies.

The dependence of the overall effect of a predictor variable on its range suggests that we might normalize the partial regression coefficients by multiplying them by their ranges or, more typically, their respective standard deviations. This effectively makes the unit of measurement either the range or the standard deviation, respectively. The *standardized regression coefficients* are also traditionally normalized to include the standard deviation of Y_i according to the following formula:

$$b_j^* = b_j \left[\frac{s_{X_j}}{s_Y} \right]$$

Some books and computer programs unfortunately use β_j or "beta weights" to refer to standardized regression coefficients. We prefer b_j^* to avoid confusion with the unknown values of the model coefficients in our models and to emphasize that standardized regression coefficients are simple transformations of the regression coefficients b_j.

Unfortunately, the standardized regression coefficients do not solve the problem of making inferences about relative importance. First, they still depend on the particular range and distribution of the predictor in the particular dataset. If this distribution were to change for any reason—for example, if the researcher were to include a wider range of cases—then the standardized regression coefficients could change dramatically even if the underlying relationship between the predictor and the data remained the same (i.e., the partial regression coefficient is unchanged). Such dependence on the vagaries of a particular set of observations hardly provides a good basis for inferring relative importance. Second, relative importance is clouded by redundancy. A regression coefficient (standardized or not) might be low, not because its associated predictor variable has no relationship with Y, but because that predictor variable is partially redundant with other predictors. Third, the fact that one estimate is larger than another estimate does not constitute a statistical test. Researchers generally understand that they cannot simply look at two means and conclude that the larger differs from the smaller. A statistical test is required to determine whether the difference between means is significant. In the same way, one cannot conclude that two standardized coefficients differ simply because one is somewhat larger than the other. Thus, standardized regression coefficients do not really allow us, as they were intended, to make relative comparisons of predictors. The fundamental problem is that we cannot frame a question about relative importance in terms of a Model A versus Model C comparison because both models would necessarily have the same predictors. Hence, our machinery for statistical inference cannot address the question of relative importance.

In sum, relative importance of predictor variables is a slippery concept. Any statement about relative importance must be accompanied by many caveats that it applies to this particular range of variables in this particular situation. As a result, in our opinion, they are seldom useful. Although it is tempting to compare the importance of predictors in multiple regression, it is almost always best not to do so.

Automatic Model Building

Many regression programs have procedures for automatically building models. One commonly used procedure is *stepwise regression*, which, as its name implies, builds a model step by step. On each step a predictor variable may be added or deleted from the model. On the first step the variable with the largest F might be added (so long as it exceeds a minimum threshold). On subsequent steps, the variable not in the model that has the largest F is added to the model *or* a variable in the model whose F has fallen below the threshold is removed. The stepwise procedure terminates when all the variables have F values above the threshold and all the variables not in the model have F values below the threshold. There are variations of stepwise regression (e.g., forward selection, backward selection) and other, distinct strategies (e.g., LASSO, best-subset) for automatic model building that exist. Machine learning and artificial intelligence are also rapidly changing the landscape of alternatives.

These variants seem to promise the "best" model with virtually no effort by the researcher or data analyst. However, the allure is deceptive—even when a strategy can find the model that maximizes PRE. We do not recommend the use of any of the automatic model-building procedures for three reasons. First, an unfocused search through many possible models (sometimes pejoratively referred to as a "fishing expedition") increases the likelihood of capitalizing on chance (i.e., making Type I errors) and thereby finding a model that represents only a spurious relationship. We generated 51 random variables with 100 observations for each. We randomly designated one variable as Y and then used stepwise regression to build a model with the remaining 50 variables as potential predictor variables. The estimated final model was

$$\hat{Y}_i = 0.03 + 0.25X_{11i} - 0.29X_{21i} - 0.35X_{23i} + 0.26X_{31i} - 0.21X_{44i}$$

with the test of the overall model against the simple model, which was the true model, having PRE = .31, $F_{5,94} = 8.49$, $p < .0001$. All five partial regression coefficients are significant, with the minimum being PRE = .04, $F_{1,94} = 5.78$, $p = 0.018$. Hence, the stepwise regression procedure produced six Type I errors. Given that there are about 50 statistical tests at each step and there is ample opportunity to make Type I errors, it is not surprising that six actual Type I errors were made.

Second, as we learned in this chapter, the interpretation of coefficients and the meaning of the questions being asked depend on what other variables are included in the model. Testing whether there is a relationship between age and weight is different from testing whether there is a relationship between age and weight after controlling for height. It seems unwise to let an automatic algorithm determine the questions we do and do not ask of our data.

Third, but related to both the first and second points, this book focuses on using data to test hypotheses, not on explaining all variation in the outcome measure. It is our experience and strong belief that better models and a better understanding of one's data result from focused data analysis guided by substantive theory. For an interesting illustration of the superiority of substantively guided data analysis over automatic model building, see Henderson and Velleman (1981). They state a fundamental principle of their philosophy of data analysis, a principle with which we agree: "The data analyst knows more than the computer." Failure to use that knowledge produces inadequate and often flawed data analyses.

Statistical Power in Multiple Regression

The discussion of statistical power in the context of simple regression, using a single predictor variable, generalizes quite readily to multiple regression. If we have an estimate of the true proportional reduction in error, η^2, for whatever model comparison is of interest, then we can estimate the statistical power for a given number of observations. The usual sources of estimates of the true proportional reduction in error are appropriate here: past studies, particularly meta-analyses or estimates computed from expected parameter values.

The interesting complexities of power in multiple regression arise from the presence of the other predictors in the model. As we have already discussed, the denominator of the confidence interval for the coefficient for a particular predictor includes the tolerance of that predictor. As that tolerance decreases, the confidence interval for its regression coefficient increases in width. Accordingly, as the tolerance of a predictor decreases (i.e., as it becomes more redundant with the other predictors in the model), tests of its regression coefficient become less powerful.

At the same time, other predictors may also enhance the power of tests of a regression coefficient in multiple regression. Recall that in the numerator of the confidence interval is the mean square error (MSE) for the augmented model. As MSE decreases in size (i.e., as the predictors in the augmented model do a better job of predicting Y_i), the width of the confidence interval decreases. Factors that reduce the mean square error thus increase the power of tests of the regression coefficients included in the model. The power of any model comparison is enhanced whenever nonredundant and error-reducing predictor variables are added to both the augmented and compact models. When control variables are redundant with the focal variables (the variables we want to test) there is a tradeoff: The gain in power from reducing MSE may be offset by a reduction in power due to redundancy. It is difficult to estimate this tradeoff in power. However, the implication is that to improve power one should try to add significant, nonredundant predictors to the model, assuming that their presence makes substantive sense. Of course, the presence of predictors in the compact and augmented models that do not significantly relate to Y will result in less power, because the degrees of freedom for error for augmented and compact models will be reduced, thus increasing MSE.

Summary

In this chapter we have considered models with multiple continuous predictors, traditionally known as multiple regression models. Such models permit complex questions about the simultaneous roles of multiple, partially redundant predictors in explaining variation in the dependent variable. The major issue that complicates interpretation of parameter estimates in such models is that of redundancy among predictors. Accordingly, we spent a considerable amount of time developing notions of what it means to interpret a partial regression coefficient and, more particularly, what it means to examine the effect of one predictor controlling for others that are included in a model.

Our model comparison approach to statistical inference in such models is identical to the approach of earlier chapters, where we compare an augmented model to a compact one. In the context of multiple-predictor models, the number of possible model comparisons that might be of interest is considerably larger than in the case of simpler models considered in earlier chapters. We illustrated a number of possible model comparisons that might be of interest, many of which are typically conducted by standard multiple regression computer programs. The most frequently reported of these are tests of the overall model and tests of individual partial regression coefficients. However, there are many other possibilities that are not routinely output in standard regression programs. The advantage of the model comparison approach is that whenever one is dealing with augmented and compact models, these can be separately estimated and then their error sums of squares compared. The range of questions we can thus ask of our data extends considerably beyond those that the computer programmers thought we would be interested in when they wrote the standard multiple regression programs.

8 Moderated and Nonlinear Regression Models

In Chapter 7, we considered in some detail models with multiple predictor variables. In this chapter we extend these models by discussing the addition of predictors that are themselves products of other predictors. Such models, as we will show, permit us to extend dramatically the range of questions that we can ask of our data. In the models of Chapter 7, our questions about individual predictor variables had the following generic form: "Is one predictor useful over and above or controlling for the set of other predictors in the model?" In this chapter we will ask a set of rather different questions about predictors. The questions considered in this chapter take the following generic form: "Does the utility of one variable as a predictor depend on the value of another variable or even of that particular variable itself?"

At this point, it may not be clear why this second sort of generic question is likely to be of interest. It probably is not even clear exactly what this second question means. To clarify things, we start this chapter with an extended example, using new data. We first recapitulate some of the major points of Chapter 7 in a somewhat different language, once again clarifying the sorts of questions about models that we asked there. We then show how the questions we will ask of our data with the models developed in this chapter differ from and extend the questions we asked earlier. Our hope is that this initial section will give an understanding of the second generic question mentioned earlier and will further the understanding of the sorts of questions addressed earlier in Chapter 7.

The models we consider in this chapter go by a variety of names. Occasionally they are known as *moderator regression* models. They are also known as *polynomial, interactive*, *nonadditive*, or *nonlinear* models. Regardless of the name we give them, the goal in this chapter is to discuss the estimation, interpretation, and tests of models that include, as predictors, products of other predictors.

Let us turn to the data that we will use to illustrate things. These are hypothetical data, supposedly gathered from 80 individuals who are all runners. Three variables have been measured. The first variable (Time) is each runner's most recent 5-kilometer race time, measured in minutes. Next, we have each runner's age (Age). The final variable is the average number of miles per week that each runner runs when in training for a race (Miles). The data are presented in Figure 8.1.

Simple and Additive Models Predicting Race Times

Our interest in these data will focus on how the age of the runner and the amount of training relate to race times in the 5-kilometer race. Let us start by examining how these two variables relate to race time individually, using the single predictor models developed in Chapter 6. If we regress Time on Age, the following simple regression equation results:

$$\hat{Time} = 14.955 + .217 Age$$

DOI: 10.4324/9781003438335-13

Time	*Age*	*Miles*	*Time*	*Age*	*Miles*
24.91	29	10	29.68	22	10
21.82	25	20	20.36	24	30
21.54	27	40	18.07	22	40
23.03	25	50	28.78	39	10
25.35	37	20	18.80	39	30
22.84	31	40	14.87	39	50
30.16	43	10	32.19	47	20
27.00	44	20	19.46	40	30
16.42	46	40	35.32	58	10
24.87	53	20	20.83	59	30
24.95	58	40	22.69	27	20
18.32	30	30	15.69	24	40
16.66	27	50	30.89	32	10
25.08	36	20	23.11	34	30
16.53	32	40	19.17	35	50
35.29	45	10	26.09	41	20
26.41	45	30	21.81	50	40
22.48	48	50	37.75	51	10
34.38	55	10	30.30	51	20
19.00	53	30	22.95	51	40
21.86	24	10	20.14	20	20
22.11	24	30	15.23	24	40
17.45	25	40	18.53	22	50
24.58	34	20	24.96	35	30
19.97	36	40	19.21	31	50
34.34	45	10	25.07	41	20
25.22	49	30	20.99	43	40
20.52	44	50	29.94	59	10
29.98	55	20	23.08	59	30
21.62	57	40	21.52	53	50
19.35	29	20	21.38	27	30
20.45	24	40	19.52	28	50
16.07	28	50	27.00	37	10
23.06	38	20	17.61	39	30
20.94	31	40	15.19	39	50
28.77	42	10	26.03	45	20
29.86	46	30	21.51	42	40
22.96	46	50	34.53	58	10
25.62	56	20	23.68	56	30
24.98	55	40	23.50	53	50

Figure 8.1 Runners' data

The sum of squared errors for this model equals 1714.720. The sum of squared errors for a single-parameter compact model in which we predict the mean time for each case equals 2211.091. Testing the null hypothesis to determine whether Age is a significant predictor of race time yields a value of PRE of .224, which converts to an F of 22.58 with 1 and 78 degrees of freedom. These values indicate that indeed Age is a significant predictor of Time.

This simple regression result is graphed in Figure 8.2. As the positive slope makes clear in the graph, our best prediction based on these data is that race times increase with age. Since this

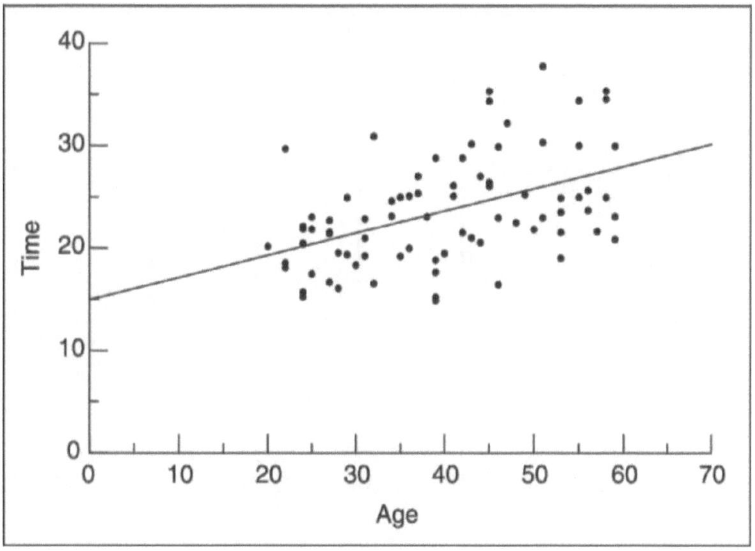

Figure 8.2 Time–Age simple relationship

is a simple regression model and no other predictor variables have been included, we are unable to answer questions about what other variables may be related to Time or what would happen to the Time–Age relationship if we controlled for some other variable, such as Miles. It may be that Age and Miles are redundant or confounded in these data, such that older runners train with fewer miles. If this were the case, then if we included Miles in the model, we might reach different conclusions about how Age was related to Time once Miles was controlled.

The simple relationship between Time and Miles is captured by the least-squares simple regression equation:

$$\hat{Time} = 31.911 - .280 Miles$$

The sum of squared errors from this model equals 1029.013. To determine whether Miles and Time are significantly related, we want to compare the sum of squared errors from this model with the sum of squared errors from the compact single-parameter model. This compact model is the same one we used to test whether Age is significantly related to Time. Its sum of squared errors equals 2211.091. The value of PRE that we get when comparing these two models equals .535, which converts to an F statistic of 89.60 with 1 and 78 degrees of freedom. Again, this value of F is significant, and we conclude that Miles and Time are related. The simple relationship is graphed in Figure 8.3.

When we regress Time on both Miles and Age, the following multiple regression equation results:

$$\hat{Time} = 24.605 + .167 Age - .257 Miles$$

The sum of squared errors for this three-parameter model equals 741.195, which gives a PRE of .665 when compared to the single-parameter compact model that includes neither predictor. The overall

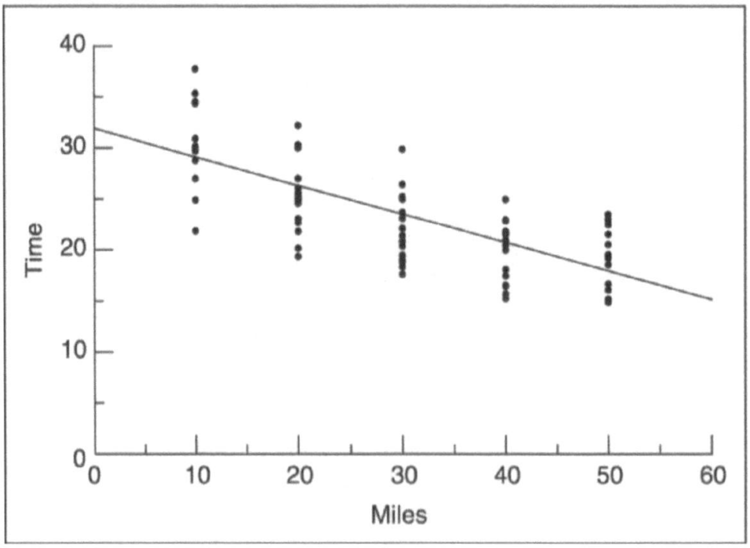

Figure 8.3 Time–Miles simple relationship

Source	*SS*	*df*	*MS*	*F*	*PRE*	*p*
Model	1469.90	2	734.95	76.35	.665	<.001
Age	288.13	1	288.13	29.90	.280	<.001
Miles	974.56	1	974.56	101.20	.568	<.001
Error	741.19	77	9.63			

Figure 8.4 Multiple regression source table

F statistic to compare this three-parameter model with the single-parameter model equals 76.35 with 2 and 77 degrees of freedom.

We can also compare this three-parameter model with each of the simple regression models to determine whether each predictor variable significantly increases the accuracy of our predictions when controlling for the other predictor variable. The value of PRE when comparing compact and augmented models without and with Age equals .280, which converts to an *F* statistic of 29.92 having 1 and 77 degrees of freedom. The value of PRE when comparing compact and augmented models without and with Miles equals .568, which converts to an *F* statistic of 101.20 with 1 and 77 degrees of freedom. Hence, we conclude that each of these predictor variables significantly increases the accuracy of our predictions of Time even when the other predictor is controlled. The source table in Figure 8.4 summarizes the results of the inference tests we have just conducted using this multiple regression model.

It is instructive to graph this multiple regression equation much like we did in the case of the simple equations. In Figure 8.5, we have graphed the prediction plane that this model generates in a three-dimensional space, where the vertical axis is Time, and Miles and Age are the two horizontal axes. As this figure makes clear, a specific predicted value of Time is made by the model at each pair of values for Miles and Age. The individual data points could also be included in this graph, and errors of prediction would be simply the vertical distances between each data point and the plane of prediction.

"Simple" Re-expressions of the Additive Model

To make the presentation of this model a little simpler, we could also take slices of this prediction plane, treating Age, for instance, as the predictor on the horizontal axis and graphing Time–Age prediction lines at different levels of Miles. If we group terms in the multiple regression equation a little differently, we get the re-expression

$$\hat{Time} = (24.605 - .257Miles) + .165Age$$

We can think of this re-expression as telling us about the "simple" relationship between Time and Age, allowing the intercept in the simple regression to change as Miles changes. In other words, suppose we wanted to predict race time conditional on age for someone who trained 20 miles per week. We would substitute 20 for Miles in the above re-expression and get

$$\hat{Time} = (24.605 - .257(20)) + .165Age = 19.465 + .165Age$$

For a runner who trained 30 miles per week, our "simple" predicted relationship between Time and Age would be

$$\hat{Time} = (24.605 - .257(30)) + .165Age = 16.895 + .165Age$$

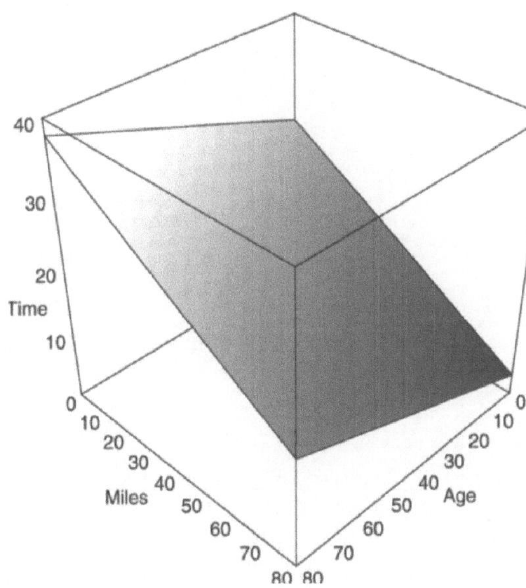

Figure 8.5 Prediction plane plot from multiple regression predicting Time from Age and Miles

We call these re-expressions "simple" relationships, meaning that we are acting as if we are examining a "simple" regression model between Age and Time at particular values of the other variable, Miles.

If we graphed the Time–Age relationship from this multiple regression equation where we also use Miles as a predictor, we would get many different "simple" regression prediction functions between Time and Age, one for each of the possible levels of Miles. A few such lines are graphed in Figure 8.6. Importantly, these "simple" relationships all have the same slopes: They are parallel lines that differ only in their intercepts. Accordingly, this multiple regression equation suggests that the slope or "simple effect" of Age on Time is the same regardless of the level of Miles. As Age increases, we predict the same change in Time, regardless of the level of Miles.

Since Miles is in theory a continuous variable, our plot of the Time–Age relationship within levels of Miles contains an infinite number of lines, one for each of the possible levels of Miles. All of these "simple" prediction lines combine to form the prediction plane of Figure 8.5.

Now let us turn the equation around and graph the Time–Miles "simple" relationship at different Age levels. Re-expressing our multiple regression equation, we get

$$\hat{Time} = (24.605 + .165Age) - .257Miles$$

For representative values of Age, we have graphed this function in Figure 8.7, putting Miles on the horizontal axis this time. Again, we see that as Age changes, the intercept for the Time–Miles "simple" relationship changes but the slope does not. In other words, the "simple effect" of Miles on Time (i.e., its slope) is assumed to remain the same regardless of Age.

Interactions Between Predictor Variables

The assumption that we are making in this multiple regression equation, and that these graphs make clear, is known as the *additive* assumption. In slightly more abstract language, if we regress

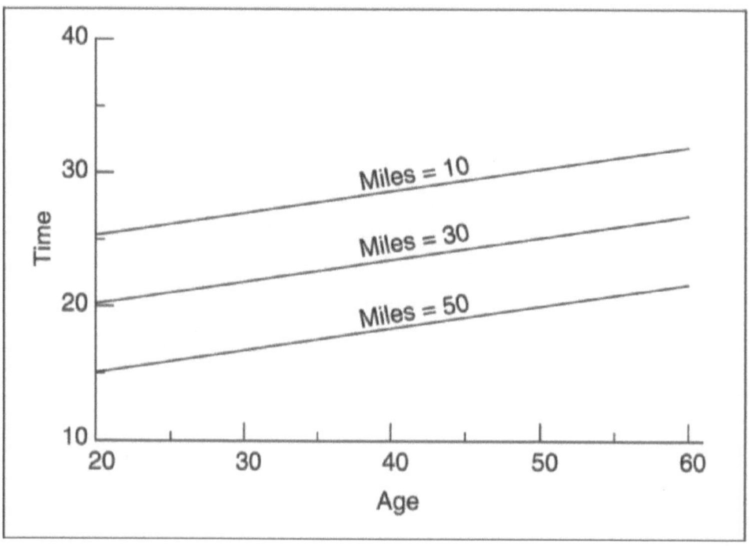

Figure 8.6 Time–Age "simple" relationships

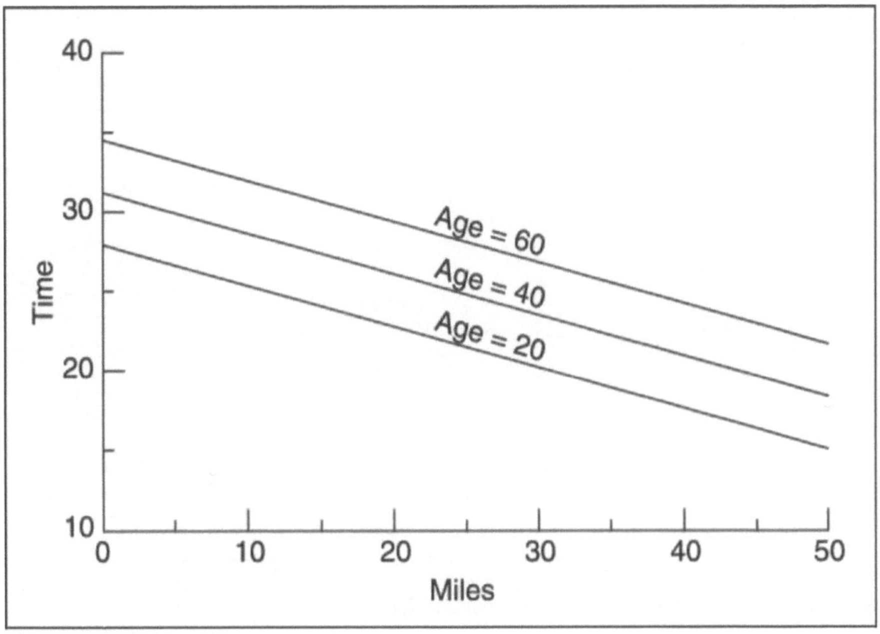

Figure 8.7 Time–Miles "simple" relationships

some variable Y_i on predictors X_{1i} and X_{2i}, we are assuming that the contribution of each predictor to the best predicted values, $\hat{Y}_i$, does not depend on the value of the other predictor. The two contributions are thus additive: Regardless of the level of X_{1i}, the relationship (i.e., slope) between Y_i and X_{2i} does not change. Likewise, regardless of the level of X_{2i}, the relationship between Y_i and X_{1i} is presumed not to change. We can simply "add" up the "effects" of two predictors in deriving $\hat{Y}_i$: In figuring out what the contribution of each predictor variable is to these $\hat{Y}_i$ values, we do not need to worry about the level or value of the other predictor.

How reasonable is this additive assumption? It depends on the substantive domain under examination. For the variables at hand, it may not be a very good assumption. For instance, it seems reasonable to suggest that the age difference in race times might be greater when dealing with people who do not run very much than when dealing with people who run a lot. To put this hypothesis another way, training a lot may reduce the relationship between Age and Time, such that differences in race times associated with age may be smaller with more training. This hypothesis is graphed in Figure 8.8. Notice that the slope of the prediction function linking Age and Time is hypothesized to be steeper when Miles is relatively low than when it is relatively high. In other words, the relationship between Age and Time is hypothesized to be stronger when Miles is low than when it is high.

We are suggesting that rather than Age and Miles combining *additively* in giving us our best predictions of Time, they may *interact* in affecting time. An *interaction* between two variables, say X_{1i} and X_{2i}, is said to occur whenever the slope or the "simple effect" of one of them, say X_{1i}, on the dependent variable Y_i varies with or depends on the value of the other variable, X_{2i}. In the case at hand, we are hypothesizing that Age and Miles interact in affecting Time. This means that we suspect that the relationship between Age and Time depends on or varies with the number of miles run.

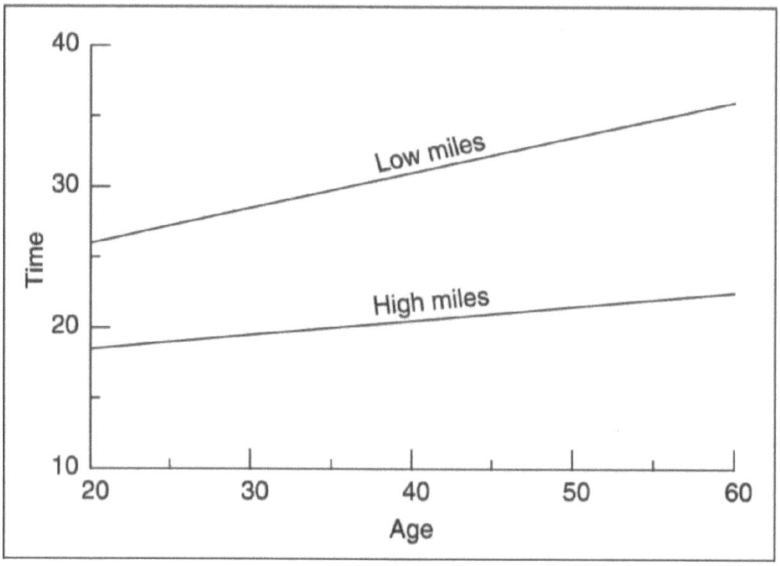

Figure 8.8 Hypothesized "simple" Time–Age relationships

Whether there is an interaction between two variables in predicting a third is an issue that is totally independent of whether the two predictor variables are correlated or redundant with each other. To suggest that Age and Miles may interact in predicting Time is to suggest that the "simple" relationship between Age and Time may depend on or vary with Miles. This may be true regardless of whether Age and Miles are themselves correlated or redundant predictor variables.

To test this hypothesis about the interaction between Age and Miles, we want to compare our three-parameter additive model with one in which the "simple" slope of one predictor variable is allowed to change as the other predictor variable changes. We saw earlier that the simple additive model allows the "simple" intercept of one predictor variable to change as the other predictor variable changes. However, the "simple" slope does not. The "simple" slope of Age, for instance, was invariant across changes in Miles. To allow for changing slopes, we need to build a model that allows both the intercept and the slope in the "simple" re-expressions to vary. In the following, we show how this can be done.

First, consider the additive model:

$$\hat{Time} = b_0 + b_1 Age + b_2 Miless$$

and its two "simple" re-expressions, the first focusing on the Time–Age "simple" relationship at different levels of Miles and the second focusing on the Time–Miles "simple" relationship at different levels of Age:

$$\hat{Time} = (b_0 + b_2 Miles) + b_1 Age$$

$$\hat{Time} = (b_0 + b_1 Age) + b_2 Miles$$

Now we would like to modify these "simple" expressions so that not only the "simple" intercepts vary as a function of the other variable, but so do the slopes. In other words, we want the "simple" re-expressions to have the following forms:

$$\hat{Time} = (b_0 + b_2 Miles) + (b_1 + b_3 Miles) Age$$

$$\hat{Time} = (b_0 + b_1 Age) + (b_2 + b_3 Age) Miles$$

In the first of these two, examining the "simple" relationship between Time and Age at different levels of Miles, the "*simple*" intercept (b_0 + b_2Miles), varies as a function of Miles, and now so does the "simple" slope (b_1 + b_3Miles). That is, as Miles changes value, the "simple" slope for Age varies, to the extent that b_3 differs from zero. Likewise, in the second of the previous "simple" expressions, focusing on the Time–Miles "simple" relationship, both the "simple" intercept (b_0 + b_1Age) and the "simple" slope (b_2 + b_3Age), are allowed to vary as a function of Age.

These two "simple" expressions are symmetrical and can be equivalently written as

$$\hat{Time} = b_0 + b_1 Age + b_2 Miles + b_3 (Age \times Miles)$$

This then indicates how we go about estimating models in which we allow for interactions, that is, models in which the "simple" slope for each variable is allowed to vary as a function of the value of the other variable. We simply form a new variable that is the product of the two predictors and estimate a model that includes this product variable as an additional predictor, in addition to the two component variables as predictors themselves. And the symmetry of these two "simple" effect expressions and their equivalence to the model that includes the product predictor means that interactions can be interpreted in either of two ways. If the product is a useful predictor, over and above its component variables, then that can be interpreted either as "the simple slope for Age depends on the value of Miles" or as "the simple slope for Miles depends on the value of Age."

Accordingly, in the following model,

$$\hat{Time} = b_0 + b_1 Age + b_2 Miles + b_3 (Age \times Miles)$$

two different, but equally correct, interpretations of the regression coefficients are possible, depending on which "simple" relationship we focus on. If we think about this model in the context of the "simple" Time–Age relationship,

$$\hat{Time} = (b_0 + b_2 Miles) + (b_1 + b_3 Miles) Age$$

the following interpretations follow:

b_0: the intercept for the "simple" Time–Age relationship when Miles equals zero.
b_1: the slope for the "simple" Time–Age relationship when Miles equals zero.
b_2: the change in the intercept for the "simple" Time–Age relationship as Miles increases by one unit.
b_3: the change in the slope for the "simple" Time–Age relationship as Miles increases by one unit.

On the other hand, if we think about this model in terms of the "simple" Time–Miles relationship,

$$\hat{Time} = (b_0 + b_1 Age) + (b_2 + b_3 Age) Miles$$

then the following equivalent interpretations follow:

b_0: the intercept for the "simple" Time–Miles relationship when Age equals zero.
b_1: the change in the intercept for the "simple" Time–Miles relationship as Age increases by one unit.
b_2: the slope for the "simple" Time–Miles relationship when Age equals zero.
b_3: the change in the slope for the "simple" Time–Miles relationship as Age increases by one unit.

Although these two sets of interpretations sound rather different, in fact they are fundamentally equivalent. In any specific context, one set of interpretations, and accordingly one "simple" relationship, is likely to be more compelling than the alternative. But both sets remain viable regardless of such preferences.

Estimation and Interpretation of the Interactive Model

Let us now illustrate this interactive model, and the resulting possible interpretations, with the data at hand. We have already estimated the three parameters of the linear additive model, including both Age and Miles as predictors:

$$\hat{Time} = 24.605 + .167Age - .257Miles$$

Now let us estimate the four parameters of the interactive model. To do this we need to compute a new predictor variable that is the product of Age and Miles. We call this variable AgeMiles. Its value, for each observation, is the product of runners' value on Age multiplied by their respective values on Miles. We then regress Time on Age, Miles, and this new predictor variable, AgeMiles:

$$\hat{Time} = 18.899 + .308Age - .069Miles - .005AgeMiles$$

The sum of squared errors for this model equals 697.625.

Let us compare this estimated model to the results that we have already given for the additive model. We can do this in two steps: compare the models to ask whether the interactive model does significantly better than the simpler additive one; and compare the estimated coefficients of the two models, and their interpretations.

To compare the quality of predictions of the two models, we treat the additive one as Model C and the interactive one as Model A:

$$\text{MODEL A: Time} = \beta_0 + \beta_1\text{Age} + \beta_2\text{Miles} + \beta_3\text{AgeMiles} + \varepsilon_i$$
$$\text{MODEL C: Time} = \beta_0 + \beta_1\text{Age} + \beta_2\text{Miles} + \varepsilon_i$$

This model comparison provides a test of the null hypothesis that the parameter associated with the additional predictor (the AgeMiles product) equals zero:

$$H_0: \beta_3 = 0$$

The additive model has a sum of squared errors of 741.195 with 77 degrees of freedom for error, having estimated three parameters. We have just seen that the interactive model has a sum of squared errors of 697.625, with 76 degrees of freedom. Thus, the comparison yields an SSR of 43.570 (i.e., 741.195–697.625) with PA – PC = 1. The values of PRE and *F* that result from this comparison are thus

$$\text{PRE} = \frac{43.570}{741.195} = .059$$

$$F_{1,76} = \frac{.059/1}{(1-.059)/76} = \frac{43.570/1}{697.625/76} = 4.75$$

The values of both *F* and *PRE* exceed their critical values. Accordingly, the interactive model, including the product predictor, does a significantly better job of predicting the data than the simpler additive model. In other words, the product predictor is a useful predictor over and above the two variables that are its components, that is, over and above Age and Miles.

To help us understand what this conclusion means, and to understand exactly how this model differs from the additive one, it is important to compare the models in terms of the interpretations we can give the parameter estimates. We will do this in the context of the models' "simple" re-expressions, as defined earlier. In general, we strongly encourage interpretations of parameter estimates in the context of "simple" re-expressions, particularly in the case of interactive models.

We have already interpreted the various parameter estimates in the additive model twice: once as we examined the "simple" Time–Age relationship within levels of Miles, and once as we examined the "simple" Time–Miles relationship within levels of Age. As we saw there, when we focused on the "simple" Time–Age relationship, the estimated coefficient for Miles, –.257, was the change in the "simple" intercept as Miles increases, while the estimated coefficient for Age, .167, was the "simple" slope for Age, invariant across levels of Miles. When we focused on the "simple" Time–Miles relationship, the estimated coefficient for Age (.167) was the change in the "simple" intercept as Age increases, while the estimated coefficient for Miles (–.257) was the "simple" slope for Miles, invariant across levels of Age.

Parallel interpretations can be given for the interactive model. First, looking at the "simple" Time–Age relationship within levels of Miles, we re-express the estimated interactive model as follows:

$$\hat{Time} = (18.899 - .069Miles) + (.308 - .005Miles)Age$$

Now let us evaluate this "simple" re-expression at various levels of Miles. We start with the value of zero. Obviously, we must be cautious here in talking about the "simple" Time–Age relationship at this value because no observation in our dataset had a Miles value of zero. Nevertheless, based on the model, we can substitute the value of zero for Miles in the earlier re-expression. Our predicted Time–Age "simple" relationship for this value of Miles is

$$\hat{Time} = 18.899 + .308Age$$

Accordingly, when interpreted from the point of view of the "simple" Time–Age relationship, the intercept in the interactive model, 18.899, gives us our best estimate of the intercept for the "simple" Time–Age relationship when Miles equals zero. The coefficient for Age in the interactive model, .308, tells us the estimated "simple" slope for the Time–Age relationship when Miles equals zero. Note that this interpretation is quite different from the interpretation given for the regression coefficient for Age in the additive model. There the regression coefficient for

Age was interpreted as the slope for the Time–Age relationship within levels of Miles, regardless of the actual value of Miles. In other words, there the "simple" slope did not change as Miles changed.

Now, however, in the interactive model, the "simple" slope for Age changes as the value of Miles changes. Suppose we were interested in the "simple" relationship between Time and Age for runners who average 30 miles per week. The prediction function for such runners is

$$\hat{Time} = \left(18.899 - .069(30)\right) + \left(.308 - .005(30)\right) Age = 16.829 + .158 Age$$

Clearly this "simple" relationship between Time and Age is not what it was for the prediction function when Miles was set equal to zero. As Miles has increased, both the "simple" intercept and the "simple" slope for Age have decreased. For each unit change in Miles, the "simple" intercept has decreased by .069 units. And for each unit change in Miles, the "simple" slope has decreased by .005.

From this perspective, focusing on the "simple" Time–Age relationship in this interactive model, we can now interpret all four parameter estimates. First, the intercept in the model, 18.899, is the "simple" intercept when Miles equals zero. The estimated coefficient for Miles, –.069, is the change in the "simple" intercept as Miles increases by one unit. The estimated coefficient for Age, .308, is the "simple" slope for Age when Miles equals zero. And the coefficient for the product predictor, –.005, is the change in the "simple" slope as Miles increases by one unit.

In Figure 8.9, we have graphed a few of the "simple" Time–Age prediction functions for representative values of Miles. Clearly, both the intercepts and the slopes of these "simple" prediction functions vary as a function of Miles—the intercept by –.069 per unit increase in Miles and the slope by –.005 per unit increase in Miles. In comparison to the graph of this same simple relationship from the additive model (contained in Figure 8.6), we now are allowing nonparallel "simple" prediction functions. Our earlier statistical conclusion that the product predictor variable significantly improved the quality of the predictions made by the model is equivalent to the conclusion that the graphed "simple" prediction functions in Figure 8.9 are significantly nonparallel.

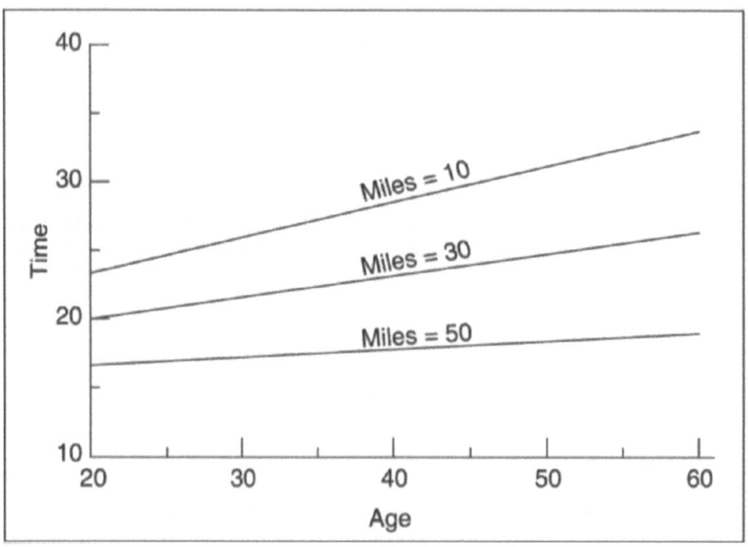

Figure 8.9 "Simple" Time–Age relationships from interactive model

Substantively, the interaction represented by these plotted lines amounts to the conclusion that while 5-kilometer race times may be slower for older than for younger runners, that age difference is less pronounced with more training.

We now turn to the alternative "simple" re-expression of this interactive model, focusing on the "simple" Time–Miles relationship at different levels of Age and interpreting the estimated regression coefficients from this perspective. The re-expressed interactive model is

$$\hat{Time} = (18.899 + .308Age) + (-.069 - .005Age)Miles$$

The value of the estimated intercept, 18.899, is the "simple" intercept when Age equals zero. The estimated coefficient for the Age predictor is the change in the "simple" intercept as Age increases by one unit. The estimated coefficient for the Miles predictor, –.069, is the "simple" slope for Miles when Age equals zero. And finally, the estimated coefficient for the product predictor, –.005, is the change in the "simple" slope for Miles as Age increases by one unit.

Again, in Figure 8.10, we have graphed a few representative prediction functions from this "simple" re-expression. Once again, our earlier statistical test, comparing the additive model as the compact one with the interactive one as the augmented one, is in essence asking whether these "simple" prediction functions are significantly nonparallel. Substantively, our conclusion from this graph of the "simple" prediction functions focuses on the effect of training and how that varies as a function of age. While people who train more run faster 5-kilometer races, that training benefit is greater among those who are older than among those who are younger.

As the aforementioned makes clear, each estimated regression coefficient in an interactive model has multiple interpretations (which in a deep sense are statistically equivalent). So the coefficient associated with the Age predictor can be interpreted either as the "simple" slope for Age when Miles equals zero or as the change in the intercept for the "simple" Time–Miles relationship as Age increases. The estimated coefficient for the Miles predictor in a parallel manner can be interpreted either as the "simple" slope for Miles when Age equals zero or as the change in the intercept for

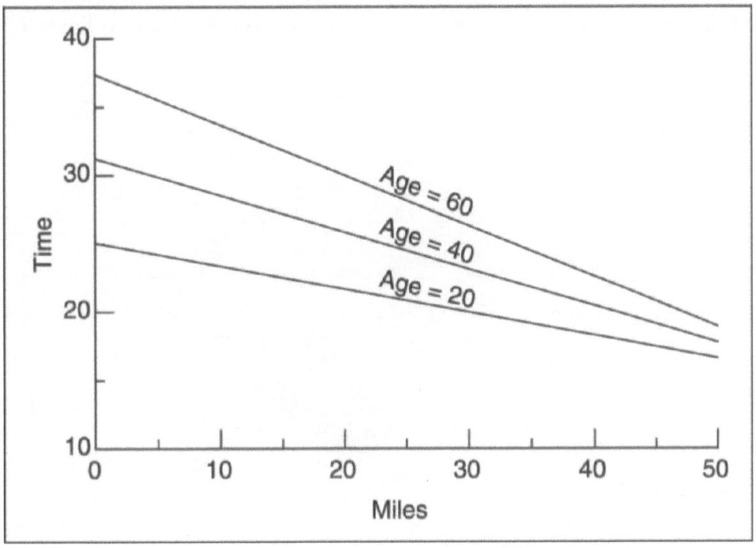

Figure 8.10 "Simple" Time–Miles relationships from interactive model

the "simple" Time–Age relationship as Miles increases. And finally, the estimated coefficient associated with the product predictor can be interpreted either as the change in the "simple" slope for Age as Miles increases or as the change in the "simple" slope for Miles as Age increases.

While the choice of interpretation for the coefficient associated with the product predictor is largely arbitrary, we have a preference about one of the two interpretations for the coefficients associated with the variables that are components of the product. In general, we find it easier to think in terms of "simple" slopes rather than "simple" intercepts. Accordingly, in the case at hand, we prefer to interpret the estimated coefficient for Age as its "simple" slope when Miles equals zero, and the estimated coefficient for Miles as its "simple" slope when Age equals zero. Both interpretations for the estimated coefficient for the product predictor refer to "simple" slopes, either how the Age "simple" slope varies as a function of Miles or how the Miles "simple" slope varies as a function of Age. Here, the preference of one over the other of these depends entirely on theoretical assumptions about which of the two component variables is in some sense primary and which one modifies or moderates its "effect." In the case at hand, we have a slight preference for saying that the age difference in running times decreases as one trains more. But someone else might prefer to say that the training difference in running times is greater among older runners. Both are equally viable interpretations of the significant interaction in these data.

Figure 8.11 presents a three-dimensional graph of the prediction plane generated by this interactive model. As the graphs of the "simple" relations in Figures 8.9 and 8.10 make clear, if we take slices of this prediction plane parallel to one of the horizontal axes, the slope of the resulting prediction lines change. Thus, as we move up the Age horizontal axis, the slope of the plane becomes more and more negative. Compared to the prediction plane generated by the additive model (graphed in Figure 8.5), this prediction plane seems to twist as we move across the values of one of the horizontal axes.

Tests of Other Parameters in the Interactive Model

We have already tested the null hypothesis that the value of β_3 equals zero when we compared the augmented interactive model with the compact additive one. This test of whether the parameter associated with the product predictor departs from zero is formally a test of the interaction: Does

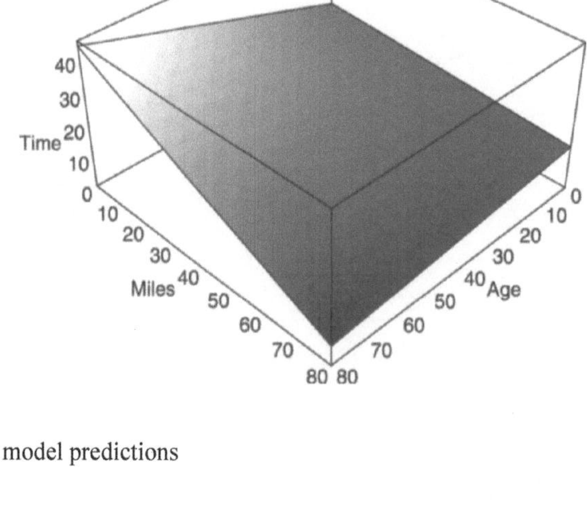

Figure 8.11 Interactive model predictions

the "effect" of Age depend on Miles or, equivalently, does the "effect" of Miles depend on Age? Importantly, this model comparison must include the two component predictors in both Models A and C. A test of a product predictor is a test of the interaction only when both component predictor variables are included in the models.

With multiple predictors in the interactive model, there are of course other Model Cs with which the interactive model might be compared, other than the additive one that leaves out the product predictor. For instance, suppose we constructed a compact model in which the parameter associated with the Age predictor was set equal to zero and compared such a model with the full interactive one. This comparison would be testing the null hypothesis that the component variable Age is not a useful predictor over and above Miles and the product predictor. Models A and C for this comparison are estimated as follows:

$$\text{Model A}: \hat{Time} = 18.899 + .308Age - .069Mile - .005AgeMiles$$

$$\text{Model C}: \hat{Time} = 31.542 - .416Miles - .004AgeMiles$$

The augmented interactive model has a sum of squared errors of 697.625, as we saw previously, while this compact model has a sum of squared errors of 869.153. Their comparison yields a PRE value of .197 and an F statistic of 18.688 with 1 and 76 degrees of freedom. Clearly, we must reject the null hypothesis that the parameter associated with Age in the interactive model has a value of zero.

We can also test the null hypothesis that the parameter associated with Miles in the interactive model equals zero. Estimating a compact model with only Age and AgeMiles as predictors, we get

$$\text{Model C}: \hat{Time} = 16.784 - .357Age - .006AgeMiles$$

with a sum of squared errors of 702.953. Again, comparing this model with the augmented interactive model, we get PRE = .008 and $F = 0.581$ with 1 and 76 degrees of freedom. We thus cannot reject the null hypothesis that the parameter associated with Miles in the interactive model equals zero.

At the very start of this chapter, we tested null hypotheses about the parameters associated with both Age and Miles in the additive model that did not include the product term. Testing whether the regression coefficient for Age in the additive model differed from zero, we obtained a PRE value of .280 and an F of 29.899 with 1 and 77 degrees of freedom. Testing whether the regression coefficient for Miles in the additive model differed from zero, we obtained a PRE value of .568 and an F of 101.143 with 1 and 77 degrees of freedom. Clearly these tests of the Age and Miles parameters in the additive model do not give the same results as the tests of the parallel parameters associated with these variables in the interactive model. For the test of the Age parameter, the resulting PRE value in the additive model was .280, while in the interactive model it equaled .197. For the tests of the Miles parameter, the resulting PRE value in the additive model was .568, while in the interactive model it equaled only .008. Why this difference?

In a general sense, we have already discussed the reason for this difference. As shown in Chapter 7, the interpretation of a partial regression coefficient depends on what other predictors are included in the model. Since the interactive model includes a different set of predictors than the additive one, we would expect the partial regression coefficient for a given predictor variable to differ between the two models.

In this case, a more specific and substantive answer to this question is extremely important, although in fact this more specific answer is just a restatement of the general one given in Chapter 7 and in the previous paragraph. The partial regression coefficients for Age and Miles have quite different interpretations in the interactive model than they do in the additive one, as we have repeatedly stressed in interpreting the interactive model. Hence, testing whether they significantly differ from zero in the two models amounts to testing entirely different substantive hypotheses. In the additive model, the Age parameter is interpreted as the slope for Age within levels of Miles or holding Miles constant. But that slope is presumed not to change in the additive model as Miles changes. Hence, we can interpret the slope as something like the average Time–Age slope within levels of Miles. On the other hand, the regression coefficient for Age in the interactive model is the slope for Age *when and only when* Miles equals zero.

Therefore, in the additive model, when we test whether the Age parameter equals zero, we are testing whether Time and Age are significantly related on average within levels of Miles or holding Miles constant. On the other hand, in the interactive model, we are testing whether Time and Age are significantly related *when Miles equals zero*. Similarly, when we test whether the regression coefficient for Miles equals zero in the additive and interactive models, we are testing entirely different substantive null hypotheses. In the additive case, we are testing whether the slope for Miles on average, within levels of Age or holding Age constant, differs significantly from zero. On the other hand, in the interactive model, the test of whether the Miles parameter differs from zero amounts to testing whether Time and Miles are significantly related *when Age equals zero*.

In the present case, of course, asking questions about "simple" slopes of one predictor variable when the other one equals zero is totally uninteresting and uninformative since our data do not include any observations with zero values on these variables. Given that we have already found that the interactive model does a better job than the additive one, and therefore that the "simple" slope of one variable changes as a function of the other variable, it seems unlikely that we would be interested in testing a "simple" slope at values outside the range of those included in the data.

In our experience advising others and reviewing articles that report results from multiple regression models, misinterpreting regression coefficients associated with predictor variables that are components of interactions occurs frequently. In general, when a product predictor is included in a model, along with its components, the coefficients associated with each component estimate the "simple" effect of that component when the other component with which it is multiplied to make the product equals zero. Since it is frequently the case that values of zero are not of theoretical interest and may not even be possible values of the data, it makes no sense to interpret such regression coefficients. Certainly, they do not represent "overall" or "net effects" of predictors. Indeed, the search for some estimate of the "overall effect" of a component predictor variable is generally misguided. The very nature of a significant interaction between one variable and another means that the "effect" of either one of them "depends"; there is no constant or overall "effect." Rather, the magnitude (and/or direction) of the effect depends on the level of the other component variable.

Deviating or Centering Component Predictors

It is possible to obtain and test more meaningful "simple" slopes of component predictor variables by transformations that change the zero values of the predictors. Recall that in Chapter 6, we showed that by deviating a predictor from its mean, or centering it, the intercept in simple regression models came to equal the predicted value at the mean level of the predictor, which equivalently is the mean of Y_i. It then became possible to compare augmented and compact models to ask potentially interesting questions about the mean of Y_i. In a similar vein, it often makes sense in interactive models to center the two component variables to make their regression coefficients of greater interest. Importantly, one must center the predictor variables before computing the product. In other words, the product variable that is included as a predictor in the interactive model should be the product of the two centered predictors. If such an interaction model is estimated, then the regression coefficients for each component predictor variable will equal the "simple" slope for that predictor when the other centered predictor equals zero, that is, at the mean level of the other predictor. Admittedly, no overall or "net" effect of each component predictor is readily available in the context of a significant interaction. Yet one may well want to ask questions about the average "simple" effect of a component predictor, that is, its "simple" effect at the average level of the other component predictor variable.

Let us illustrate with the data at hand. In the dataset, the mean Age is 39.662 and the mean Miles trained is 29.875. We compute new centered or mean-deviated versions of both variables, Age_0 and $Miles_0$, subtracting the means of the variables from each observation's raw values. The zero subscript is used to indicate that each variable, in its centered form, now has a mean of zero. We now compute a new product predictor variable, taking the product of the two centered components, Age_0Miles_0.[1] We then estimate the interactive model again, this time using the centered predictor and their product:

$$\hat{Time} = 23.431 + .166Age_0 - .258Miles_0 - .005Age_0Miles_0$$

In a deep sense, this is exactly the same interactive model that we estimated previously, without the centered variables. We have simply relocated the zero values on the two horizontal axes of Figure 8.11, putting zero at the mean values of both Age and Miles. The data have not otherwise changed nor have the model's predicted values, that is, the prediction plane in Figure 8.11. Thus, the sum of squared errors for this model is identical to what it was for the estimated interactive model with all variables in their uncentered form, that is, 697.625.

But now all of the estimated coefficients, except for that associated with the product predictor, have changed. And this is because, within the context of this interactive model, they are now estimating different things. The intercept, 23.431, is the predicted value when all three predictors equal zero, which are the mean values of Age_0 and $Miles_0$. The regression coefficient associated with Age_0, .166, is the "simple" slope for Age when $Miles_0$ equals zero, that is, for runners who run the average number of miles in the dataset. Similarly, the regression coefficient associated with $Miles_0$, –.258, is the "simple" slope for Miles when Age_0 equals zero, that is, for runners who are of the average age in the dataset. Since this interactive model is one in which each of these "simple" slopes changes as a function of the other variable, these are different values from the simple slopes estimated in the model with the uncentered predictors. And the degree to which they depart from those uncentered "simple" slopes is a function of the regression coefficient associated with the product predictor. The coefficient associated with the predictor variable is therefore the one estimated coefficient that remains unchanged in the model: It continues to indicate the extent to which the "simple" slope of Age changes as Miles increases or, equivalently, the extent to which the "simple" slope of Miles changes as Age increases. Graphically, the regression coefficients associated with the predictors that are components of the product in the interactive model (Age and Miles in the original model *or* Age_0 and $Miles_0$ in the centered model) are slopes of the prediction plane at particular values along the two horizontal axes (i.e., where the values equal zero), while the coefficient associated with the product predictor is the change in these simple slopes, that is, the degree to which the predicted plane twists as values along a horizontal axis change.

Tests of the regression coefficients associated with the component predictor variables in the centered interactive model are considerably more informative than those same tests in the uncentered version of the model. Comparing the centered interactive model with a compact one that fixes the coefficient for Age_0 at zero yields an SSR of 283.351, a PRE of .289, and an F of 30.91 with 1 and 76 degrees of freedom. Comparing this same interactive model with a compact one that fixes the $Miles_0$ coefficient at zero yields an SSR of 976.934, a PRE of .583, and an F of 106.50 with 1 and 76 degrees of freedom. Conclusions from these tests suggest that within the confines of this interactive model (i.e., allowing "simple" slopes to change) the effect of each component predictor variable is significant at the mean level of the other predictor.

Note that the statistical results of these "simple" tests are similar, although not identical, to the tests of each of the predictors in the three-parameter additive model reported at the start of this chapter. There the statistical test for the additive effect of Age yielded a PRE of .280 and an F of 29.92 with 1 and 77 degrees of freedom, and the test of the additive effect of Miles yielded a PRE of .568 and $F = 101.20$ with 1 and 77 degrees of freedom. Although these results are similar, they are fundamentally addressing different questions. In the context of the additive model, we are asking about the "effect" of each variable controlling for the other. In the context of the centered interactive model, we are asking about the "simple effect" of each variable at the average value of the other, allowing those "simple effects" to vary as the other predictor changes values.

Since the regression coefficient associated with the product predictor has not changed as a function of centering its components, and since in a deep sense this centered version of the interactive model is identical to the uncentered version, it has to be the case that the Model A/Model C comparison to test whether the parameter associated with the product equals zero yields the same result in the centered version as in the uncentered case already examined, that is, PRE = .059 and $F = 4.75$ with 1 and 76 degrees of freedom. In other words, regardless of whether we center Age and Miles, the test of the regression coefficient for their product is always a test of whether the "simple" slope of one variable changes as the other variable changes in value, so *long as the two variables involved in the product term, Age and Miles or Age_0 and $Miles_0$, are included as separate predictors in the augmented model*. The italics are to emphasize that the regression coefficient for a product of two predictor variables estimates the effect of the interaction between those variables only when the regression coefficient is a partial one, partialing out or controlling for the component variables that were multiplied together to yield that product. As long as the components are included as separate predictors in the model, a test of whether the regression coefficient for the product departs from zero will be invariant across all linear transformations of the component variables (Cohen, 1978).

The utility of a centering transformation for both Age and Miles to increase the interpretability of their regression coefficients generalizes to other sorts of deviation transformations in order to test hypotheses about particular "simple" relationships. Suppose, for instance, in the interactive model, allowing for different "simple" Time–Age relationships at various levels of Miles, we were interested in determining whether Time and Age were significantly related for runners who averaged 50 miles a week. Since the coefficient for Age in the interactive model estimates the "simple" Time–Age slope when the other predictor with which Age interacts equals zero, we could transform Miles by subtracting 50 from the value of Miles for each observation, resulting in a deviation transformation defined as

$$Miles_{50} = Miles - 50$$

Now we regress Time on Age_0, $Miles_{50}$, and their product Age_0Miles_{50}:

$$\hat{Time} = 18.244 + .070Age_0 - .258Miles_{50} - .005Age_0Miles_{50}$$

Notice that while the coefficient associated with Miles_{50} has not changed as a result of its transformation (it still equals the same value as it did when both predictors were centered), the coefficient for Age_0 has changed considerably and now is no longer significant: PRE = .022 and $F = 1.69$ with 1 and 76 degrees of freedom. The change occurs because the coefficient associated with Age_0 now estimates the "simple effect" of Age for runners who train an average of 50 miles per week (i.e., when Miles_{50} equals zero). This estimated "simple effect" is not significant in these data: We cannot conclude that there is an age difference in 5-kilometer race times among runners who train 50 miles per week.

Note the curious but correct implication of this transformation exercise: A linear transformation of one component predictor variable in an interactive model (and a recomputation of the product predictor) does not affect the regression coefficient associated with that component predictor; rather, it affects the regression coefficient associated with the other component predictor. This fact results from the definition of these component coefficients as "simple" slopes for each component when the other equals zero. If the zero value of one is altered, the "simple" slope for the other changes, assuming that an interaction between the two exists.

One set of commonly used and recommended transformations, to present the results of an interactive model, is to estimate and test "simple" slopes of one component variable at three values of the other: its mean, one standard deviation above its mean, and one standard deviation below its mean. Although there is nothing sacred about such values, and others might be more informative in a particular context, getting into the habit of estimating and testing different "simple" slopes of variables that are components of interactions is good practice.

Power Considerations in Interactive Models

In Chapter 7, we presented the general formula for the confidence interval of a regression coefficient in multiple regression. That formula, reproduced from Chapter 7, is

$$\text{CI} = b_{p.12\ldots p-1} \pm \sqrt{\frac{F_{crit}(\text{MSE})}{s_p^2(n-1)(1-R_p^2)}}$$

where $b_{p.12\ldots p-1}$ is the regression coefficient for the pth predictor variable, F_{crit} is the critical value of the F statistic with 1 and $n -$ PA degrees of freedom, MSE is the mean squared error of the model, $s_p^{\,2}$ is the variance of the pth predictor, and $1 - R_p^{\,2}$ is its tolerance. This formula is the appropriate one regardless of the nature of the predictor variable. Thus, it provides the confidence intervals for the regression coefficients of products and their components in interactive models, as well as for the coefficients of additive models.

The same considerations affect the power of tests in interactive models as in additive models. Namely, factors that tend to increase the size of the confidence interval tend to decrease the power of the test of the null hypothesis that the estimated parameter equals some a priori value. Factors that decrease the size of the confidence interval increase power. Accordingly, as the mean square error of the model decreases, as the variance for the pth predictor increases, and as the tolerance of the pth predictor increases, the inferential test of its associated parameter is more powerful.

Within the context of interactive models, however, there is an interesting twist to the power issue because the product predictor variable is computed from the other predictor variables in the model and thus in some sense depends on them. Consider the following two variables and their product:

X_1	X_2	X_1X_2
3	2	6
1	3	3
4	5	20
2	1	2
5	4	20

If we were to regress some data variable Y_i on these three predictors, the confidence interval for the regression coefficient for the product variable would be a function of the critical value of F, the mean square error from the model, the variance of the product predictor, its tolerance (i.e., $1-R_p^2$), and the sample size. Let us think about these last three terms, all in the denominator of the formula for the confidence interval. Multiplying the variance of the product by $n-1$ gives the sum of squares of the product around its mean. The tolerance of the product is obtained by regressing it on its two components and subtracting the resulting R^2 from 1. In this case, the total sum of squares of the product around its mean equals 328.80 and the tolerance of the product equals 0.0096563. (Notice that the product and its components are extremely redundant here; also we are reporting tolerance to a very high degree of accuracy to avoid rounding errors.) If we multiply these two together we get 3.175, which is the sum of squared errors resulting from the model where the product is regressed on its two components. This is the value of the denominator in the formula for the confidence interval.

Now consider what happens if we center both X_1 and X_2, subtracting their mean values (3 in both cases), and then compute the product of the transformed variables:

X_{10}	X_{20}	$X_{10}X_{20}$
0	−1	0
−2	0	0
1	2	2
−1	−2	2
2	1	2

Now the variance of the product variable equals 1.2, $n-1$ still equals 4, and the tolerance of the product equals .6615. As a function of the transformations of X_{10} and X_{20}, their product is considerably less variable and relatively less redundant (linearly) with its two components. Even so, by multiplying these three terms together, the product in the denominator of the confidence interval still equals 3.175 (and this number is again the error sum of squares from a model in which the product of the two centered variables is regressed on its two centered components). This is exactly as it should be since inferential tests of a product variable's regression coefficient are not affected by linear deviation transformations of its components, as we have seen. Thus, the power of the test of the interaction or product variable is unaffected by deviation transformations of its components, *as long as those components are included as predictors in the model.*

What is interesting about this result, however, is that deviation transformations do affect the tolerance of the product variable or its linear redundancy with its components. Yet the width of the confidence interval of its regression coefficient is unaffected since the change in tolerance is exactly compensated for by the change in the variance of the product variable. This is an interesting exception to the general rule given in the last chapter that redundancy among predictors tends to reduce power. When a predictor is a product of other predictors, then various transformations can radically affect its redundancy with its components. Yet such transformations have no effect on the power of the test of the product's regression coefficient, as long as the component variables are included as predictors in the model.

For the running data, we compute the confidence interval for the regression coefficient of the product predictor for the interactive model, first with the component predictors in their original metrics, and then with them centered. In both models, the critical value of F, with 1 and 76 degrees of freedom, is 3.96 and the MSE from both models equals 9.18.

In the original uncentered model, the variance of the product, AgeMiles, equals 380,387.09 and its tolerance ($1 - R^2$ when regressed on Age and Miles) equals .064. Accordingly, the confidence interval for the coefficient of the product predictor is

$$\mathrm{CI}_{b_3} = .005 \pm \sqrt{\frac{3.96 \times 9.18}{380{,}387.09 \times 79 \times .064}}$$

$$= .005\sqrt{\frac{36.35}{1{,}917{,}017.84}}$$

$$= .005 \pm .004$$

And when the two component predictors are centered and these and their product are the predictors in the interactive model, the confidence interval for the coefficient for the product equals:

$$\mathrm{CI}_{b_3} = .005 \pm \sqrt{\frac{3.96 \times 9.18}{24{,}276.01 \times 79 \times .999}}$$

$$= .005 \pm \sqrt{\frac{36.35}{1{,}917{,}017.84}}$$

$$= .005 \pm .004$$

The lesson from all of this is that while a product variable may be quite redundant, in a linear manner, with its components, and while transformations of its components may reduce that redundancy, such transformations will have no effect on the statistical inference of the product variable's regression coefficient as long as the component variables are included in the model.

A General Procedure for the Derivation of "Simple" Slopes

As we have emphasized, we find it easiest to interpret parameter estimates in models with product predictors by thinking in terms of "simple" slopes. Up to this point we have relied on a simple heuristic to derive these "simple" slopes: We have simply rearranged terms in the model. With more complex models, this approach does not work. In this section we present a more general procedure for deriving "simple" slopes, relying on a notion in calculus: the derivative or partial derivative of a function with respect to some variable.

A derivative *is* a “simple” slope, in the way we have been using it, in so many words. We can express any estimated multiple regression equation in the following generic terms:

$$\hat{Y}_i = F(X_1, X_2, X_3, \ldots, X_p)$$

meaning simply that our predicted values are some function of a set of predictor variables. The “simple” slope associated with one of these variables, say X_j, is the partial derivative of the function with respect to that variable:

$$\frac{\Delta\hat{Y}}{\Delta X_j}$$

This standard notation for a derivative means that it is a slope, in the sense that the “Δ” indicates a change or difference: We are asking about the “simple” difference in the predicted values as a function of a one-unit difference in a particular predictor, X_{ji}.

Even those who are unfamiliar with calculus can calculate partial derivatives of the sorts of functions addressed in this book. Three simple rules suffice:

1 The partial derivative of a sum, with respect to X_j, equals the sum of the partial derivatives of the components of that sum.
2 The partial derivative of aX_j^m with respect to X is amX_j^{m-1}, where a can be either a constant or another variable (or some product of both).
3 The partial derivative of a component of a sum, with respect to X_j, where that component does not contain X_j, is zero.

Let us take the interactive model that we have been dealing with so far in this chapter and apply these rules to derive the “simple” slopes of both Age and Miles at particular levels of the other variable.

The function is

$$\hat{Time} = b_0 + b_1 Age + b_2 Miles + b_3 AgeMiles$$

To evaluate the “simple” slope for Age, we take the partial derivative of all four components of this expression with respect to Age. According to the aforementioned third rule, the partial derivative of the first component, b_0, is zero, as is that of the third. And the partial derivative of b_1Age is b_1 according to the second rule, and that of b_3AgeMiles is b_3Miles, also according to the second rule. This yields the following partial derivative for Age:

$$\frac{\Delta\hat{Time}}{\Delta Age} = b_1 + b_3 Miles$$

This derivation of the Age “simple” slope is, of course, identical to that discussed earlier: b_1 is the “simple” slope when Miles equals zero and b_3 is the increase in that “simple” slope as Miles increases. Unsurprisingly, a parallel process yields the “simple” slope for Miles identified earlier, that is, $b_2 + b_3 Age$.

Consider now a model, which we have yet to explicate, involving a product of a predictor with itself:

$$\hat{Y}_i = b_0 + b_1X_i + b_2X_i^2$$

Let us take the derivative of this with respect to X, thus obtaining its "simple" slope:

$$\frac{\Delta \hat{Y}_i}{\Delta X_i} = b_1 + 2b_2 X_i$$

What does this mean? This is the "simple" slope at a particular value of X_i. It says that the "simple" slope depends on the value of X_i itself. At higher values of X_i, the "simple" slope is something different than at lower values. This defines a curvilinear relationship between X_i and Y_i, such that the "simple" slope varies as a function of X_i itself. The "simple" slope is thus the slope of the line (actually curve) at a particular value of X_i, with that slope changing continuously as X_i increases. If it helps, think about this "simple" slope (i.e., the derivative) as the slope of a line tangent to the curve that results from the plot of the predicted values from the estimated model:

$$\hat{Y}_i = b_0 + b_1 X_i + b_2 X_i^2$$

We will return with a data example to illustrate this shortly.

Finally, consider a considerably more complicated model:

$$\hat{Y}_i = b_0 + b_1 X_{1i} + b_2 X_{2i}^2 + b_3 X_{1i}^2 + b_4 X_{1i} X_{2i}$$

Again, we will worry about the specifics of such models later. For now, we simply derive the "simple" slopes of both X_1 and X_2:

$$\frac{\Delta \hat{Y}_i}{\Delta X_{1i}} = b_1 + 2b_3 X_{1i} + b_4 X_{2i}$$

$$\frac{\Delta \hat{Y}_i}{\Delta X_{2i}} = b_2 + b_4 X_{1i}$$

Thus, according to this model the "simple" slope of X_1 depends both on its value and on the value of X_2, while the "simple" slope of X_2 depends only on X_1.

Powers of Predictor Variables

A generalization of what has been learned about interactions between two predictor variables can be used to understand models that include powers of a predictor variable. Suppose, for instance, with the data at hand, we suspect that the relationship between Time and Miles depends on the level of Miles. We suspect that when one does not train very much, the relationship between training and race times might be a lot stronger than it is when one trains a great deal. In other words, adding 5 miles more per week to one's training regimen is likely to make much more of a difference to one's race time if one's base were 10 miles per week than if one's base were 50 miles per week. In other words, we are hypothesizing that the "simple" Time–Miles relationship changes as Miles (rather than Age) changes. In essence, we are hypothesizing an interaction of Miles with itself in predicting Time.

We use the same functional form to incorporate such an interaction into our model; namely, we include the product of the two variables that are hypothesized to interact as a separate predictor in the regression model. But this time, because it is Miles that is thought to interact with itself, we include a product predictor that is Miles × Miles (or $Miles^2$):

$$\hat{Time} = b_0 + b_1 Miles + b_2 Miles^2$$

Given the present data, this model is estimated as

$$\hat{Time} = 37.479 - .754 Miles + .008 Miles^2$$

with a sum of squared errors of 894.190.

To interpret this model and what its parameters mean, we do two things: We derive and plot predicted values of Time as a function of Miles, and we interpret the parameters by focusing on the "simple" slope for Miles. Because the maximum value of Miles in our dataset is 50, we derive predicted values of Time (as a function of Miles) that vary between 0 and 50. For instance, when Miles equals zero, the predicted value of Time equals

$$37.479 - .754(0) + .008(0^2) = 37.479$$

And when Miles equals 10, the predicted value of Time equals

$$37.479 - .754(10) + .008(10^2) = 30.739$$

Once we have a full set of predicted values we plot them, as in Figure 8.12. As is apparent, these predicted values, as a function of Miles, no longer lie along a straight line. Although in a formal sense our model is still a "linear" one, the prediction function is now a curve. And, as we expected, the shape of the curve is such that the slope is decidedly steeper at lower levels of Miles than it is at higher levels: Training one additional mile per week has more of a benefit in reducing race times if one is at a low training base than if one is at a high training base.

To interpret the coefficients of the model, we can start with the intercept, 37.479. This number represents what it always has, namely the predicted value when the predictors both equal zero. In this case, it is the predicted race time for someone who trains zero miles per week.

To interpret the two slope estimates in this model, we derive the "simple" slope for Miles, using the procedure explicated in the last section to get the derivative of this function with respect to Miles:

$$\frac{\Delta \hat{Time}}{\Delta Miles} = -.754 + (2).008 Miles$$

This Miles "simple" slope obviously varies as a function of Miles. It amounts to the slope of lines that are at a tangent to the prediction curve in Figure 8.12 at various points along the Miles horizontal axis. So, for instance, when Miles equals zero, the "simple" slope of the curve equals –.754. When talking about someone who runs 10 miles per week, the "simple" slope of the curve is $-.754 + (2).008(10) = -.594$. At the point of the curve where Miles equals 20, the "simple" slope is $-.754 + (2).008(20) = -.434$. And by the time we get up to someone who runs 40 miles per week, the "simple" slope is $-.754 + (2).008(40) = -.114$.

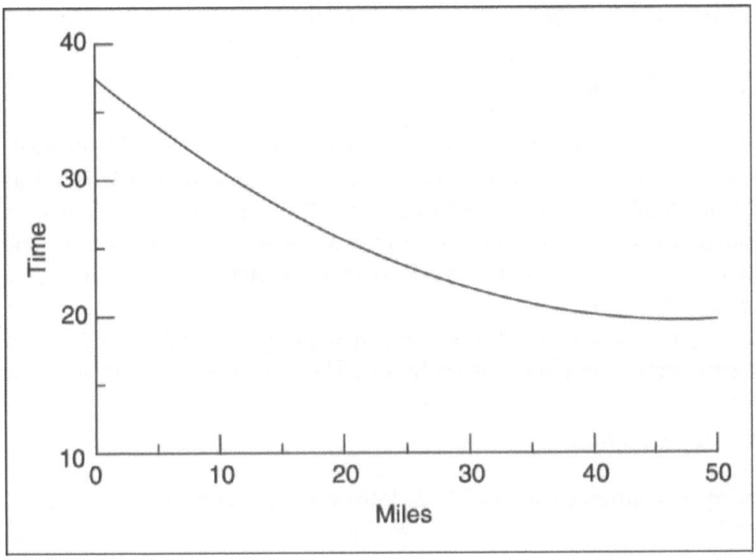

Figure 8.12 Time–Miles nonlinear model

Let us be clear about what these "simple" slopes are and are not. They are not predicted values of Time. Predicted values are what we plotted when we derived the graph in Figure 8.12. These "simple" slopes are rather the slopes of the curve at each different point along the Miles axis. They tell us the expected benefit of running one more mile per week on race times, given that one already is training by running a given number of miles each week. It may be difficult to think about the slope of this curve at particular points as the curve is obviously changing its slope continuously. Accordingly, think about the slope of lines that are at a tangent to the curve at different values of Miles. The slopes of such lines, at particular values of Miles, are the "simple" slopes.

From the point of view of the "simple" slope, the coefficient associated with Miles, –.754, is the "simple" slope of the curve at only one point, namely when Miles equals zero. In the plot of Figure 8.12, it is the slope of the curve at the intercept, when the curve is steepest. And as Miles increases, the "simple" slope becomes less and less negative. For each one-unit increase in Miles, the slope becomes less negative by $2 \times .008$. The interpretation for the regression coefficient associated with Miles^2: .008 is thus half the change in the "simple" Miles slope as Miles increases by one unit. If you wish, it indexes the degree of nonlinearity in the curve plotted in Figure 8.12.

We can ask whether this model, with its changing "simple" slope and its quadratic prediction function, does a better job than the simpler linear model with Miles as the only predictor. Our model comparison is thus

MODEL A: $\text{Time} = \beta_0 + \beta_1\text{Miles} + \beta_2\text{Miles}^2 + \varepsilon$
MODEL C: $\text{Time} = \beta_0 + \beta_1\text{Miles} + \varepsilon$

and the null hypothesis is H_0: $\beta_2 = 0$. In words, the null hypothesis is that the Time–Miles relationship is simply a linear one; there is no diminishing marginal return for additional miles of training on 5-kilometer race times.

At the very start of this chapter, we provided the estimates for Model C:

$$\hat{Time} = 31.911 - .280Miles$$

with SSE(C) = 1029.013. Recall that SSE(A) = 894.190. Thus, the value of SSR for this comparison is 134.823, which converts to a PRE of .131 and an F of 11.61 with 1 and 77 degrees of freedom. Thus, Model A is to be preferred to Model C and we can conclude that a quadratic prediction function is a better fit to the data than a linear one. There is thus evidence that the benefit of training one more mile per week, on 5-kilometer race times, is less the higher the training base is.

We could also, of course, compare this augmented nonlinear model to a compact one in which β_1 is fixed at zero, regressing Time only on Miles^2. The estimates for such a model are

$$\hat{Time} = 28.101 - .004Miles^2$$

with a sum of squared errors equal to 1222.243; the comparison generates a PRE of .268 and an F of 28.25 with 1 and 77 degrees of freedom. We thus conclude that the parameter associated with Miles in this augmented model is not equal to zero. But what exactly have we tested here? Have we tested the overall linear effect of Miles on Time? The answer is clearly "no." What we have just tested is not an "overall" linear effect of Miles, but rather the "simple" effect of Miles when and only when Miles equals zero. Based on these data and allowing Miles to be nonlinearly related to Time, we can conclude that running one more mile per week lowers race times when one is currently running zero miles per week. But because the effect of running more miles depends on the number of miles one already runs per week, this test does not tell us about the benefits of running one more mile given any other base per week. Additionally, because the value of zero on Miles lies outside the range of values included in the dataset, we probably should not have confidence in this prediction even if it is statistically significant.

As with the interactive model discussed previously, the interpretation of slopes of component predictor variables is a bit easier if those components are mean-deviated or centered and then the product predictor is recomputed using this centered variable. When we center Miles and regress Time on Miles$_0$ and Miles$_0$ (i.e., the centered variable and its square), we get the following parameter estimates:

$$\hat{Time} = 22.054 - .279Miles_0 + .008Miles_0^2$$

Again, in a deep sense, this model is the same as that estimated before, prior to centering the predictor. All we have done is move the zero point on the horizontal axis of Figure 8.12, so that it is now at the mean value of Miles, 29.875. And at that point, the "simple" slope of Miles in this model is –.279. Thus, the coefficient for Miles$_0$ in this centered model is interpreted as the simple slope when Miles$_0$ equals zero, which of course is at the mean of Miles.[2]

A comparison of this model with a Model C that fixes β_1 at zero tests whether the "simple" effect of Miles is different from zero when dealing with runners who train the average number of miles per week (average in the dataset). The estimated Model C is

$$\hat{Time} = 22.021 + .008Miles_0^2$$

with a sum of squared errors of 2070.265. This comparison gives a PRE of .568 and an F of 101.264, with 1 and 77 degrees of freedom. At the average level of Miles in the dataset, there is a significant Time–Miles relationship: Running one more mile per week leads to a lower 5-kilometer race time.

Note that this analysis is subtly different in both its details and its conclusion from asking the simple regression question of whether there is a significant overall linear relationship between Time and Miles. Here we are asking about a significant "simple" linear relationship only when Miles equals its mean, in the context of an overall quadratic model. When we ask whether Miles is related to Time in a simple regression model, we are assuming a linear and thus constant Time–Miles relationship.

We have just provided tests of "simple" slopes both when Miles equals zero and when Miles equals its mean. We could re-estimate this nonlinear model with Miles deviated from any other particular value where the Miles "simple" slope was of interest. Suppose, for instance, that someone ran 40 miles a week and wanted to know whether the present data suggest a significant effect of running one more mile, given this base. To test the "simple" slope of Miles when Miles equals 40, within the confines of this nonlinear model, we would simply re-estimate it with a new deviated Miles predictor, deviating it from 40, and its square:

$$\hat{Time} = 20.042 - .118Miles_{40} + .008Miles_{40}^2$$

Again, in a deep sense, this is the same nonlinear prediction function, making the same predictions. But now the coefficient associated with Miles_{40} estimates the "simple" slope for Miles when it equals 40. And a comparison of this augmented model with a Model C that omits the Miles_{40} predictor provides a test of that "simple" slope. The resulting PRE is .057, which converts to an F of 4.62 with 1 and 77 degrees of freedom. We can conclude that there is a significant "simple" effect of Miles when Miles equals 40, allowing a nonlinear relationship between Time and Miles.

Just as we noticed when discussing interactive models, the coefficient for Miles^2 in all three versions of the quadratic model we have estimated (with Miles in its raw metric, with Miles centered, and with Miles deviated from 40) remained the same. And Model A/Model C tests of its significance yield the same result in every case. Again, assuming that the component predictor is included in the model, the test of the coefficient of the product predictor is invariant across deviation transformations of its component. It continues to examine whether the "simple" slope changes as Miles changes value.

More Complex Product Models

We have found evidence in our data for both an interaction between Age and Miles and a quadratic effect of Miles in predicting race times. We have examined these two nonlinear effects separately, rather than putting them both in a single model, for didactic purposes. Let us now look at a more complete model that incorporates both the Age by Miles interaction and the quadratric Miles term. The estimated model is

$$\hat{Time} = 25.578 + .273Age - .565Miles + .008Miles^2 - .004AgeMiles$$

The sum of squared errors for this model equals 579.724. If we compare this model with a model that omits the Miles^2 predictor, the resulting PRE equals .169 and the F, with 1 and 75 degrees of freedom, equals 15.30. Hence, the nonlinearity in the Time–Miles relationship is still found even when we control for Age and the interaction of Age with Miles. However, when we compare this model with one that deletes only the AgeMiles interaction term, the resulting PRE (.039) and F (3.06) are not significant. Hence, although we found that Age and Miles reliably interact in predicting Time when $Miles^2$ is not controlled, that effect ceases to be significant in this more complete model.

Nevertheless, let us continue to interpret the various coefficients of this model in order to clarify the interpretation of more complex nonlinear functional forms. As before, we want to do two things to interpret this model. First, we derive predicted values of Time for all combinations of Age and Miles and we plot these, this time in the three-dimensional graph of Figure 8.13. And second, to interpret the regression coefficients of the model, in the context of this graph of the predicted values, we derive the "simple" slopes of both Age and Miles according to this model:

$$\frac{\Delta \hat{Time}}{\Delta Age} = .273 - .004 Miles$$

$$\frac{\Delta \hat{Time}}{\Delta Miles} = -.565 + (2).008 Miles - .004 Age$$

First, let us interpret the coefficient associated with Age in the model, .273. According to the aforementioned "simple" slope derived for Age, it is that "simple" slope of Age when and only when Miles equals zero. In the three-dimensional plot of Figure 8.13, we can think about taking slices of the prediction plane parallel to the Age axis. Such slices are prediction lines as a function of Age, at particular values of Miles. One of these slices is found when Miles equals zero. And the slope of that slice (or line) is .273. Substantively, we might say that with no training, a 1-year increase in Age is predicted to result in slower 5-kilometer race times by .273 minutes.

The coefficient associated with the Miles predictor, –.565, can be interpreted in the context of the derived "simple" slope for Miles, earlier. There we can see that it is the "simple" slope of Miles when and only when *both* Age and Miles equal zero. Graphically, it is the slope of the plane, with respect to Miles, at the point in the plot where both Age and Miles equal zero. If we met a runner who was zero years old and who ran zero miles per week, we would predict that running one more mile per week would result in a faster 5-kilometer race time by .565 minutes! Obviously, neither this coefficient nor the one associated with the Age predictor focuses on a "simple" slope

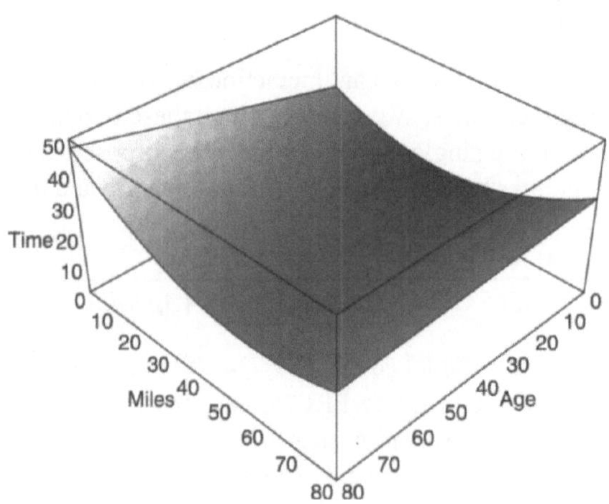

Figure 8.13 Predictions of model with Miles squared and Miles by Age interaction

of interest, given the values in our dataset. We will revisit them, with centered predictor variables later once we have interpreted the other two slope coefficients in this model.

The coefficient for $Miles^2$ in the model, .008, is half the change in the "simple" slope for Miles as Miles increases. Again, looking at the prediction plane in Figure 8.13, taking slices parallel to the Miles axis we get a series of curves. This coefficient tells us about the degree to which these curves depart from linearity. The Miles "simple" slope is not constant; it changes as Miles increases, becoming less negative. This interpretation is similar to the interpretation given earlier for the coefficient for $Miles^2$ in the model that included neither Age nor AgeMiles. The difference is that here we are estimating the degree of nonlinearity in the Time–Miles "simple" relationship, controlling for Age and its interaction with Miles.

Finally, the coefficient associated with the AgeMiles product, –.004, can be interpreted in the context of either of the derived "simple" slopes. From the point of view of Age, it is the change in the Age "simple" slope as Miles increases. In Figure 8.13, taking slices of the prediction plane parallel to the Age axis we can see that the "simple" slope for Age is not invariant. It seems steeper (i.e., more positive: older runners have longer race times) when Miles is relatively low than when the value of Miles is higher. From the point of view of the Miles "simple" slope, it is the change in that "simple" slope as Age increases. Taking slices of the prediction plane parallel to the Miles axis, in general the predictions are going down as Miles increases (i.e., faster race times with more miles training). But the extent to which this is true seems greater among older runners than among younger ones. The linear effect of Miles is greater among older runners than among younger ones.

If we were to center Age and Miles and regress Time on the centered predictors and their product terms, it is an interesting exercise to figure out which of the coefficients in the model would change and which would not. Again, all we would be doing is changing where the value of zero is found on both the Age and Miles axes. Any coefficient that makes reference to the value of zero of either variable would thus change. Accordingly, when we re-estimate this model, having centered the two variables and recomputed the product terms, we get the following:

$$\hat{Time} = 22.044 + .167Age_0 - .257Miles_0 + .008Miles_0^2 - .004Age_0Miles_0$$

The coefficients associated with Age_0 and with $Miles_0$ now have different values. The former is the "simple" slope for Age when Miles equals its mean value; the latter is the "simple" slope for Miles when both Miles and Age equal their mean values. The coefficients associated with $Miles_0$ and Age_0Miles_0 have not changed, for they make no reference to zero values on the two axes. Rather, these coefficients indicate *changes* in the "simple" slopes; they are not "simple" slopes when something else equals zero. As before this "centered" model has the same sum of squared errors as the "uncentered" one estimated previously.

Although the data that we have been using throughout this chapter are not consistent with any more complex nonlinear functional forms than those already explored, we could extend this model to a more complex one, should our theories suggest that more complex forms are plausible. There is nothing inherently more difficult about interpreting more complex nonlinear models, so long as the basic principles that we have explained are kept in mind. In general, we want to generate predicted values as a function of the possible values of the predictors, and then graph these. To interpret the resulting coefficients, it is easiest to do so in the context of derived "simple" slopes. Doing this makes clear the general principle that the coefficient associated with any predictor variable that is a component of one or more product predictors included in the model estimates the "simple" slope of that variable when and only when the other component(s) of those products equal zero. So in the example just discussed, the estimated Age coefficient is the "simple" slope for Age when Miles equals zero because Age is a component of the AgeMiles product that is included in the model. And the estimated coefficient for Miles is the "simple" slope for Miles when both Age and Miles equal zero because Miles is a component of both the Miles^2 and AgeMiles products that are included in the model.

Additionally, if the interpretation of a coefficient is in terms of a "simple" effect at a zero level of some other component variable, then transformations that affect that zero level will alter the obtained coefficient. Only the coefficients of predictors in models that are not themselves components of product predictors will be unaffected by such deviation transformations.

We close the chapter with one last model that includes the product of three different predictor variables:

$$\hat{Y} = b_0 + b_1X_{1i} + b_2X_{2i} + b_3X_{3i} + b_4X_{1i}X_{2i} + b_5X_{1i}X_{3i} + b_6X_{2i}X_{3i} + b_7X_{1i}X_{2i}X_{3i}$$

As before, whenever we have products in a model, we want to make sure that we include all the components of those products as predictors. And in this case, all components of the triple product $X_{1i}X_{2i}X_{3i}$ include not only the single predictor variables, e.g., X_{1i}, but also the products of two of the three predictors, e.g., $X_{1i}X_{2i}$. To aid interpretation of the resulting coefficients, we will assume that all variables have been centered and the included product predictors have been computed from their centered components.

With a model such as this, a single plot is not possible because only three dimensions are visually available. Instead, we could generate separate three-dimensional plots at various levels of the fourth variable. For instance, we might put X_{1i} and X_{2i} on the horizontal axis and plot predicted values of Y_i in one graph at one value of X_{3i}, creating additional graphs for different values of X_{3i}.

To interpret the various parameter estimates, we generate the "simple" effect expressions:

$$\frac{\Delta\hat{Y}_i}{\Delta X_{1i}} = b_1 + b_4X_{2i} + b_5X_{3i} + b_7X_{2i}X_{3i}$$

$$\frac{\Delta\hat{Y}_i}{\Delta X_{2i}} = b_2 + b_4X_{1i} + b_6X_{3i} + b_7X_{1i}X_{3i}$$

$$\frac{\Delta\hat{Y}_i}{\Delta X_{3i}} = b_3 + b_5X_{1i} + b_6X_{2i} + b_7X_{1i}X_{3i}$$

These "simple" slopes inform us first of all about the estimated coefficients attached to the three predictor variables by themselves, b_1, b_2, and b_3. These are the "simple" effects of each variable at the mean (zero value given centering) of both of the other predictors. Each of the estimated coefficients associated with the products involving two variables, b_4, b_5, and b_6, can be interpreted in the context of two different "simple" slopes. These estimate "simple" two-way interactions between the variables involved in the product at the mean level (zero level) of the third variable. For example, b_4 can be interpreted either as the change in the "simple" slope of X_{1i} as X_{2i} increases when X_{3i} equals the mean, or as the change in the "simple" slope of X_{2i} as X_{1i} increases when X_{3i} equals the mean. And finally, there are three possible interpretations for the coefficient associated with the triple product, b_7. These three possible interpretations are perhaps best shown if we derive "simple" slopes for each of the three two-way products:

$$\frac{\Delta \hat{Y}_i}{\Delta X_{1i} X_{2i}} = b_4 + b_7 X_{3i}$$

$$\frac{\Delta \hat{Y}_i}{\Delta X_{1i} X_{3i}} = b_5 + b_7 X_{2i}$$

$$\frac{\Delta \hat{Y}_i}{\Delta X_{2i} X_{3i}} = b_6 + b_7 X_{1i}$$

These are derived in exactly the same way as we explicated for the "simple" slopes of single predictor variables, but now we apply those rules to a product predictor. In this context, b_7 can be interpreted in the following three equivalent ways:

1 The change in the effect of the X_1X_2 interaction as X_3 increases.
2 The change in the effect of the X_1X_3 interaction as X_2 increases.
3 The change in the effect of the X_2X_3 interaction as X_1 increases.

These interpretations imply that one has already interpreted the "simple" two-way interactions at the mean level of the third variable, as previously discussed. Thus, the process of understanding a model like this, with a triple interaction, happens in steps. First, one attempts to make sense of the "simple" effects of single variables at the mean level of the others, then the two-way interactions at the mean of the third (i.e., how the "simple" effect of one variable depends on the level of a second, at the mean of the third), and finally one interprets the three-way interaction by talking about how a "simple" two-way interaction is affected by changes in the third variable.

As we have discussed, with a two-way interaction there are two equally viable interpretations ("the simple effect of X_{1i} depends on the value of X_{2i}" or "the simple effect of X_{2i} depends on the value of X_{1i}"), but one of these is likely to be more compelling than the other. Similarly, with a three-way interaction and its three possible interpretations, one will generally emerge as more compelling than the other two.

One final word of caution is in order about the sorts of models we have been exploring in this chapter. With three variables as predictors in a model, there are three two-way interactions and one three-way interaction that might be estimated. With four predictor variables, there are potentially six two-way interactions, four three-way interactions, and one four-way interaction. And with increasing numbers of predictor variables, the possible interactions go up geometrically. And this count does not include the possibilities that arise from including quadratic terms and higher order powers of predictors (and the interactions of these with other predictors).

Although we have explicated the general approach for testing and interpreting models involving product predictors, in general the inclusion of such predictors should be strongly dependent on compelling theoretical expectations. We do not encourage the inclusion of product predictors in models unless there are such theoretical expectations. Building more and more complex models is not our goal, even if we have the tools to interpret them.

Summary

The models that have been considered in this chapter considerably extend the range of questions that can be asked of data. They do so by allowing the slopes of predictor variables to vary as a function of the values of those variables and of other variables. In the most frequently encountered case, these models thus allow statistical interactions between predictor variables—examining whether the "effect" of one variable depends on the value of another. These models also allow nonlinear effects, examining for instance whether the "effect" of one variable changes as its own values change.

In general, these interactive and nonlinear effects are estimated by including products of predictors as additional predictor variables, making sure that all components of those products are included as predictors as well. Interpreting such models can be difficult and we have recommended a two-stage procedure, first graphing predicted values and then deriving "simple" slopes and interpreting parameter estimates in the context of such simple slopes. By focusing on these "simple" slopes, it becomes clear that the parameter estimate associated with a predictor that is a component of some product predictor estimates the "simple" effect of that predictor when and only when the other components of the product equal zero. To make these estimates generally more interpretable, we have recommended centering or mean-deviating variables prior to computations of the product predictors.

While models such as those that have been considered in this chapter (allowing interactions and nonlinear effects) dramatically increase the range of questions that one can ask of data, their complexity can become overwhelming. We therefore encourage analysts to examine such effects only when there are good theoretical reasons for them. Adding products willy-nilly to models will only result in Type I errors and overly complex models.

Notes

1 While this product variable is the product of two centered variables, it will not necessarily be centered itself. What is important is that the product be computed from the two centered variables, not that the product itself be centered.

2 This value can be computed directly from the expression for the "simple" slope of Miles given earlier from the estimated nonlinear model prior to centering. There the "simple" slope was calculated as $-.754 + (2).008\text{Miles}$. If we substitute the mean level of Miles, 29.875, in this expression we get $-.754 + (2).008(29.875) = -.276$, which is the estimated coefficient for Miles_0 in the centered model (within rounding error).

9 One-Way ANOVA

Models With a Single Categorical Predictor

To this point we have relied on regression models where the predictor variables have been treated as continuous variables. Our purpose in this chapter and the following two is to examine our basic approach to data analysis when predictors are categorical variables. In the language of traditional statistics books, earlier chapters concerned multiple regression. The present chapter concerns one-way analysis of variance (ANOVA) models or, equivalently, models with a single categorical predictor. In the next chapter we consider models having multiple categorical predictors, or higher order ANOVA models. Chapter 11 is devoted to models in which some predictors are categorical and some are continuous. Such models have been traditionally labeled analysis of covariance models. By integrating these into a common approach, we will not only explore these traditional topics but also consider others that considerably extend the sorts of questions that we are able to ask of our data, in the context of categorical predictor variables. Throughout we will continue to use our basic approach to statistical inference, testing null hypotheses by the comparison of augmented and compact models.

The Case of a Categorical Predictor With Two Levels

Figure 9.1 contains hypothetical data from a study in which the impact of a SAT training course was evaluated. Twenty high school seniors were randomly assigned to either take the 2-week training course, designed to improve SAT performance, or to a control no-course condition. As we can see, 10 students wound up in each of the two groups. At the end of the 2-week period, all 20 students took the SAT test, and their performance was recorded. What we would like to do is examine whether the course made a difference in subsequent performance. Our question thus is whether we can reliably predict subsequent SAT performance as a function of whether a student was in the Course group or the No Course control group.

If we are to tell the computer to specify a model in which SAT performance is predicted by which of the two groups a student was in, we need some way of coding or numerically representing the group variable. This variable is categorical rather than numerical or continuous, meaning that students in the two groups differ in whether they have taken the course, but no automatic numerical representation of that difference is implied. Such variables require some coding scheme to represent them numerically, so that they can be used as predictor variables in models. It turns out that any numerical representation of this group variable would do, as long as we use that numerical representation consistently. By consistent use, we mean that if a given value on the variable that represents group (Course versus No Course) numeric ally is assigned to one group, then every student in that group has that same value on the variable, and no student in the other group has that value.

To illustrate, suppose we created a variable X_i to represent group numerically, arbitrarily assigning the value of -1 to students in the No Course group and $+1$ to students in the Course group.

DOI: 10.4324/9781003438335-14

Student	*Group*	*SAT*
1	Course	580
2	Course	560
3	Course	660
4	Course	620
5	Course	600
6	Course	580
7	Course	590
8	Course	640
9	Course	620
10	Course	600
11	No Course	580
12	No Course	530
13	No Course	590
14	No Course	550
15	No Course	610
16	No Course	590
17	No Course	600
18	No Course	530
19	No Course	590
20	No Course	600

Figure 9.1 Two-group SAT data

Because every student is in one group or the other, all students have values of either –1 or +1 on variable X_i. Notice that our purpose in creating this variable is simply to differentiate numerically between the two groups. Because group is a nominal or categorical variable, no rank order or interval information needs to be preserved in our coding scheme. We could just as easily have given the value of –1 to the Course group and +1 to the No Course group. Similarly, we could have given the value of 203 to students in the Course group and the value of –20.5 to students in the No Course group. The point is that the values that represent the categorical variable are arbitrarily defined, but they must be consistently used.

Throughout the rest of the book, we will use a convention for coding nominal predictors known as *contrast codes*. Contrast codes are simply one of the possible arbitrary coding schemes for numerically representing categorical predictors. Two conditions define contrast codes and differentiate them from other coding schemes. For now, we will only define one of these two conditions. The other is only relevant when the categorical variable has more than two categories and will be given later in this chapter. Let us define a value on a contrast-coded categorical variable X_i as λ_k ("lambda$_k$"), where the subscript k refers to the level of the categorical variable being coded. In this case, k refers to the two levels of the group variable: Course versus No Course. Across levels of k, or across all categories of the variable, a contrast code is one where

$$\sum_k \lambda_k = 0$$

Notice that we are summing here across levels or categories rather than across individual observations. In other words, the values of the contrast-coded variable sum to zero across the two categories, not across the individual observations in those two categories.

In our example, the values of +1 for students in the Course group and –1 for students in the No Course group constitute values of a contrast-coded variable because the sum of these two values across the two categories equals zero. Another valid contrast-coded variable would use values of +0.5 for the Course group and –0.5 for the No Course group. Notice, however, that the values of 203 for the Course group and –20.5 for the No Course group do not meet the condition for a contrast-coded variable. Note also that in this example the values of a contrast-coded variable, say +1 and –1, sum to zero not only across the two categories but also across the 20 students in those two categories. This will be the case when a contrast-coded variable is used and when there are an equal number of observations in the two groups or categories. Had we had more students in one of the two groups than in the other, then the sum of the values across the two groups would be zero, but the sum of the values across all the students would not be.

In the following sections, we will use two different contrast-coded predictors to predict SAT performance with the data presented in Figure 9.1. We will first use the values of +1 and –1, and then we will use the values of +0.5 and –0.5. These possible values of a contrast-coded predictor are simply two from an infinite number of such values that might be used. At a later point, we will also briefly discuss the estimation of models using coding conventions other than contrast codes.

Model Estimation and Inference With a Contrast-Coded Predictor

We start by estimating a model in which we predict SAT performance with a contrast-coded predictor having values of +1 for students in the Course group and –1 for students in the No Course group. SAT is thus our Y variable and our predictor, X, is the contrast-coded predictor. Our model is the simple regression model with a single predictor variable of Chapter 5. We will want to compare this model, making predictions of SAT conditional on X_i, with a compact one in which we predict the same value for every student regardless of whether they were in the Course group or the No Course group:

MODEL A: $Y_i = \beta_0 + \beta_1 X_i + \varepsilon_i$

MODEL C: $Y_i = \beta_0 + \varepsilon_i$

Assuming that $\beta_1 \neq 0$, the augmented model makes predictions of SAT performance conditional on whether students were in the Course group or the No Course group. On the other hand, the compact model makes the same prediction for all students, regardless of their group.

The least-squares estimates for these models are

MODEL A: $\hat{Y}_i = 591 + 14X_i$

MODEL C: $\hat{Y}_i = 591$

Both models, as well as the data on which they are based, are graphed in Figure 9.2. The horizontal prediction function is Model C, while the prediction line that makes differential predictions is Model A. The sums of squared errors of these two models are 16,060 for Model A and 19,980 for Model C.

The calculations of these two sums of squares are shown in Figure 9.3, where we derive for each observation the predicted value from each model, the residual, and the squared residual. The sums of these squared residuals, given in the last row of Figure 9.3, are the sums of squared errors for the two models.

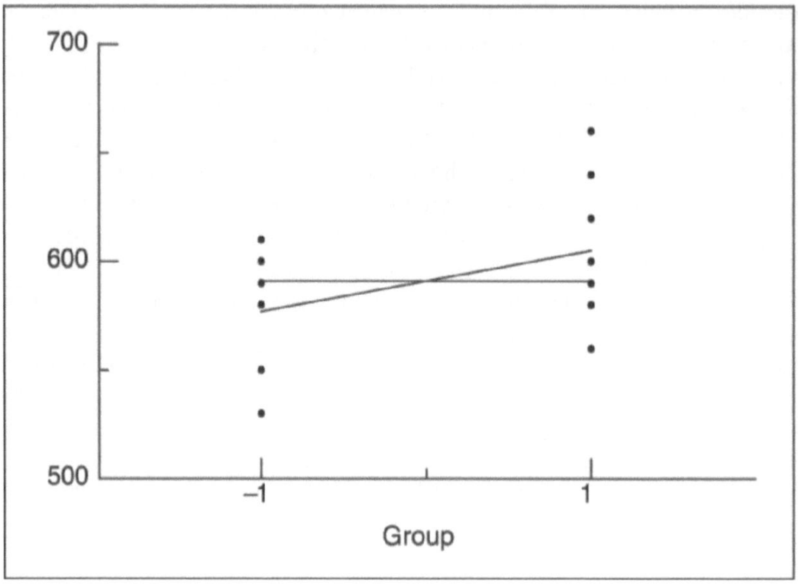

Figure 9.2 Models A and C for the SAT data

				Compact			Augmented		
Student	*Group*	X_i	*SAT* (Y_i)	$\hat{Y}_i$	e_i	e_i^2	$\hat{Y}_i$	e_i	e_i^2
1	Course	1	580	591	–11	121	605	–25	625
2	Course	1	560	591	–31	961	605	–45	2025
3	Course	1	660	591	69	4761	605	55	3025
4	Course	1	620	591	29	841	605	15	225
5	Course	1	600	591	9	81	605	–5	25
6	Course	1	580	591	–11	121	605	–25	625
7	Course	1	590	591	–1	1	605	–15	225
8	Course	1	640	591	49	2401	605	35	1225
9	Course	1	620	591	29	841	605	15	225
10	Course	1	600	591	9	81	605	–5	25
11	No Course	–1	580	591	–11	121	577	3	9
12	No Course	–1	530	591	–61	3721	577	–47	2209
13	No Course	–1	590	591	–1	1	577	13	169
14	No Course	–1	550	591	–41	1681	577	–27	729
15	No Course	–1	610	591	19	361	577	33	1089
16	No Course	–1	590	591	–1	1	577	13	169
17	No Course	–1	600	591	9	81	577	23	529
18	No Course	–1	530	591	–61	3721	577	–47	2209
19	No Course	–1	590	591	–1	1	577	13	169
20	No Course	–1	600	591	9	81	577	23	529
					SSE(C) = 19,980			SSE(A) = 16,060	

Figure 9.3 Calculation of SSE for Models C and A

The comparison of these two models, asking whether the predictions conditional on the contrast-coded predictor do a better job than the unconditional predictions, yields

$$\text{PRE} = \frac{19980 - 16060}{19980} = \frac{3920}{19980} = .196$$

Model A contains two parameters and Model C one, hence PA – PC equals 1 and n – PA equals 18. Accordingly, we can compute the F statistic for this comparison either from the computed value of PRE or from the values of the sums of squares:

$$F_{1,18} = \frac{PRE/(PA - PC)}{(1 - PRE)/(n - PA)} = \frac{.196/1}{(1 - .196)/18} = 4.39$$

$$F_{1,18} = \frac{SSR/(PA - PC)}{SSE(A)/(n - PA)} = \frac{3920/1}{16060/18} = 4.39$$

These statistics fall just short of their critical values, with 1 and 18 degrees of freedom, with α set at .05. Hence, we are unable to conclude that the conditional predictions of SAT are significantly better than the unconditional one made by Model C.

So far, once we have coded our categorical predictor, there is nothing different about this simple regression model from those simple models using continuous predictors that were discussed in Chapter 6. The estimation of the model parameters and the calculations of PRE and F proceed just as before. Because all of the assumptions underlying the use of the critical values of PRE and F involve assumptions about the distribution of Y_i, or really of ε_i, and because the categorical nature of X_i has no effect on the distribution of ε_i, none of the assumptions underlying this analysis have been impacted by the use of the categorical predictor.

What has changed somewhat, however, is the interpretation of the estimated parameters of the model and, accordingly, the interpretation of the statistical inference results. It is not the case that the old interpretations (those developed in Chapter 6) are incorrect, for the value of the intercept in the augmented model, 591, is still the predicted value of Y_i when X_i equals zero. The coefficient for X_i, 14, is still a slope—the amount by which the predicted value of Y_i changes for each unit increase in X_i. And PRE and F still tell us about the reduction in errors of prediction. Rather, when we have categorical predictors, new interpretations of these statistics become possible.

To understand these new interpretations, it is helpful to consider the predicted values that the augmented model makes. These are contained in Figure 9.3. If we are dealing with a student in the Course group, the predicted value from the augmented model is

$$\hat{Y}_{+1} = 591 + 14(+1) = 605$$

And if we are dealing with a student in the No Course group, the predicted value from the augmented model is

$$\hat{Y}_{-1} = 591 + 14(-1) = 577$$

These predicted values turn out to be the mean SAT scores of the 10 students in each of the two groups. That is, 605 is the mean SAT score of those students who received the course, and 577 is the mean SAT score of those students who did not. And given our use of +1 and –1 as the values of the contrast-coded predictor, the slope associated with that predictor equals half the difference between the means of the two groups:

$$\begin{aligned}\bar{Y}_C - \bar{Y}_{NC} &= \hat{Y}_{+1} - \hat{Y}_{-1} = 605 - 577 \\ &= (591 + 14(+1)) - (591 + 14(-1)) \\ &= 14(+1) - 14(-1) \\ &= 2(14)\end{aligned}$$

In general, the least-squares parameter estimate or slope associated with a contrast-coded predictor is given by

$$b = \frac{\sum_k \lambda_k \bar{Y}_k}{\sum_k \lambda_k^2} \tag{9.1}$$

which, in the example at hand, is evaluated as

$$b_1 = \frac{(-1)577 + (+1)605}{(-1)^2 + (+1)^2} = \frac{28}{2} = 14$$

Equation 9.1 is a very general and useful formula for interpreting the slope associated with any contrast-coded predictor. The numerator represents a comparison between group means, in this case the difference between the mean for the Course group and the mean for the No Course group, and the denominator is a scaling factor dependent on the specific values used for the contrast-coded predictor. The important conceptual point is that the regression coefficient associated with any contrast-coded predictor tells us about the difference between group means. The direction of that difference is determined by which group is coded with a positive value and which group is coded with a negative value.

The estimated intercept of 591 equals, as always, the predicted value of Y_i when X_i equals zero. Because X_i equals zero halfway between the two values of +1 and –1 that code the two groups, the estimated value of the intercept is equal to the average of the two group means. This result is made clear by the graph of the model in Figure 9.2. It can also be shown algebraically as follows:

$$\hat{Y}_{+1} = \bar{Y}_C = 591 + 14(+1)$$

$$\hat{Y}_{-1} = \bar{Y}_{NC} = 591 + 14(-1)$$

Adding these two equalities gives

$$\bar{Y}_C + \bar{Y}_{NC} = (591 + 14(+1) + (591 + 14(-1))$$

$$\bar{Y}_C + \bar{Y}_{NC} = (2)591$$

$$\bar{Y}_C + \bar{Y}_{NC} = 591/2$$

Notice that this interpretation of the intercept in the augmented model, including the contrast-coded predictor, is not the same as the interpretation of the intercept in the compact model, the one making unconditional predictions. The intercept in the compact model is estimated as the mean of all the observations, what we might call the grand mean, $\bar{Y}$, defined as

$$b_{0C} = \bar{Y} = \frac{\sum_i Y_i}{n}$$

On the other hand, the intercept in the augmented model is the mean of the group means, defined as

$$b_{0A} = \frac{\sum_k \bar{Y}_k}{m}$$

where m is the total number of groups, in this case 2.

In the dataset that we have been using, the values of these two intercepts, one from the compact model and one from the augmented, are identical, i.e., 591, because there are an equal number of students in the two groups. In general, however, they estimate different things. The intercept in the compact model, the one making unconditional predictions, is the mean of all the observations. The intercept in the augmented model, the one making predictions conditional on group, is the mean of the group means.

One last result when using contrast-coded predictors is important to know. Equation 9.1 represents the regression coefficient for any contrast-coded predictor as a comparison among group means. Similarly, the SSR associated with a contrast-coded predictor can be represented in terms of group means. As always, the SSR associated with a predictor is the difference between the SSE(A) and SSE(C) for compact and augmented models with and without that predictor. And, as always, that SSR equals

$$\text{SSR} = \sum_i \left(\hat{Y}_{iA} - \hat{Y}_{iC}\right)^2$$

In the case of a categorical predictor, as we have seen, the predicted values for the augmented model in this expression are the group or category means, $\bar{Y}_k$, and the predicted value from the compact model is the grand mean of all the observations, $\bar{Y}$. As a result, the SSR associated with a contrast-coded predictor can be expressed as a function of the category means (and the number of observations in each group, n_k) as follows:

$$\text{SSR} = \frac{\left(\sum_k \lambda_k \bar{Y}_k\right)^2}{\sum_k \left(\lambda_k^2 / n_k\right)} \tag{9.2}$$

In the case at hand, we have seen that SSE(C) equals 19,980 and SSE(A) equals 16,060, resulting in an SSR associated with the contrast-coded predictor of 3920. That is equivalently computed as

$$\text{SSR}_X = \frac{\left((-1)577 + (+1)605\right)^2}{\left((-1)^2/10\right) + \left((+1)^2/10\right)} = 3920$$

Estimation With Alternative Values for the Contrast-Coded Predictor

If we had defined the values of the contrast-coded predictor as +0.5 for the Course group and –0.5 for the No Course group, rather than +1 and –1, the estimated model would be

$$\text{MODEL A: } \hat{Y}_i = 591 + 28X_i'$$

where X_i' is the new contrast-coded predictor. Importantly, this model makes exactly the same predictions for students in the two groups, i.e., their respective group means:

$$\hat{Y}_{+0.5} = 591 + 28(+0.5) = 605 = \bar{Y}_C$$

$$\hat{Y}_{-0.5} = 591 + 28(-0.5) = 577 = \bar{Y}_{NC}$$

Accordingly, in a deep sense, it is the same augmented model with the same sum of squared errors. The slope for the contrast-coded predictor now equals 28 instead of 14, as there is now a one-unit difference on X_i'—*that separates the two groups* (between –0.5 and +0.5) rather than the two-unit difference that separated them on X_i (between +1 and –1). From Equation 9.1, that slope now equals the difference between the two group means:

$$b_{1'} = \frac{\sum_k \lambda_k \bar{Y}_k}{\sum_k \lambda_k^2} = \frac{(-0.5)577 + (+0.5)605}{(-0.5)^2 + (+0.5)^2} = \frac{0.5(605 - 577)}{0.5} = 605 - 577 = 28$$

Of course, because this Model A is in a deep sense the same as the one where the contrast-coded predictor had values of +1 and –1, the model comparison of it with the compact model, making predictions that were not conditional on group, yields the same SSR, PRE and *F* statistics. Recomputing the SSR for this comparison, using these new codes and Equation 9.2, gives us

$$\text{SSR}_{X'} = \frac{\left((-0.5)577 + (+0.5)605\right)^2}{\left((-0.5)^2/10\right) + \left((+0.5)^2/10\right)} = 3920$$

Equivalence With ANOVA and Two-Group* t*-Test

The slope in the augmented model, making conditional predictions, informs us about the difference between the two group means (regardless of the values of λ_k used to construct the contrast-coded predictor). Therefore, the comparison of this augmented model with the compact one making unconditional predictions asks both whether the parameter associated with the contrast-coded predictor departs from zero and whether the two group means differ from each other. In other words, the model comparison we have examined addresses the following equivalent null hypotheses:

$$H_0: \beta_1 = 0$$

$$H_0: \mu_C = \mu_{NC}$$

where μ_C and μ_{NC} are the true but unknown means of the two groups.

In more traditional statistical textbooks, the test of a null hypothesis about the difference between two group means is usually conducted by computing a two-group ANOVA or a two-group independent samples *t*-test. It is therefore important to show that our model comparison and its associated PRE and *F* statistics are identical to those generated by these more traditional approaches.

In Figure 9.4, we give the ANOVA source table for the model comparison we have just conducted, using the formulas developed in Chapter 6 and earlier. We also provide the source table using the numeric values generated by our data example.

The formula for the SSR for the model comparison that we have given before is

$$\text{SSR} = \sum_i \left(\hat{Y}_{iA} - \hat{Y}_{iC}\right)^2$$

summing across all individual observations. In the present case, $\hat{Y}_C$ is the mean of all of the observations, $\bar{Y}$, and the predicted values from the augmented model, $\hat{Y}_A$, are the two group means $\bar{Y}_C$ and $\bar{Y}_{NC}$. Let us generically represent these group means as $\bar{Y}_k$, thereby indicating the group mean for the *k*th group. Then we can write the previous expression for the sum of squares reduced as

$$\text{SSR} = \sum_i \left(\hat{Y}_{iA} - \hat{Y}_{iC}\right)^2 = \sum_i \left(\bar{Y}_k - \bar{Y}\right)^2 = \sum_k n_k \left(\bar{Y}_k - \bar{Y}\right)^2$$

where n_k is the number of observations in the *k*th group.

We can also re-express the formula for SSE(A) and SSE(C) in terms of means because those are the predicted values from each model:

$$\text{SSE}(\text{A}) = \sum_i \left(Y_i - \hat{Y}_{iA}\right)^2 = \sum_i \left(Y_i - \bar{Y}_k\right)^2$$

$$\text{SSE}(\text{C}) = \sum_i \left(Y_i - \hat{Y}_{iC}\right)^2 = \sum_i \left(Y_i - \bar{Y}\right)^2$$

Accordingly, the formulas in the source table that we have used all along for summarizing our computations of PRE and *F* (given in the top half of Figure 9.4) can, in this case, be equivalently written with the formulas used to compute an analysis of variance to compare group means in traditional statistics textbooks. This revised version of the source table is given in Figure 9.5.

Source	*SS*	*df*	*MS*	*F*	*p*	*PRE*
Reduction	SSR	PA – PC	MSR	$\frac{\text{MSR}}{\text{MSE(A)}}$		$\frac{\text{SSR}}{\text{SSE(C)}}$
Error	SSE(A)	n – PA	MSE(A)			
Total	SSE(C)	n – 1	MSE(C)			

Source	*SS*	*df*	*MS*	*F*	*p*	*PRE*
Reduction	3920	1	3920.00	4.39	< .10	.196
Error	16060	18	892.22			
Total	19980	19	1051.58			

Figure 9.4 ANOVA source tables

Source	*SS*	*df*	*MS*	*F*	*p*	*PRE*
Between	$\sum_k n_k (\bar{Y}_k - \bar{Y})^2$	1	MSB	$\frac{\text{MSB}}{\text{MSW}}$		$\frac{\text{SSB}}{\text{SST}}$
Within	$\sum_i (Y_i - \bar{Y}_k)^2$	n – 2	MSW			
Total	$\sum_i (Y_i - \bar{Y})^2$	$n - 1$	MST			

Figure 9.5 ANOVA source table

The names given to the rows in this version of the source table have been changed to reflect those traditionally used in analysis of variance. So the sum of squares reduced and its mean square are traditionally called the sum of squares and mean square *between groups*, and the sum of squares and mean square from Model A are traditionally called the sum of squares and mean square *within groups*. But fundamentally and algebraically this is the same source table, yielding the exact same F and PRE, as the one we are more used to, given in Figure 9.4.

The square root of F is the t statistic (because n – PA =1), with $n - 2$ degrees of freedom, that is traditionally called the two-group independent samples t-test.

Thus, our integrated approach to statistical inference, resting on model comparisons estimated with any least-squares multiple regression program, yields the same results as the cookbook recipes given in more traditional statistical textbooks. Our model comparison in this case, testing whether the regression coefficient for a single contrast-coded predictor departs from zero, is exactly equivalent to a two-sample t-test for examining whether the means of two groups differ from each other.

Confidence Interval for the Slope of a Contrast-Coded Predictor

The formula that we gave in Chapter 7 for the confidence interval for a regression coefficient continues to be applicable in the situation where predictor variables are contrast coded. The confidence interval for the slope associated with any predictor variable was given there as

$$\text{CI} = b \pm \sqrt{\frac{F_{crit;1,n-p-1;\alpha}\text{MSE}}{(n-1)s_{X_p}^2\left(1-R_p^2\right)}}$$

Because the slope of a contrast-coded predictor informs us about the magnitude of the difference between group means, its confidence interval informs us about the confidence interval associated with that mean difference. To see this, let us take the case where the contrast-coded predictor used values of –.5 for the No Course group and +.5 for the Course group. The resulting slope in this case equals 28 and the confidence interval for that slope is calculated as

$$\text{CI}_{b_{1'}} = 28 \pm \sqrt{\frac{4.41(892.22)}{(19)0.263(1.00)}}$$

$$= 28 \pm 28.06$$

where 4.41 is the critical F value with 1 and 18 degrees of freedom, 892.22 is the mean square error from our Model A, $n - 1$ equals 19, the variance of the 20 individuals on the contrast-coded predictor is 0.263, and its tolerance ($1 - R_p^2$) is of course 1 because it is the only predictor in the model. This confidence interval can also be written as

$$-0.06 \leq \beta_1 \leq 56.06$$

Because the parameter that we estimated with this contrast-coded predictor is an estimate of the true mean difference between the two groups, this confidence interval can be equivalently expressed as

$$-0.06 \leq \mu_C - \mu_{NC} \leq 56.06$$

When the contrast-coded predictor used the values of –1 and +1 rather than –0.5 and +0.5, the estimated slope was half the difference in the group means (i.e., 14) and its confidence interval is computed as

$$\text{CI}_{b_1} = 14 \pm \sqrt{\frac{4.41(892.22)}{(19)1.052(1.00)}}$$

$$= 14 \pm 14.03$$

The term that is different in this interval (other than the value of the estimated slope itself) is the variance of the predictor (coded +1 and –1), which is 1.052 rather than 0.263.

In this case, the confidence interval is given equivalently as

$$-.03 \leq \beta_1 \leq 28.03$$

$$-.03 \leq \frac{\mu_C - \mu_{NC}}{2} \leq 28.03$$

Thus, it continues to tell us about the confidence interval for the mean difference, except with these codes of course it is the confidence interval for half the mean difference. If we multiply this expression by 2, we get the confidence interval for the mean difference.

Categorical Predictors With More Than Two Levels

Suppose a developmental psychologist is interested in the effects of feedback about performance on subsequent motivation to do a task. She hypothesizes that subsequent motivation will decline if children are told that they earlier failed at the task. To test this hypothesis, she randomly assigns children to three conditions: In one condition they are told that they failed on the task; in a second condition they are given no feedback; and in a third condition they are told they succeeded. The experimenter then monitors the number of tasks they subsequently complete, after the differential feedback has been given. Twenty-four children are run in total, eight in each of the three conditions. The hypothetical raw data are given in Figure 9.6.

Contrast Codes for Multilevel Categorical Predictors

To examine the effects of feedback on the number of tasks subsequently completed, we need to derive a coding scheme to represent the three levels of the categorical feedback variable. We might

	Failure	*No Feedback*	*Success*
	2	4	4
	2	3	6
	2	4	5
	3	5	4
	4	5	6
	4	2	4
	3	4	3
	4	3	3
$\bar{Y}_k$	3.000	3.750	4.375

Figure 9.6 Hypothetical experimental data for three conditions (values represent number of tasks each child completes)

think that a single variable that codes all three conditions would be appropriate, giving observations from the Success condition a higher value on the variable than observations for the No Feedback condition who in turn receive a higher value than observations from the Failure condition. We could then see if such a coded variable would be predictive of Y_i. The problem with using a single variable to code the three levels of this categorical variable is that with such a coding scheme we are assuming that the categories can be ordered in an a priori manner and that the relationship between the values of the single predictor variable and Y_i is a linear one. Although we may have a reason for expecting that Y_i should be lower in the Failure condition than in the other two, we do not have any reason for assigning particular values to the groups, expecting linear predictions as a function of those particular values. In other words, a single predictor variable that codes the three conditions with particular values does not make much sense, given that we are dealing with a categorical variable whose levels do not differ in a neat linear way.

To ask whether we can predict Y_i as a function of some categorical variable having in general m levels or groups is equivalent to asking whether there are differences among the m group means ($\bar{Y}_k$) across those levels (with k varying from 1 to m). To answer this question, we need to employ $m - 1$ contrast-coded predictor variables in our model. We will then be able to ask about mean differences among the groups, allowing for all possible orderings of those means. To define these $m - 1$ contrast-coded predictors, it is now time to introduce the second defining condition for contrast codes. The first defining condition, you will recall, was that the sum of the λ values across the groups or levels of the categorical variable must equal zero:

$$\sum_k \lambda_k = 0$$

When we use more than one contrast-coded predictor to code a categorical variable having more than two levels, the second condition that must be met is that across levels of the categorical variable all pairs of contrast codes must be orthogonal to each other. Given that the first condition is met, this second condition of orthogonality is met whenever the sum (across k or the levels of the categorical variable) of the products of the λ values from pairs of contrast codes equals zero.

In our example, we have three levels of the categorical variable. We will therefore use two contrast codes to code it. Each value of λ now has two subscripts, the first one designating which contrast code we are talking about and the second one designating the level of the categorical variable (k). The condition of orthogonality is met when

$$\sum_k \lambda_{1k}\lambda_{2k} = 0$$

To make this second defining condition of contrast codes more understandable, let us illustrate codes that do and do not meet it for the example at hand. In Figure 9.7, two sets of codes, with two codes in each set, are given for coding the three levels of the categorical predictor variable: Failure, No Feedback, and Success.

Each of the four codes meets the first defining condition for a contrast code because the sum of the λ values for each code, computed across the three levels of the categorical variable, equals zero. The second defining condition, however, is only met by the codes in Set A. If we multiply the value of λ_{1k} by the value of λ_{2k} at each of the levels of the categorical predictor variable and then add up the resulting three products, we get a sum of 0 from Set A (i.e., $0 + (-1) + 1 = 0$) and a sum of 1 from Set B (i.e., $0 + 0 + 1 = 1$). Accordingly, only the codes in Set A can legitimately be called contrast codes.

This second defining condition means that a given code cannot be defined as a contrast code in isolation. We could not, for instance, look at the code for λ_{1k} in Set A and identify it as a contrast code, unless we looked at the other code or codes with which it is used in combination to code the categorical predictor variable. For instance, if we changed the values of λ_{2k} in Set A to be –1, –1, and 2 for Failure, No Feedback, and Success, respectively, then the codes in Set A would no longer be contrast codes, even though we had not changed the values of λ for the first code. This set of codes would no longer be contrast codes because the sum of the products of the λ values across the category levels would no longer equal zero.

If our categorical predictor variable had four levels, as in Figure 9.8, we would need three contrast codes to code it completely. The second defining condition for contrast codes would be met in such a case if the sums of the products of the λ values for all possible pairs of codes equaled zero. Let us define three contrast codes with values of λ_{1k}, λ_{2k}, and λ_{3k}. We then have three pairs of codes, and for each of these pairs the sum of the products of the λ values must equal zero. For codes 1 and 2, the sum of the products of the λ_{jk} values equals $(-3)0 + 1(-2) + 1(1) + 1(1) = 0$. For codes 1 and 3, the sum of the products of the λ_{jk} values equals $(-3)0 + 1(0) + 1(-1) + 1(1) = 0$. And for codes 2 and 3, the sum of the products of the λ_{jk} values equals $0(0) + 0(-2) + 1(-1) + 1(1) = 0$.

	Failure	*No Feedback*	*Success*
Set A			
λ_{1k}	–2	1	1
λ_{2k}	0	–1	1
Set B			
λ_{1k}	–1	0	1
λ_{2k}	0	–1	1

Figure 9.7 Sets of codes for a three-level categorical predictor

	Level 1	*Level 2*	*Level 3*	*Level 4*
λ_{1k}	−3	1	1	1
λ_{2k}	0	−2	1	1
λ_{3k}	0	0	−1	1

Figure 9.8 Codes for a four-level categorical predictor

With a categorical predictor having three levels, then, we need two contrast codes and a single sum of products of λ values must equal zero. With a categorical predictor having four levels, we need three contrast codes. Those three codes result in three possible pairs of codes, and hence three sums of products of λ values must equal zero. In general, with a categorical variable having m levels, we need $m - 1$ contrast codes to code it. From these $m - 1$ contrast codes, there are $(m - 1)(m - 2)/2$ pairs of codes. This many sums of products of λ values must equal zero to meet the second defining condition of contrast codes.

For any given categorical predictor, there are an infinite number of sets of contrast codes that could be used. The choice of codes to be used should be guided by some theoretical or substantive notions about how the groups defined by the categorical predictor variable are expected to differ on the dependent variable. For instance, in the illustration at hand, we expected children in the Failure condition to have lower scores than children in the other two conditions. As we saw in the case of a categorical predictor with only two levels, the regression coefficient for a contrast code tells us about the relative mean difference between observations having different values on the contrast code. It thus makes sense to derive codes that allow us to examine predictions about mean differences on the dependent variable. In other words, given that we want to see whether children in the Failure condition have lower scores than children in the other two conditions, the first contrast code we gave in Set A of Figure 9.7 is one that we may well choose to examine.

As was the case with a single contrast-coded predictor that codes a categorical variable with two levels, the regression coefficient associated with a contrast-coded predictor in the case of a categorical variable with more than two levels tells us about mean differences among the various groups or levels of the categorical variable, according to Equation 9.1 (see further for a recapitulation of the formula). But this will be the case *only* if a complete set of $m - 1$ contrast-coded predictors is included in the model and *only* if the contrast codes used meet the orthogonality condition that we just defined. At a later point in this chapter we will discuss estimation in the presence of nonorthogonally coded predictors. For now, the important point is that slopes tell us about coded mean differences among the categories only with a complete set of codes and only with orthogonality.

Based on theoretical considerations, then, we are interested in the comparison that is made by the first contrast code of Set A in Figure 9.7. With three levels of our categorical variable and one code chosen, the second code is constrained to be one that compares the means in the No Feedback and Success conditions, like the second code in Set A. In general, with m levels of a categorical variable and $m - 1$ contrast codes, the final code is constrained once the first $m - 2$ codes have been defined, in order to meet the orthogonality condition.

We can use these two codes to define two predictor variables, X_{1i} based on the codes −2, 1, 1 (for Failure, No Feedback, and Success respectively) and X_{2i} based on the codes 0, −1, 1, and then estimate a multiple regression model in which these are used as simultaneous predictors of Y_i. If we do this, we can exactly specify the mean differences estimated by the two resulting slopes using Equation 9.1 in the context of a model with a full set of orthogonal contrast-coded predictors:

$$b_1 = \frac{\sum_k \lambda_{1k}\bar{Y}_k}{\sum_k \lambda_{1k}^2} = \frac{(-2)\bar{Y}_F + (1)\bar{Y}_{NF} + (1)\bar{Y}_S}{(-2)^2 + (1)^2 + (1)^2} = \frac{(\bar{Y}_{NF} + \bar{Y}_S) - 2\bar{Y}_F}{6} = \frac{\left(\frac{\bar{Y}_{NF} + \bar{Y}_S}{2}\right) - \bar{Y}_F}{3}$$

$$b_2 = \frac{\sum_k \lambda_{2k}\bar{Y}_k}{\sum_k \lambda_{2k}^2} = \frac{(0)\bar{Y}_F + (-1)\bar{Y}_{NF} + (1)\bar{Y}_S}{(0)^2 + (-1)^2 + (1)^2} = \frac{\bar{Y}_S - \bar{Y}_{NF}}{2}$$

In general, such slopes will inform us about differences among category means following the codes used, with the numerator of the aforementioned expressions representing the mean difference, and the denominator representing a scaling factor. Notice that group means for levels of the categorical variable that are coded with a zero value of λ on a particular contrast-coded predictor drop out of the numerator of the slope and thus do not figure in the comparison that is made (i.e., the group mean for the Failure condition does not play a role in the slope of the second contrast-coded variable).

To show the impact of the scaling factor in the denominator of the slope expression, had we used fractional values for λ($-⅔$, ⅓, ⅓ for X_{1i}', and 0, $-½$, ½ for X_{2i}') rather than those defined above, then the following would be the values of the slopes:

$$b_{1'} = \frac{\sum_k \lambda_{1'k}\bar{Y}_k}{\sum_k \lambda_{1'k}^2} = \frac{\left(-\frac{2}{3}\right)\bar{Y}_F + \left(\frac{1}{3}\right)\bar{Y}_{NF} + \left(\frac{1}{3}\right)\bar{Y}_S}{\left(-\frac{2}{3}\right)^2 + \left(\frac{1}{3}\right)^2 + \left(\frac{1}{3}\right)^2} = \left(\frac{\bar{Y}_{NF} + \bar{Y}_S}{2}\right) - \bar{Y}_F$$

$$b_{2'} = \frac{\sum_k \lambda_{2'k}\bar{Y}_k}{\sum_k \lambda_{2'k}^2} = \frac{(0)\bar{Y}_F + \left(-\frac{1}{2}\right)\bar{Y}_{NF} + \left(\frac{1}{2}\right)\bar{Y}_S}{(0)^2 + \left(-\frac{1}{2}\right)^2 + \left(\frac{1}{2}\right)^2} = \bar{Y}_S - \bar{Y}_{NF}$$

The advantage of such fractional codes is that their slopes will equal the mean differences rather than fractions of mean differences.

Without practice, it may seem difficult to come up with a set of orthogonal contrast codes, particularly when dealing with a categorical variable having more than three or so levels. Our advice is that one should initially create codes that represent mean comparisons one would like to make theoretically and then derive the remainder of the codes so that orthogonality is preserved. One way to do this, once one or more initial codes have been defined, is to construct further contrast codes that compare category means that were tied (or received the same value of λ) on already used code(s). With practice, deriving orthogonal codes becomes a relatively easy task.

In the absence of any motivated comparisons, one can always use a convention called Helmert codes, regardless of the number of levels. A simple algorithm generates such codes. If there are m levels of the categorical variable, one defines the first of $m - 1$ contrast codes by assigning the value of $m - 1$ to the first level and the value of -1 to each of the remaining $m - 1$ levels. For the second contrast code, the first level is given the value of 0, the second level is given the value of $m - 2$, and all remaining levels are given the value of -1. For the third contrast code, the first two levels of the categorical predictor are assigned values of 0, the third level is given the value of $m - 3$, and the remaining levels are given the value of -1. One proceeds in this manner to define all $m - 1$ contrast codes, with the last one having values of 0 for all levels of the predictor variable except for the last two. These last two levels have values of 1 and -1. The resulting code values are presented in Figure 9.9.

Estimation and Inference With Multilevel Categorical Predictors

Using the data in Figure 9.6, we estimated the parameters of the following multiple regression model, with X_{1i} and X_{2i} as contrast-coded predictors, given the values of λ defined by Set A in Figure 9.7:

$$\text{MODEL A: } Y_i = \beta_0 + \beta_1 X_{1i} + \beta_2 X_{2i} + \varepsilon_i$$

The parameter estimates are

$$\text{MODEL A: } \hat{Y}_i = 3.7083 + .3542X_{1i} + .3125X_{2i}$$

and the sum of squared errors is 23.375.

Unsurprisingly, the predicted values from this model are the means of the three categories (given in Figure 9.6) of the categorical independent variable:

$$\hat{Y}_F = 3.7083 + .3542(-2) + .3125(0) = 3.000$$
$$\hat{Y}_{NF} = 3.7083 + .3542(1) + .3125(-1) = 3.750$$
$$\hat{Y}_S = 3.7083 + .3542(1) + .3125(1) = 4.375$$

As we have said before, a model with a categorical independent variable will make predictions of the group or category level means whenever a complete set of $m - 1$ codes is used as predictors.

	Category level						
Code	*1*	*2*	*3*	...	*m–2*	*m–1*	*m*
λ_{1k}	m–1	–1	–1	...	–1	–1	–1
λ_{2k}	0	m–2	–1	...	–1	–1	–1
λ_{3k}	0	0	m–3	...	–1	–1	–1
⋮	⋮	⋮	⋮	...	⋮	⋮	⋮
λ_{m-2k}	0	0	0	...	2	–1	–1
λ_{m-1k}	0	0	0	...	0	1	–1

Figure 9.9 Helmert contrast codes

We have already discussed the interpretations of the two parameter estimates associated with the contrast-coded predictors in terms of the category means. Let us revisit these interpretations now that we have the numerical estimates:

$$b_1 = \frac{\sum_k \lambda_{1k}\bar{Y}_k}{\sum_k \lambda_{1k}^2} = \frac{\left(\frac{\bar{Y}_{NF}+\bar{Y}_S}{2}\right)-\bar{Y}_F}{3} = \frac{\left(\frac{3.750+4.375}{2}\right)-3.000}{3} = 0.3542$$

$$b_2 = \frac{\sum_k \lambda_{2k}\bar{Y}_k}{\sum_k \lambda_{2k}^2} = \frac{\bar{Y}_S-\bar{Y}_{NF}}{2} = \frac{4.375-3.750}{2} = 0.3125$$

And just as we found with a categorical predictor with two levels, the estimated intercept in this model equals the mean of the three category means:

$$b_0 = \frac{\sum_k \bar{Y}_k}{m} = \frac{3.750+4.375+3.000}{3} = 3.7083$$

Although these interpretations for the regression coefficients in models with contrast-coded predictors are typically the most useful, interpretations we gave earlier for parameter estimates in multiple regression models continue to be entirely appropriate. Thus, slopes of a predictor can be interpreted as differences in $\hat{Y}_i$ values as the predictor increases by one unit, holding constant other predictors. In the case of the slope for X_{1i}, as we move from a score of –2 (for the Failure condition) to a score of +1 (for the No Feedback and Success conditions) the predicted values go from the mean of the Failure condition (3.000) to the means in the No Feedback and Success conditions (3.750 and 4.375). Thus, for a three-unit increase in X_{1i}, we go from a predicted value of 3.000 to one of 4.0625, meaning that the increase in predicted values for a one-unit increase in X_{1i} is .3542. And for X_{2i}, as we go from a score of –1 to 1, the predicted value goes from 3.75 to 4.375. Accordingly, per unit increase in X_{2i}, we predict a .3125 increase in Y_i. And finally, the intercept equals the predicted value when both contrast-coded predictors equal zero. When do these predictors equal zero? From the first condition used to define contrast codes, the mean of each contrast code, across categories, equals zero. Accordingly, the intercept is the predicted value for the average of the categories.

There are, of course, many Model Cs with which we can compare this model to test various null hypotheses. One obvious comparison is with the single-parameter simplest model, estimating just the intercept:

MODEL C: $Y_i = \beta_0 + \varepsilon_i$

and predicting the grand mean, $\bar{Y}$, for all the observations. Since, in this example, each of the levels of the categorical variable has the same number of observations, the overall grand mean of the 24 observations is the same as the mean of the category means. Hence, it is the case that the estimated parameter in this Model C is identical to the intercept in the three-parameter Model A with which we are comparing it:

MODEL C: $\hat{Y}_i = 3.7083$

This estimated Model C has a sum of squared errors of 30.958.

What exactly is the null hypothesis that is tested by this model comparison? It is that the two predictors have slopes of zero, that is, that using them as predictors does nothing to improve the quality of our predictions:

$$H_0\text{: } \beta_1 = \beta_2 = 0$$

But this null hypothesis can also be expressed in terms of the equality of the category means, since Model C predicts the grand mean, $\bar{Y}$, for every observation and Model A makes predictions that are conditional on category membership, predicting the category mean, $\bar{Y}_k$, for each observation. Accordingly, the null hypothesis can equivalently be expressed as

$$H_0\text{: } \mu_{\text{F}} = \mu_{\text{NF}} = \mu_{\text{S}}$$

where these are the true but unknown means of the three levels of the categorical independent variable. The comparison of these two models yields the following values of PRE and F:

$$\text{PRE} = \frac{30.958 - 23.375}{30.958} = .245$$

$$F_{2,21} = \frac{\text{PRE}/(\text{PA}-\text{PC})}{(1-\text{PRE})/(n-\text{PA})} = \frac{.245/2}{(1-.245)/21} = 3.406$$

$$F_{2,21} = \frac{\text{SSR}/(\text{PA}-\text{PC})}{\text{SSE(A)}/(n-\text{PA})} = \frac{7.582/2}{23.375/21} = 3.406$$

And these come just short of beating the critical values for 2 and 21 degrees of freedom. Hence, we cannot reject the null hypothesis that there are no mean differences among these three categories or conditions.

This conclusion does not mean that we should accept the null hypothesis of no mean differences. And in this case, since we clearly had an expectation that the mean in the Failure condition would be less than the mean in the other two conditions, we should certainly proceed to directly test that hypothesis, which is the comparison made by the first contrast-coded predictor. Within the analysis of variance tradition, it is sometimes maintained that one should not test specific focused comparisons among category means unless the overall multiple-degree-of-freedom test that we have just conducted—that there are no mean difference among the categories—is rejected. We strongly disagree with this point of view. For reasons we explained earlier, we generally prefer model comparisons where PA – PC = 1. One of the advantages of a regression-based approach to traditional analysis of variance procedures is that one must construct individual one-degree-of-freedom comparisons or contrasts among group means. Many traditional ANOVA programs automatically provide only the omnibus, multiple-degree-of-freedom test, and this, we think, is a disservice.

As we have seen, the regression coefficients for X_{1i} and X_{2i} estimate particular differences among category means, the first comparing the Failure mean with the average of the No Feedback and Success means, and the second comparing the No Feedback and Success means. Hence, model comparisons that test whether these two parameters depart from zero are equivalently tests of mean differences among the three categories. Specifically, one model comparison is whether the parameter associated with X_{1i} equals zero:

$$\text{MODEL A: } Y_i = \beta_0 + \beta_1 X_{1i} + \beta_2 X_{2i} + \varepsilon_i$$

$$\text{MODEL C1: } Y_i = \beta_0 + \beta_2 X_{2i} + \varepsilon_i$$

with the following equivalent null hypotheses:

$$H_0: \beta_1 = 0$$

$$H_0: \mu_F = (\mu_{NF} + \mu_S)/2$$

And the other model comparison tests whether the parameter associated with X_{2i} equals zero:

$$\text{MODEL A: } Y_i = \beta_0 + \beta_1 X_{1i} + \beta_2 X_{2i} + \varepsilon_i$$

$$\text{MODEL C2: } Y_i = \beta_0 + \beta_1 X_{1i} + \varepsilon_i$$

with the following equivalent null hypotheses:

$$H_0: \beta_2 = 0$$

$$H_0: \mu_{NF} = \mu_S$$

Model A for both of these comparisons is the same three-parameter augmented model that we estimated earlier, with a sum of squared errors of 23.375. Model C1 for the first comparison is estimated as follows:

$$\text{MODEL C1: } \hat{Y}_i = 3.7083 + .3125X_{2i} \qquad \text{SSE(C1)} = 29.396$$

And Model C2 for the second comparison is estimated as follows:

$$\text{MODEL C2: } \hat{Y}_i = 3.7083 + .3542X_{1i} \qquad \text{SSE(C2)} = 24.937$$

The estimated intercepts and slopes in these models are unchanged from what they were in the Model A with both predictors because of the conjunction of two conditions: First, we have employed contrast-coded predictors, which by definition are orthogonal at the level of the three categories. Second, we have an equal number of observations in each of the three conditions. As a result of these two conditions, the contrast-coded predictors are uncorrelated with each other across the 24 individual observations. Their tolerance in Model A is 1.00.

The first model comparison, asking whether β_1 differs from zero, yields the following PRE and F statistics:

$$\text{PRE} = \frac{29.396 - 23.375}{29.396} = .205$$

$$F_{1,21} = 5.41$$

This F statistic exceeds the critical value of F with α at .05. Hence, we conclude that β_1 differs significantly from zero. Equivalently, we conclude that the mean value of Y_i in the Failure condition is significantly different from the average of the mean values in the Success and No Feedback conditions. Since the sample mean in the Failure condition is less than the average of the other two sample means, we conclude that Failure feedback in this study decreases subsequent performance relative to Success and No Feedback.

The test of the second null hypothesis, that β_2 equals zero, yields the following PRE and F statistics:

$$\text{PRE} = \frac{24.937 - 23.375}{24.937} = .063$$

$$F_{1,21} = 1.40$$

Because this F does not exceed its critical value, we conclude that β_2 does not differ significantly from zero. Equivalently, we cannot conclude that the mean performance under the Success condition is different from that under No Feedback.

Earlier in this chapter we introduced Equation 9.2 to compute the SSR associated with a contrast-coded predictor in terms of the category means (see the following for a recapitulation). This expression for the SSR of a contrast-coded predictor continues to apply in the case of categorical variables with more than two levels, as long as a full set of $m - 1$ contrast-coded predictors is included in Model A. Thus, in the present case, we have seen that the SSR for the Model A/Model C1 comparison that tested whether β_1 equaled zero was equal to

$$\text{SSR}_{X_1} = \text{SSE}(\text{C1}) - \text{SSE}(\text{A}) = 29.396 - 23.375 = 6.021$$

This can be obtained equivalently in terms of the category means as

$$\text{SSR}_{X_1} = \frac{\left(\sum_k \lambda_{1k}\bar{Y}_k\right)^2}{\sum_k \left(\lambda_{1k}^2 / n_k\right)} = \frac{\left((-2)3.000 + (1)3.750 + (1)4.375\right)^2}{(-2)^2/8 + (+1)^2/8 + (+1)^2/8} = 6.021$$

Likewise, we saw that the SSR for the Model A/Model C2 comparison that tests whether β_2 equals zero yields

$$\text{SSR}_{X_2} = \text{SSE(C2)} - \text{SSE(A)} = 24.937 – 23.375 = 1.562$$

This can be obtained equivalently in terms of the category means as

$$\text{SSR}_{X_2} = \frac{\left(\sum_k \lambda_{2k}\bar{Y}_k\right)^2}{\sum_k \left(\lambda_{2k}^2 / n_k\right)} = \frac{\left((0)3.000 + (-1)3.750 + (1)4.375\right)^2}{(0)^2/8 + (-1)^2/8 + (+1)^2/8} = 1.562$$

We have now done three different tests comparing the augmented model, which includes both contrast-coded predictors, with three different compact ones. The results of these three tests are presented in Figure 9.10. Notice that we have given labels, in parentheses, for each of these tests to indicate the questions they examine in terms of the group means. The two-degree-of-freedom test, done first, comparing Model A to a Model C that predicted the grand mean for all observations, was an omnibus test of any group mean differences. The second, comparing models with and without X_{1i} as a predictor, examined whether the mean in the Failure condition differed from the average of the two in the other conditions. And the third, comparing models with and without X_{2i} as a predictor, examined whether the means in the No Feedback and Success conditions differed. We want to emphasize again that even though the two-degree-of-freedom test was not significant, we found a significant mean difference when we tested the more focused contrast question represented by X_{1i}. As always, we strongly encourage focused PA – PC = 1 model comparisons.

As the sums of squares in this source table show, the SSRs for the individual predictor variables sum to the SSR for the first model comparison, where the overall augmented model was compared to a compact single-parameter model, predicting the grand mean for all observations. As we saw in Chapter 7, this will be the case whenever predictors are completely nonredundant, with tolerances of 1.0. In the present case, this results from the conjunction of two conditions: the use of contrast-coded predictors, which are by definition orthogonal at the level of the groups or categories; and the fact that each category contains the same number of observations.

A Quick Look at Alternative Contrast Codes

Earlier we stated that there were many, many possible sets of contrast codes that could be used to code a categorical predictor. Let us examine the same data that we have been focusing on using a different set of codes. Suppose we now define our contrast codes as follows:

	Failure	*No Feedback*	*Success*
λ_{1k}	−1	0	1
λ_{2k}	−1	2	−1

To differentiate these codes from the earlier set, we define Z_1 and Z_2 as contrast-coded predictors, assigning individuals the indicated values to represent category membership. We then regress Y_i on Z_{1i} and Z_{2i} with the estimated parameters

$$\hat{Y}_i = 3.7083 + .6875Z_{1i} + .0208Z_{2i}$$

The parameter estimates for the two contrast-coded predictors in this model are different from those that we estimated using the earlier set, but in a deeper sense this model is equivalent to the model

Source	*b*	*SS*	*df*	*MS*	*F*	*PRE*
Model (between conditions)		7.583	2	3.792	3.406	.245
X_1 (Failure vs. No Feedback, Success)	.3542	6.021	1	6.021	5.409	.205
X_2 (No Feedback vs. Success)	.3125	1.562	1	1.562	1.403	.063
Error		23.375	21	1.113		
Total		30.958	23			

Figure 9.10 Summary source table

we developed under the old set of codes. Substituting for the values of Z_{1i} and Z_{2i}, we see that the group or condition means continue to be the predictions made by the model for all observations:

$$\hat{Y}_F = 3.7083 + .6875(-1) + .0208(-1) = 3.000$$

$$\hat{Y}_{NF} = 3.7083 + .6875(0) + .0208(+2) = 3.750$$

$$\hat{Y}_S = 3.7083 + .6875(+1) + .0208(-1) = 4.375$$

Because this model makes the same predictions for all observations as the model with the previous set of contrast-coded predictors, the sum of squared errors is identical to what it was before, that is, 23.375.

The regression coefficients for the contrast-coded predictors have changed because the new contrast codes make comparisons among condition means that are different from the comparisons made by the old set of codes. The contrast-coded predictor Z_{1i} now compares the means in the Success and Failure conditions. The value of its regression coefficient equals half the difference between these two group means. The second contrast-coded predictor, Z_{2i}, compares the mean in the No Feedback condition with the average of the means of the two conditions in which some kind of feedback was provided. Its regression coefficient equals one-third of the difference between the mean in the No Feedback condition and the average of the other two means. These values for the regression coefficients are easily derived using Equation 9.1. They also follow immediately once we realize the comparisons made by the contrasts and the number of units that separate observations in the various conditions on Z_{1i} and Z_{2i}. The intercept has not changed in value as a result of the new set of codes. It still equals the mean of the three condition means, as it will whenever a full set of contrast-coded predictors is used.

Because the change in codes has not changed the predicted values or the sum of squared errors for this model, a test of the null hypothesis that both β_1 and β_2 equal zero produces the same values of PRE and F as it did under the old set of codes. The compact single-parameter model is

$$\text{MODEL C: } \hat{Y}_i = 3.7083$$

with a sum of squared errors of 30.958. PRE continues to equal .245, which converts to an F of 3.406 with 2 and 21 degrees of freedom. Thus, a test of the omnibus null hypothesis—that all of the condition means equal one another—reaches the same conclusion regardless of our choice of contrast codes.

Single-degree-of-freedom tests of whether β_1 or β_2 equals zero, however, reach different conclusions than they did before. These regression coefficients now estimate different comparisons between the condition means from those estimated with the earlier set of contrast codes. A test of whether β_1 equals zero now tests whether the means in the Failure and Success conditions are equal. The compact model for this test is

$$\text{MODEL C1: } \hat{Y}_i = 3.7083 + .0208Z_{2i} \qquad \text{SSE(C1)} = 30.938$$

Accordingly,

$$\text{PRE} = \frac{30.938 - 23.375}{30.938} = .245$$

which converts to an F statistic of 6.793 with 1 and 21 degrees of freedom. Because .244 exceeds the critical value, our test reveals both that β_1 is significantly greater than zero and that the mean in the Success condition is significantly greater than the mean in the Failure condition. As before, the sum of squares reduced associated with this contrast-coded predictor can be computed as a function of the relevant condition means using Equation 9.2:

$$\text{SSR}_{Z_1} = \frac{\left((-1)3.000 + (0)3.750 + (1)4.375\right)^2}{(-1)^2/8 + (0)^2/8 + (1)^2/8} = 7.562$$

The test of whether β_2 differs from zero is a test of whether the No Feedback mean is significantly different from the average of the Failure and Success means. The compact model for this test is

MODEL C2: $\hat{Y}_i = 3.7083 + .6875Z_{1i}$ $\qquad$ SSE(C2) = 23.396

The resulting PRE equals .001, which converts to an F of 0.019 with 1 and 21 degrees of freedom. Thus, the No Feedback mean does not significantly differ from the average of the means in the other two conditions. As before, the SSR for this comparison can be directly calculated using Equation 9.2:

$$\text{SSR}_{Z_2} = \frac{\left((-1)3.000 + (2)3.750 + (-1)4.375\right)^2}{(-1)^2/8 + (2)^2/8 + (-1)^2/8} = 0.021$$

Because these two new contrast-coded predictors are nonredundant, just as were the earlier two, the two SSRs explained by each predictor over and above the other can be added to equal the SSR explained by them both as a set. The three tests we have just conducted are summarized in Figure 9.11. Note that the only changes in this source table compared to the one using the earlier set of contrast codes (Figure 9.10) occur in the two rows of the table testing the specific comparisons made by the coefficients associated with Z_{1i} and Z_{2i}. All we have done is divide up the sum of squares between conditions (7.583) differently here, focusing on a different set of contrasts. As we have said before, with m category levels, there are only $m - 1$ orthogonal contrasts or comparisons that can be used. And whether we use these codes or those we discussed earlier is only a matter of theoretical preference.

Source	*b*	*SS*	*df*	*MS*	*F*	*PRE*
Model (between conditions)		7.583	2	3.792	3.406	.245
Z_1 (Failure vs. Success)	.6875	7.562	1	7.562	6.793	.244
Z_2 (No Feedback vs. Failure, Success)	.0208	0.021	1	0.021	0.019	.001
Error		23.375	21	1.113		
Total		30.958	23			

Figure 9.11 Summary source table

Problems and Pitfalls in Using Nonorthogonal Codes

Suppose we were interested in asking the following two questions of these data, coded by the following set of codes:

	Failure	*No Feedback*	*Success*
λ_{1k}	−2	+1	+1
λ_{2k}	−1	0	+1

The first question is whether the Failure mean differs from the average of the means in the other two conditions. This was the first of the codes that we used in our first set, used to create the predictor X_{1i}. The second is whether the means in the Failure and Success conditions differ. This was the first of the codes that we used in our second set, used to create the predictor Z_{1i}. As we have seen, these are perfectly legitimate questions that we might want to ask of these data, but they are not orthogonal questions about the mean differences, as revealed by the fact that the second condition for contrast codes is not met by these two codes if we combine them into a single model:

$$\sum_k \lambda_{1k}\lambda_{2k} = (-2)(-1) + (+1)(0) + (+1)(+1) = +3$$

It is for this reason that we have not called them "contrast codes" when we think about them as a set. They are perfectly valid questions to ask of the data, but they are nonorthogonal.

If we were interested in these two questions, we might be tempted to use the two resulting predictors, X_{1i} and Z_{1i}, created with these codes to simultaneously predict Y_i, even though the codes themselves are not orthogonal. The resulting model would be

$$\hat{Y}_i = 3.7083 + .0417X_{1i} + .625Z_{1i}$$

This model, even with these nonorthogonal codes, is the same in a deep sense as the earlier model in that it makes the same predictions of the condition means for every observation:

$$\hat{Y}_F = 3.7083 + .0417(-2) + .625(-1) = 3.000$$

$$\hat{Y}_{NF} = 3.7083 + .0417(+1) + .625(0) = 3.750$$

$$\hat{Y}_S = 3.7083 + .0417(+1) + .6250(+1) = 4.375$$

And, as a result, it has the same sum of squared errors, 23.375. A model comparison between it, as Model A, and the single-parameter Model C, making a constant prediction for all observations, thus still provides the omnibus two-degree-of-freedom test about whether there are any differences among the category means. The resulting test of the overall model is summarized in the first row of the source table given in Figure 9.12 (PRE = .245, $F(2,21) = 3.406$). Both this row and the final two rows of the table are identical to what they were in the earlier source tables that we presented from these data. However, when we estimate the model in a regression program and examine the regression coefficients for the individual predictor variables and their respective SSRs, as given in the source table of Figure 9.12, they are not the values that we might expect.

Source	*b*	*SS*	*df*	*MS*	*F*	*PRE*
Model (between conditions)		7.583	2	3.792	3.406	.245
X_1	.0417	0.021	1	0.021	0.019	.001
Z_1	.6250	1.562	1	1.562	1.403	.063
Error		23.375	21	1.113		
Total		30.958	23			

Figure 9.12 Summary source table using nonorthogonal codes

The coefficient for X_{1i}, when it was embedded in an orthogonal set of contrast-coded predictors (i.e., with X_{2i}), equaled .3542, which was shown to be

$$\frac{\left(\frac{\bar{Y}_{NF}+\bar{Y}_S}{2}\right)-\bar{Y}_F}{3}$$

and its associated SSR was 6.021, which with Equation 9.2 was shown to equal

$$\frac{\left((-2)\bar{Y}_F+(+1)\bar{Y}_{NF}+(+1)\bar{Y}_S\right)^2}{(-2)^2/8+(+1)^2/8+(+1)^2/8}$$

Now, however, when it is used to predict Y_i along with the nonorthogonal predictor Z_{1i}, its estimated coefficient equals .0417 and its SSR equals 0.021.

Similarly, the coefficient for Z_{1i} in this model equals .625, whereas its coefficient when used in the orthogonal set with Z_{2i} equaled .687, which was half the difference between $\bar{Y}_S$ and $\bar{Y}_F$. And now its SSR equals 1.562, whereas earlier, when used in the orthogonal set with Z_{2i}, its SSR was 7.562, which was shown to equal

$$\frac{\left((-1)\bar{Y}_F+(0)\bar{Y}_{NF}+(+1)\bar{Y}_S\right)^2}{(-1)^2/8+(0)^2/8+(+1)^2/8}$$

The important point is that predictors that code the levels of a categorical independent variable will not yield coefficients that equal the expected mean differences and their associated SSRs unless a full set of orthogonal contrast-coded predictors is used. In other words, the important Equations 9.1 and 9.2 (shown next) will *not* hold, unless a full set of contrast-coded predictors is included in the augmented model:

$$b=\frac{\sum_k \lambda_k \bar{Y}_k}{\sum_k \lambda_k^2}$$

$$\text{SSR}=\frac{\left(\sum_k \lambda_k \bar{Y}_k\right)^2}{\sum_k\left(\lambda_k^2/n_k\right)}$$

Serious interpretative errors can ensue if one thinks that a given categorical predictor codes a particular mean difference when it is embedded in a nonorthogonal set. Orthogonality here depends on meeting the second defining condition of contrast codes—that the sum of the products of their coded values across category levels equals zero. As we will show, unequal numbers of observations can result in redundant contrast-coded predictors, redundant across observations, but this creates no interpretative problems as long as the codes are orthogonal.

So how might one proceed if one were interested in testing the nonorthogonal questions of whether the Failure mean differs from the average of the No Feedback and Success means (i.e., the question implicit in the X_{1i} codes) and of whether the Failure and Success means differ (i.e., the question implicit in the Z_{1i} codes)? One could test these sequentially by specifying two models, one using both X_{1i} and X_{2i} as predictors and the other using Z_{1i} and Z_{2i} as predictors, just as we did in the earlier sections. Alternatively, one could rely on two bits of knowledge to test the mean differences implied by these contrasts without doing both estimations. First, as we have shown, the SSE(A) for a Model A that incorporates any complete set of contrast-coded predictors will be the same regardless of the specific set of such predictors used. Second, if a given contrast-coded predictor were included in a complete set of contrast-coded predictors, its SSR would be given by following the aforementioned.

Accordingly, in the case at hand, had one simply estimated the model that included X_{1i} and X_{2i} as predictors, one would have known that the SSE(A) for a model that used any full set of contrast-coded predictors would equal 23.375 with 21 degrees of freedom. One could then calculate the SSR associated with Z_{1i} if it were embedded in a full set of contrast-coded predictors, that is

$$\mathrm{SSR}_{Z_1} = \frac{\left((-1)\bar{Y}_F + (0)\bar{Y}_{NF} + (+1)\bar{Y}_S\right)^2}{(-1)^2/8 + (0)^2/8 + (+1)^2/8} = 7.562$$

Then one could calculate the values of PRE and F that would result if one tested Z_{1i} in the context of a complete set of contrast-coded predictors:

$$\mathrm{PRE} = \frac{7.562}{23.375 + 7.562} = .244$$

$$F_{1,21} = \frac{7.562/1}{23.375/21} = 6.793$$

When doing this, one should recognize that the questions represented by these two contrast codes are not independent. The question of whether the Failure mean differs from the No Feedback and Success means is not entirely independent of the question of whether the Failure and Success means differ from each other. The answer to one is partially informative about the answer to the other.

Dummy Codes

There is another coding convention, known as dummy coding, that is widely used in the literature. Under this convention, one of the groups defined by the categorical variable is given values of zero on all $m - 1$ codes, and the other groups are given values of zero on all but one of the codes. So, for instance, in the case at hand the following codes are consistent with this convention:

	Failure	*No Feedback*	*Success*
λ_{1k}	0	+1	0
λ_{2k}	0	0	+1

Dummy codes do not meet either of the conditions that define contrast codes. As a result, it takes care to interpret exactly what is examined when one uses such codes to form predictor variables. One might think that a predictor that uses the aforementioned first codes would be comparing the No Feedback group to the other two groups and that a predictor that uses the second code above would be comparing the Success group with the other two groups, but that is not the case. Rather, the regression coefficients for these two coded predictors, if included simultaneously, would each be asking whether the group coded zero on both codes differs from the group coded with a 1 for the predictor that is being examined. So the predictor with the first code would examine the Failure–No Feedback difference and the predictor with the second code would examine the Failure–Success difference.

Because interpretive mistakes can follow from the use of dummy codes unless one is thoroughly familiar with them, we strongly recommend that researchers adopt the contrast-coding convention that we have explicated and that we will use in the remainder of this book.

Contrast Codes With Unequal Cell Sizes

Historically, the procedure of ANOVA to detect mean differences was developed for data from experimental designs in which there were equal numbers of observations in every cell or condition of the design. In this sense, it was developed as an arithmetic shortcut, based on the assumption that predictors would be nonredundant. Of course, with the wide availability of computer programs that permit the estimation of linear regression models with partially redundant predictors, the assumption of nonredundant predictors is no longer necessary and, as we have just shown, ANOVA is easily implemented within general-purpose multiple regression procedures, even with unequal numbers of observations in the various conditions.

In this section we present a new example with four levels of a categorical variable having unequal numbers of observations in each level or category. The bottom line is that, as long as estimation is done with a full set of contrast-coded predictors, all interpretations that we have previously given continue to be applicable, even though with unequal n values those predictors will be partially redundant across observations.

Let us assume that you are in a psychology department of a major U.S. university in which there are four PhD programs to which students are admitted every year: Clinical, Developmental, Experimental, and Social. The number of students who are admitted varies across the programs. Your question is whether there are mean differences in the verbal Graduate Record Examination scores (GREs) of admitted students across the four programs. The data for a given year are presented in Figure 9.13, along with the four group means and the contrast-coded predictors that we will use. We simply use the Helmert coding convention here to derive these codes, as we have no strong expectations about where mean differences might be found.

We create three contrast-coded predictors, X_{1i}, X_{2i}, and X_{3i}, using these codes. We then proceed to estimate a Model A in which these three are used to predict verbal GRE scores:

$$\hat{Y}_i = 680.875 + 13.042X_{1i} - 8.917X_{2i} - 19.250X_{3i} \qquad \text{SSE} = 21{,}595$$

	Program			
	Clinical	*Developmental*	*Experimental*	*Social*
	750	700	640	690
	730	630	660	720
	710	620	710	750
	690		620	670
	670			650
	770			
$\bar{Y}_k$	720	650	657.5	696
n_k	6	3	4	5
λ_{1k}	3	−1	−1	−1
λ_{2k}	0	2	−1	−1
λ_{3k}	0	0	1	−1

Figure 9.13 Hypothetical GRE scores, group means, and contrast codes

The model makes predictions of the category means for all observations, and its parameter estimates can be interpreted as we have done previously because we used a full set of contrast-coded predictors, even though those predictors are now partially redundant because of the unequal numbers of observations (i.e., the tolerances of all three predictors are less than 1.0).

The intercept, 680.875, is the mean of the four category means. It is not the mean of all 19 observations, which equals 687.778. The values of the three slopes equal differences among the category means, according to the Equation 9.1:

$$b_1 = \frac{\sum_k \lambda_{1k}\bar{Y}_k}{\sum_k \lambda_{1k}^2} = \frac{(3)\bar{Y}_C + (-1)\bar{Y}_D + (-1)\bar{Y}_E + (-1)\bar{Y}_S}{12} = \frac{\bar{Y}_C - \left(\frac{\bar{Y}_D + \bar{Y}_E + \bar{Y}_S}{3}\right)}{4} = 13.042$$

$$b_2 = \frac{\sum_k \lambda_{2k}\bar{Y}_k}{\sum_k \lambda_{2k}^2} = \frac{(0)\bar{Y}_C + (2)\bar{Y}_D + (-1)\bar{Y}_E + (-1)\bar{Y}_S}{6} = \frac{\bar{Y}_D - \left(\frac{\bar{Y}_E + \bar{Y}_S}{2}\right)}{3} = -8.917$$

$$b_3 = \frac{\sum_k \lambda_{3k}\bar{Y}_k}{\sum_k \lambda_{3k}^2} = \frac{(0)\bar{Y}_C + (0)\bar{Y}_D + (1)\bar{Y}_E + (-1)\bar{Y}_S}{2} = \frac{\bar{Y}_E - \bar{Y}_S}{2} = -19.250$$

Figure 9.14 presents the source table that results from a comparison of this Model A with four different Model Cs. The first row of the table is a comparison between this Model A and the single-parameter Model C that predicts the grand mean for all observations. The null hypothesis for this comparison is that the four group means are equal. The second row of the table gives the model comparison between this Model A and a Model C that omits the X_{1i} predictor. The null hypothesis here is that β_1 equals zero or, equivalently, that the mean of the Clinical group (C) is equal to the average of the means of the other three groups. The third row of the table gives the model comparison between this Model A and a Model C that omits the X_{2i} predictor. The null hypothesis here is that β_2 equals zero or, equivalently, that the mean of the Developmental group (D) is equal to the average of the means of the Experimental (E) and Social (S) groups. And the fourth row of the table gives the model comparison between this Model A and a Model C that omits the X_{3i} predictor. The null hypothesis here is that β_3 equals zero or, equivalently, that the mean of the Experimental group is equal to the mean of the Social group. All the resulting SSRs for these single-degree-of-freedom comparisons can be expressed in terms of the means, according to Equation 9.2.

In short, all interpretations and computations explicated in this chapter for the analysis of a categorical independent variable apply regardless of whether there are equal numbers of observations across the levels of the categorical variable, as long as (*once again*) a full set of contrast-coded predictors is employed. The only thing that differs in this example from those presented earlier, as a function of the unequal values of n_k, is that the sums of squares for the individual single-degree-of-freedom tests in the previous source table cannot be added up to equal the overall SSR for the model as a whole. In this case, the sum of the SSRs for the single-degree-of-freedom comparisons equals 15,625.89, while the overall test of the model yields an SSR of 14,516. This difference occurs because across observations, the predictors are now somewhat redundant. But all model comparisons and interpretations remain as they have been all along throughout the chapter.

Orthogonal Polynomial Contrast Codes

Sometimes the observations fall into discrete categories on an independent variable of interest even though the underlying variable itself can be thought of as a continuum. For instance, suppose we were interested in age differences among elementary school children in their performance on a standardized arithmetic test. We take children from three different elementary school classes, those in the fourth, fifth, and sixth grades, and give them the standardized test. We then want to know if there are class mean differences. Our conceptual independent variable of interest is the children's age, but what we measure is their year in school and observations are clearly in three distinct categories on this measured variable.

Imagine that we had data from the 14 children given in Figure 9.15. In cases like this one, there is a special set of contrast codes that are sometimes useful for assessing trends in category means. These special codes are really just regular contrast codes, but they have the special name of "orthogonal polynomials" when values on the categorical independent variable can be ordered on some underlying continuum, as they clearly can in this case. In Figure 9.16, we present orthogonal polynomial contrast codes for categorical predictors having up to five levels.

As we indicate there, these codes have names that refer to the trend in the category means, across levels of the categorical independent variable, that they examine. So, with a two-level categorical variable, we can only examine whether the means go up or down, in essence fitting a linear function to the category means. With three levels, we can fit both a linear trend to the three category means and also ask whether the mean of the middle level is higher or lower than it ought

Source	*b*	*SS*	*df*	*MS*	*F*	*PRE*
Model (between groups)		14,516.00	3	4838.70	3.14	.40
X_1 (C vs. D, E, S)	13.04	10,727.00	1	10,727.00	6.95	.33
X_2 (D vs. E, S)	−8.92	1605.00	1	1605.00	1.04	.07
X_3 (E vs. S)	−19.25	3293.89	1	3293.89	2.14	.13
Error		21,595.00	14	1542.5		
Total		36,111.00	17			

Figure 9.14 Summary source table for analysis of GRE scores

	Grade		
	Fourth	*Fifth*	*Sixth*
	68	68	80
	72	75	75
	76	68	78
	65	72	
	70	65	
		80	
		69	
$\bar{Y}_k$	70.20	71.00	77.67
n_k	5	7	3

Figure 9.15 Hypothetical scores on standardized arithmetic test

Trend				*Category*					
				1		2			
Linear				−1		1			
			1		2		3		
Linear			−1		0		1		
Quadratic			−1		2		−1		
		1		2		3		4	
Linear		−3		−1		1		3	
Quadratic		1		−1		−1		1	
Cubic		−1		3		−3		1	
	1		2		3		4		5
Linear	−2		−1		0		1		2
Quadratic	2		−1		−2		−1		2
Cubic	−1		2		0		−2		1
Quartic	1		−4		6		−4		1

Figure 9.16 Orthogonal polynomial contrast codes

to be given a simple linear ordering (the quadratic trend). With four levels, we can fit not only linear and quadratic trends, but also a cubic one, having two bends rather than one. And so forth.

With the three-level categorical variable in our data, let us use the two codes from the orthogonal polynomials to fit the linear and quadratic trends to these data. We create two contrast-coded predictors, X_{1i} and X_{2i}, using the codes specified in Figure 9.16 for a three-level categorical variable (fourth grade is category level 1, etc.). The estimated model, using these to predict the standardized test scores, is

$$\hat{Y}_i = 72.96 + 3.73X_{1i} - .98X_{2i}$$

with a sum of squared errors of 237.47. As always, this model, with a full set of contrast codes, exactly predicts the group means:

$$\bar{Y}_{4th} = 72.96 + 3.73(-1) - 0.98(-1) = 70.20$$
$$\bar{Y}_{5th} = 72.96 + 3.73(0) - 0.98(+2) = 71.00$$
$$\bar{Y}_{6th} = 72.96 + 3.73(+1) - 0.98(-1) = 77.67$$

And as always, the parameter estimates can be interpreted in terms of the group means. The intercept, 72.96, is the mean of the three means, 3.73 is half the difference between $\bar{Y}_{6th}$ and $\bar{Y}_{4th}$ and –0.98 is one-third of the difference between $\bar{Y}_{5th}$ and the average of the other two group means. The source table that summarizes the analysis of these data is given in Figure 9.17.

So far, there is nothing new about this model or its interpretations. So why do we refer to the codes we have used as orthogonal polynomials? The reason is that the slope of the first one is the slope that results if we were to fit a straight line to the three group means, going from the fourth grade to the sixth grade. And the slope of the second contrast-coded predictor estimates the degree to which the group mean for the fifth grade does not lie on that prediction line, that is, the degree to which that prediction function deviates from a straight line if it is to predict all three group means. Given the significance of the coefficient associated with X_{1i}, we can conclude that there is a linear increase in performance on the standardized test as we go from children in the fourth grade to children in the sixth grade.

An obvious question is how these results would differ from what would be obtained if we simply regressed Y_i on Grade itself, treated as a continuous variable, numerically coded as 4, 5, and 6. Such a model would be a simple regression model, asking if there is a linear relationship between Grade and test performance. It is estimated as

$$\hat{Y}_i = 55.62 + 3.38\text{Grade}_i$$

with a sum of squared errors of 268.62. A test of whether there is a linear relationship between grade and performance yields an SSR of 88.31, PRE = .25, and $F = 4.27$ with 1 and 13 degrees of freedom. In this model, the slope associated with the Grade predictor variable informs us about the degree to which predicted performance on the standardized math test increases as grade goes up by one unit, that is, from fourth to fifth and from fifth to sixth.

While these are obviously different models in a variety of ways, the important conceptual difference is that in the one using the orthogonal polynomial contrast codes we are predicting the group means and asking about differences among those group means. And in the simple linear regression model, using Grade as our predictor, we are simply fitting a linear function to all the individual observations, rather than to the group means. Given that the group sizes are very unequal, modeling the group means and modeling the individual observations yield different results.

Type I Error Rates in Testing Mean Differences

Earlier we discussed the general strategy for testing any mean difference that was of interest in designs with multiple levels (m) of a categorical independent variable. One first derives a full set of contrast codes ($m - 1$ of them) and uses them to estimate an augmented model, which predicts the category means. This model provides tests of the specific mean differences that were used as

Source	*b*	*SS*	*df*	*MS*	*F*	*PRE*
Model (between groups)		119.47	2	59.73	3.019	.34
X_1	3.73	104.506	1	104.506	5.28	.31
$X2$	–0.98	31.17	1	31.17	1.57	.12
Error		237.47	12	19.79		
Total		356.93	14			

Figure 9.17 Summary source table for analysis of arithmetic scores

codes for the predictors. Importantly, it also provides the SSE(A) and the mean square error for any model that used a full set of codes, regardless of which set was used. Then, for any additional mean comparison of interest, one calculates the SSR associated with that contrast as if it were used as a predictor in a complete set of orthogonal contrast-coded predictors. Dividing this SSR by the MSE from the estimated model yields the F statistic associated with the mean comparison of interest, and the PRE for that comparison can be calculated as

$$\text{PRE} = \frac{\text{SSR}}{\text{SSE}(\text{A}) + \text{SSR}}$$

When there are more than two or three levels of the categorical variable, the number of potential mean differences that might be tested can become very large. For instance, with only four levels, one could in theory test the means of individual groups against each other, the means of all pairs of groups against each other, the mean of each triad of groups against the mean of the remaining group, and so forth. A problem that arises in this case, where many mean differences might be tested, is that the probability of a Type I statistical error may become unacceptably large. While α may be set at .05 for any one mean comparison, across many such comparisons the probability that somewhere a Type I error has been committed can quickly become quite large. For instance, if we had four groups and we asked whether each mean differed from each other mean, that would be six contrasts tested (in addition to others included in the orthogonal set of codes used initially to generate Model A). If on each test α was set at .05, across the six tests, the probability that we would make at least one Type I error is equal to

$$1 - (1 - .05)^6 = 1 - .95^6 = .265$$

In other words, even if the null hypothesis were true and all the true group means were equal, at least one of the six mean comparisons would be significant more than a quarter of the time. In the ANOVA literature many different procedures have been developed for dealing with this issue. We focus on only two of them, and the crucial difference between these two is whether the mean comparison that is tested is *a planned comparison* or *a post-hoc comparison.*

Planned and Post-Hoc Comparisons

Planned comparisons are those that the researcher had theoretical or substantive reasons for examining *before* conducting the experiment. In other words, the researcher specified all the planned comparisons of interest before collecting or examining the data. Ideally, as many of these planned contrasts as possible would be included in the set of orthogonal contrast-coded predictor used to generate the initial Model A. Post-hoc comparisons are those that do not occur to us until *after* we have examined the data. Often when looking at the data certain comparisons that we did not anticipate appear to be interesting. It is natural to want to test those interesting, unanticipated contrasts. However, when reporting results, we must always clearly identify these *post-hoc* analyses as exploratory.

Sometimes, both planned and post-hoc comparisons require us to estimate more than one Model A in order to test all contrasts of interest. As noted earlier, one potential issue with this approach is that it inflates our Type I error rate. Thus, researchers will often seek to "correct" or adjust the α level to make it more stringent. There are several methods that analysts can use to adjust for multiple comparisons. We discuss three procedures next: *Bonferroni*, *Scheffé*, and *Games-Howell.*

To conduct the Bonferroni test for multiple comparisons, one adds up the number of comparisons that one intends to examine, both in the initial model and in the other models of theoretical interest. Let us say that number is c. To keep α at .05 across all c tests, one wants to compare each obtained F to a critical value of F, using α/c to determine the critical F. So, for instance, if there are six tests to be conducted, one would use a critical value of F at .05/6 = .0083 rather than at .05. One issue with the Bonferroni test, however, is that it is overly conservative. Thus, although Bonferroni tests may decrease Type I error, they are also likely to increase Type II error.

The Scheffé test provides a correction when a researcher wants to test many or all of the possible pairwise comparisons of cell means, which for larger designs can yield a very large number of tests. The Scheffé test involves adjusting the critical F statistic using the formula $(m - 1)\, F_{crit;m-1,n-PA;\alpha}$ to ensure that the overall probability of making at least one Type I error equals α no matter how many contrasts are evaluated. With this approach, there will be at least one contrast whose F exceeds the adjusted critical value only if the omnibus test, comparing Model A to a Model C that predicts the grand mean for all observations, is statistically significant. Though Scheffé is valuable, like Bonferroni, it is very conservative in that it increases Type II error rates.

A flexible and potentially more powerful alternative is the Games-Howell test of multiple comparisons, which is available in many statistical software programs. The Games-Howell test uses a studentized t-distribution and an adjustment to the degrees of freedom to generate a test statistic and p-value. Importantly, it takes into account unequal cell sizes and unequal variances. Thus, the Games-Howell test is more flexible than commonly used corrections such as *Tukey's Honestly Significant Difference (HSD) test*, which assumes equal variances and equal cell sizes. In addition, the Games-Howell test adequately controls Type I error without substantially decreasing power, as is the case with procedures like Bonferroni or Scheffé tests.

Power Analysis for One-Way ANOVA

Estimating Statistical Power

The advantage of consistently adhering to the model comparison approach is that all that we have learned before still applies as we consider new types of models. Thus, the methods for estimating statistical power presented in Chapter 7 for multiple regression apply unaltered to one-way ANOVA. In particular, one can use prior research to estimate potential values of η^2 (the expected proportional reduction in error) for either the omnibus test of any mean differences or specific one-degree-of-freedom contrasts. As before, one should adjust empirical estimates of η^2 based on the number of observations and parameters. This is especially important for experiments in which the number of observations is often small relative to the number of model parameters. Also, as before, one may use values from the literature of a substantive research topic. For example, if we wanted to re-do the SAT coaching study (with which we began this chapter) with a larger number of observations, we could start our power analysis by adjusting the PRE of .196 reported in that study:

$$\hat{\eta}^2 = 1 - (1 - \text{PRE})\left[\frac{n - \text{PC}}{n - \text{PA}}\right] = 1 - (1 - .196)\left[\frac{20 - 1}{20 - 2}\right] = .15$$

A quick check of the power associated with this value suggests that about 50 participants would be required to have an 80% chance of detecting an effect of this magnitude. However, as outlined in

Chapter 5, effect sizes can vary widely across studies. Thus, one would ideally identify a range of potential effect sizes for a power analysis.

Earlier we also saw how estimates of the parameter values—the regression slopes and the variance of the error—can provide estimates of the effect size for power analysis. However, in one-way ANOVA, rather than having prior ideas about the values of regression parameters, a researcher more commonly has notions about the values of the group means that determine the regression slopes. Hence, it is worth examining how we can estimate the expected effect size by beginning with expectations of the cell means. We begin with our usual definition of PRE. We have noted before that SSE(C) = SSE(A) + SSR (i.e., the error for the compact model includes all the error of the augmented model plus the error that was reduced by the addition of the extra parameters in the augmented model). Hence

$$\text{PRE} = \frac{\text{SSR}}{\text{SSE}(\text{C})} = \frac{\text{SSR}}{\text{SSE}(\text{A}) + \text{SSR}} = \frac{1}{\left(\text{SSE}(\text{A}) + \text{SSR}\right) + 1}$$

To obtain an estimate of η^2, we simply calculate SSE(A) and SSR using the true parameter values (β_0, β_1, etc., ε^2) instead of the estimated parameters (b_0, b_1, etc., s^2). For one-way ANOVA, we can start with our expectations about what the true group means, μ_k, might be, and use those in Equation 9.2.

$$\text{SSR for a contrast} = \frac{\left(\sum_k \lambda_k \mu_k\right)^2}{\sum_k \left(\lambda_k^2 / n_k\right)}$$

For a complete model, SSE(A) depends only on σ^2, the within-cell variance. Specifically,

$$\text{SSE}(\text{A}) = \left(\sum_k n_k\right)\sigma^2 = n\sigma^2$$

where n is the grand total of observations. We multiply by n instead of by $(n - \text{PA})$ because we have not estimated any parameters from data in the calculations of SSE(A). Substituting these values for SSR and SSE(A) calculated from the presumed true parameters into the formula for PRE yields the following formula for η^2:

$$\hat{\eta}^2 = \left(\frac{n\sigma^2 \sum_k \left(\lambda_k^2 / n_k\right)}{\left[\sum_k \lambda_k \mu_k\right]^2} + 1\right)^{-1}$$

If n_k, the number of observations in each group, is equal for all groups and if m is the number of groups so that $n = mn_k$, then the above formula reduces to

$$\hat{\eta}^2 = \left(\frac{m\sigma^2 \sum_k \lambda_k^2}{\left[\sum_k \lambda_k \mu_k\right]^2} + 1\right)^{-1}$$

which does not depend on either the total number of observations or the number of observations in each group. Therefore, to find a value for η^2 to use in our power calculations, we need only specify the values for σ^2 and for μ_k that we expect to obtain in our study. And, of course, we need to specify the λ_k values for the contrast code for which we want to estimate the statistical power.

As an example of this direct approach for estimating η^2, let us again consider the feedback study described earlier in this chapter. Suppose the researcher on the basis of prior research had expected values of 3, 4, and 4 for the means of the Failure, No Feedback, and Success groups, respectively, and a within-cell variance of about 1.5. Then for the {–2,1,1} contrast the expected effect size is

$$\hat{\eta}^2 = \left(\frac{3(1.5)6}{\left[(-2)3+4+4\right]^2} + 1 \right)^{-1} = .13$$

From this we estimate that about 60 observations, 20 per group, would be necessary to provide a statistical power of .8.

Power and Research Design

When planning experiments, researchers must choose how many groups to use and how many observations should be in each group. It is important to consider the consequences that such choices have for statistical power. We first consider the design implications for the omnibus test of whether there are any differences among the means and then for the power of specific contrasts.

The statistical power of the omnibus test of whether there are any mean differences among the groups is maximized when there are an equal number of observations in each group. In that case, the omnibus F statistic equals the average of all the F statistics for a set of orthogonal contrasts. This highlights a common mistake in research design: If too many groups are used, then there are many contrasts where no differences are expected, which in turn lowers the omnibus F. Sometimes using too many groups cannot be avoided. For example, biopsychological researchers must sometimes use multiple control groups (e.g., handling the animal, injecting with a drug vehicle, injecting with a placebo) in comparison to a single treatment group. If there are no differences expected among the different control groups, then the possible magnitude of the omnibus F statistic is reduced. In this case, the omnibus F is often misleading, and researchers should simply focus on the treatment versus controls contrast.

Researchers using ordered category levels over which they expect polynomial trends often use too many groups. For example, researchers expecting linear and perhaps quadratic trends sometimes err in using as many as five levels. With equal numbers of observations at each level, the omnibus F is reduced because it is the average of not only the expected linear and quadratic contrasts but also the not-expected cubic and quartic contrasts. Even the power of the separate tests of the linear and quadratic contrasts is reduced because some of the study's valuable resources—the observations—have been allocated to test for the cubic and quartic effects. At the same time, testing more polynomial trends than expected increases the chances of making Type I errors.

For specific contrasts, allocating an equal number of observations to all groups effectively gives equal importance to all contrasts. If one or two contrasts are more important to the research purpose than the other contrasts, one may want to allocate the number of observations unequally across the groups so as to maximize the statistical power of the contrasts of greater importance. Power for a contrast is maximized when the allocation of observations to groups is proportional to the absolute values of the contrast weights. For example, for the t-test where the weights are {–1, +1}, statistical power is maximized with an equal number of observations in each group. However, for the quadratic contrast for three groups with weights $\{-1, +2, -1\}$, power is maximized by allocating a quarter of the observations to each of the extreme categories and the remaining half of the observations to the middle category. The biopsychology researcher comparing the treatment group to four control groups could maximize the power of that comparison by allocating half the total observations to the treatment group and one-eighth of the observations to each of the four control groups.

Summary

In this chapter we have considered models with a single categorical predictor having two or more levels or categories. Such predictors need to be numerically coded, and we have given our strong preference that the codes to be used are orthogonal contrast codes. A full set of such codes means that there are $m - 1$ contrast-coded predictors with m category levels. Additionally, given that the sum of the codes for any predictor equals one, orthogonality is assured by codes where the sum of all pairs of cross products equals zero.

Assuming that a full set of codes is used to predict the data variable, then the predicted values from such a model will equal the category means of that data variable. Accordingly, inferences about slopes are also inferences about difference among category means. And these models are then equivalent to traditional independent sample t-tests (given $m = 2$) and one-way ANOVA models (given $m > 2$). Although one-way ANOVA has traditionally emphasized omnibus tests to examine whether there are any mean differences among the groups or categories, our strong preference is for single-degree-of-freedom model comparisons, testing specific focused contrasts that are of theoretical interest. We encourage analysts to ask about mean differences that are of interest to them, even when those differences are not themselves orthogonal. We discuss appropriate procedures for asking about such nonorthogonal differences, given the constraint that a full set of codes be orthogonal. Finally, we discuss procedures for avoiding inflated α levels when conducting many mean comparisons, whether a priori or post hoc.

10 Factorial ANOVA

Models With Multiple Categorical Predictors and Product Terms

In the previous chapter we developed models with one categorical predictor. In this chapter we expand our consideration to models with two or more categorical predictor variables. Our reasons for wanting to include more than one categorical variable as a predictor in our model are the same as those that motivated us to expand from simple regression models with one predictor to multiple regression models with two or more predictors in Chapter 7. Models in which predictions are conditional on two or more categorical variables may be required by our data and, more importantly, by the underlying process that generates the data. Just as in multiple regression, controlling for one categorical variable by including it in the model often allows us to have a better look at the effects of other categorical variables. Also, as in multiple regression, we are often interested in modeling the joint effect of two or more categorical variables. With categorical variables we will be especially interested in whether the effect of a given categorical variable depends on the levels of the other categorical variables; that is, we are interested in whether or not there is an interaction between the categorical variables analogous to the interactions of continuous variables in multiple regression considered in Chapter 8.

The generalization of models with one categorical predictor (one-way ANOVA of the previous chapter) to models with two categorical predictors (two-way ANOVA) and to models with more than two categorical predictors (q-way factorial ANOVA, where q is the number of categorical predictors) is straightforward. As we shall demonstrate, classical analysis of variance with two or more categorical predictors is nothing more than a simple one-way ANOVA with a specific, clever set of contrast codes. In other words, the only new thing to learn is how to generate the appropriate set of contrast codes; fitting the model and testing hypotheses are *exactly* the same as for one-way ANOVA in Chapter 9.

Factorial Anova as One-Way Anova

We begin by considering the hypothetical dataset in Figure 10.1. In this hypothetical experiment clinically depressed patients either receive psychotherapy (treatment) or not (control) and receive one of three drugs (A, B, or placebo). After six months each patient completes a mood questionnaire on which higher scores mean improved mood or decreased depression.

The experimental design depicted in Figure 10.1 is known as *a factorial* design because every level of one categorical variable or factor is combined with every level of the other categorical variable or factor. There are three levels of the Drug variable and two levels of the Psychotherapy variable, so there are a total of $3 \times 2 = 6$ different combinations, each defining a group or cell in the design. We often refer to this as a 3×2 *design.* In this hypothetical study, from a total of 18 patients, three patients are randomly assigned to each of the six groups.

DOI: 10.4324/9781003438335-15

	Psychotherapy					
Drug	*Treatment*			*Control*		
A	31	31	34	17	15	19
B	25	25	28	23	18	16
Placebo	17	18	16	9	10	8

Figure 10.1 Hypothetical data (mood scores) in a Drug (3) by Psychotherapy (2) experimental design

	Group 1 (A,T)	*Group 2 (B,T)*	*Group 3 (P,T)*	*Group 4 (A,C)*	*Group 5 (B,C)*	*Group 6 (P,C)*
	31	25	17	17	23	9
	31	25	18	15	18	10
	34	28	16	19	16	8
Mean	32	26	17	17	19	9

			Group					
Contrast codes		*Predictor*	*1*	*2*	*3*	*4*	*5*	*6*
λ_1	1 vs. 2, 3, 4, 5, 6	Z_1	5	−1	−1	−1	−1	−1
λ_2	2 vs. 3, 4, 5, 6	Z_2	0	4	−1	−1	−1	−1
λ_3	3 vs. 4, 5, 6	Z_3	0	0	3	−1	−1	−1
λ_4	4 vs. 5, 6	Z_4	0	0	0	2	−1	−1
λ_5	5 vs. 6	Z_5	0	0	0	0	1	−1

Figure 10.2 Hypothetical data of Figure 10.1 arrayed as a one-way design

Although it is natural to display these data in a table with three rows and two columns as in Figure 10.1, we can also display the data as a one-way layout in terms of the six groups or cells as in Figure 10.2. Seeing the data in the one-way layout makes it clear that we can use any of the sets of contrast codes developed in the previous chapter to analyze these data. To illustrate this, we will first do an analysis with one-way contrast codes that are statistically correct but that do not ask questions that are usually interesting. In this first set of contrast codes we form the first contrast λ_1 by comparing Group 1 (Drug A combined with psychotherapy Treatment: A,T) with all the other groups; the corresponding contrast-coded predictor is Z_1.[1] The contrast λ_2 compares Group 2 against all the remaining groups except Group 1, and so on. These contrast codes are also displayed in Figure 10.2. We can verify that these codes are orthogonal by checking that the sum of each set of cross products is zero. For example,

$$\sum_{k=1}^{6} \lambda_{1k}\lambda_{2k} = 5(0) - 1(4) - 1(-1) - (-1) - (-1) = 0 - 4 + 1 + 1 + 1 + 1 = 0$$

We can regress Y_i, the mood scores, on the predictor variables $Z_1, Z_2, \ldots, Z_5$ using a standard multiple regression program or, equivalently, we could use the one-way ANOVA formulas from Chapter 9. Figure 10.3 shows the actual data matrix we would use to regress Y_i on $Z_1, Z_2, \ldots, Z_5$.

Note that the mean for each contrast-coded predictor is zero because there are equal numbers of observations in each group.

We can ask whether all six group means are equal by comparing these models:

MODEL A: $Y_i = \beta_0 + \beta_1 Z_{1i} + \beta_2 Z_{2i} + \beta_3 Z_{3i} + \beta_4 Z_{4i} + \beta_5 Z_{5i} + \varepsilon_i$
MODEL C: $Y_i = \beta_0 + \varepsilon_i$
$H_0: \mu_1 = \mu_2 = \mu_3 = \mu_4 = \mu_5$

Regressing the mood scores on the five contrast-coded predictors yields the analysis in Figure 10.4. The two estimated models are

MODEL A: $\hat{Y}_i = 20 + 2.4Z_{1i} + 2.1Z_{2i} + 0.5Z_{3i} + 1Z_{4i} + 5Z_{5i}$
MODEL C: $\hat{Y}_i = 20$

The augmented model using all five predictors reduces the error of the compact model using the mean by 960/1010 = .9505. We calculate F in the usual way:

$$F_{5,12} = \frac{.9505/5}{(1-.9505)/12} = 46.1$$

This omnibus tests rejects Model C in favor of Model A because the large value of PRE (.95) is statistically surprising ($F_{5,12} = 46.1$, $p < .0001$), which implies, as it did in one-way ANOVA, that

Group		*Mood scores*	*Predictors*				
Drug	*Psychotherapy*	*Y*	Z_1	Z_2	Z_3	Z_4	Z_5
A	T	31	5	0	0	0	0
A	T	31	5	0	0	0	0
A	T	34	5	0	0	0	0
B	T	25	−1	4	0	0	0
B	T	25	−1	4	0	0	0
B	T	28	−1	4	0	0	0
P	T	17	−1	−1	3	0	0
P	T	18	−1	−1	3	0	0
P	T	16	−1	−1	3	0	0
A	C	17	−1	−1	−1	2	0
A	C	15	−1	−1	−1	2	0
A	C	19	−1	−1	−1	2	0
B	C	23	−1	−1	−1	−1	1
B	C	18	−1	−1	−1	−1	1
B	C	16	−1	−1	−1	−1	1
P	C	9	−1	−1	−1	−1	−1
P	C	10	−1	−1	−1	−1	−1
P	C	8	−1	−1	−1	−1	−1
Mean		20	0	0	0	0	0

Figure 10.3 Data matrix for analyzing mood scores using multiple regression and one-way coded predictors

Source	b_j	*SS*	*df*	*MS*	*F*	*p*	*PRE*
Between groups		960.0	5	192.0	46.1	.0001	.95
Z_1	2.4	518.4	1	518.4	124.4	.0001	.91
Z_2	2.1	264.6	1	264.6	63.5	.0001	.84
Z_3	0.5	9.0	1	9.0	2.2	.17	.15
Z_4	1.0	18.0	1	18.0	4.3	.06	.26
Z_5	5.0	150.0	1	150.0	36.0	.0001	.75
Within groups (MSE)		50.0	12	4.2			
Total		1010.0	17				

Figure 10.4 ANOVA table of the hypothetical mood scores using the model: $Y_i = \beta_0 + \beta_1 Z_{1i} + \beta_2 Z_{2i} + \beta_3 Z_{3i} + \beta_4 Z_{4i} + \beta_5 Z_{5i} + \varepsilon_i$

the means for the six groups are not all equal to one another. That is, the six groups cannot be adequately represented by the overall mean.

But which groups are different from each other? The test of the Z_1 contrast indicates that the estimate $b_1 = 2.4$ is significantly different from zero and it reduces 91% of the error remaining after the effects of all the other codes have been taken into account.[2] Remember that this test is obtained by comparing an augmented model including all the Z values with a compact model including all the Z values except Z_1. In particular,

MODEL A: $Y_i = \beta_0 + \beta_1 Z_{1i} + \beta_2 Z_{2i} + \beta_3 Z_{3i} + \beta_4 Z_{4i} + \beta_5 Z_{5i} + \varepsilon_i$
MODEL C: $Y_i = \beta_0 + \beta_2 Z_{2i} + \beta_3 Z_{3i} + \beta_4 Z_{4i} + \beta_5 Z_{5i} + \varepsilon_i$
$H_0: \beta_1 = 0$

In this case, we can reject Model C in favor of Model A and therefore reject the null hypothesis that $\beta_1 = 0$. The corresponding contrast λ_1 compares Group 1 (Drug A with psychotherapy treatment) to the average of all the other groups; thus, we can conclude that Group 1 is significantly different from the average of all other cells in the study.

We interpret the value of the coefficient the same as we always have. Specifically, the coefficient $b_1 = 2.4$ means that for a one-unit increase in Z_1 the predicted mood score increases, on average, by 2.4 points. To verify this, re-examine Figure 10.2 to see that the value of Z_1 for all the other groups is –1 but for Group 1 the value of Z_1 is 5. Thus, there is a change of six units on Z_1 between the two comparison groups. Model A therefore predicts the mean for Group 1 to be $6 \times 2.4 = 14.4$ units higher than the average in all the other groups. This is indeed the case: The mean for Group 1 is 32 and the mean for all the other groups combined is $(26 + 17 + 17 + 19 + 9)/5 = 17.6$, thus 32–17.6 = 14.4. But why is Group 1 better, on average, than all the other groups? Is it because of Drug A? Or is it because patients receiving psychotherapy score higher regardless of which drug they receive? Or is it because Drug A is especially effective for patients who also receive psychotherapy treatment? The contrast λ_1 cannot tell us. It simply indicates that there is a difference.

We can similarly interpret the results for the contrast-coded predictor Z_2. That the estimate $b_2 = 2.1$ is significantly different from zero indicates that Group 2 (Drug B administered to those receiving psychotherapy treatment) is different from the average of all the subsequent groups (i.e., excluding Group 1). The difference in means equals $5 \times 2.1 = 10.5$. But again we do not know whether the higher average mood scores of Group 2 are due to Drug B or psychotherapy treatment or their combination.

The other contrast-coded predictor whose coefficient is significantly different from zero (using $\alpha = .05$) is Z_5; the corresponding contrast λ_5 compares Groups 5 and 6 or the difference between taking Drug B versus placebo for those in the Control condition. We know in this case that the higher average score for Group 5 ($2 \times 5 = 10$) is due to Drug B relative to the placebo because all patients in this comparison are in the control group not receiving psychotherapy. But is Drug B also better than the placebo for those patients who do receive psychotherapy treatment? None of the Z contrasts help to answer that question.

A Better Set of Contrast Codes

Even though the one-way analysis of variance using the Z contrast-coded predictors is statistically correct, it fails to answer important questions we want to ask of the data. One solution would be to use the multiple comparison procedures of Chapter 9 to address some of the unanswered questions. However, a much more efficient strategy is to begin with a set of contrast codes that do ask many of the questions we would naturally want to consider. Given the two-way layout of the data (as in Figure 10.1), there are several sets of codes that are more natural than the Z codes. "Two-way" ANOVA is nothing more than one-way ANOVA using one of these natural sets of codes.

To develop a more natural set of codes, we begin by considering each of the two categorical variables separately. The strategy is to develop contrast codes identical to the ones that we would use if each categorical variable were considered alone in its own one-way ANOVA. That is, our first step is to code each categorical variable as if the other one did not exist. For the Drug variable an interesting question is whether the drugs (either A or B), on average, do better than the placebo. The contrast code λ_1 (and its corresponding contrast-coded predictor X_1) in Figure 10.5 makes precisely that comparison with the pattern (1, 1, –2). This pattern is repeated for each level of the other categorical variable (in this case, Psychotherapy). The contrast code λ_2 (and its corresponding contrast-coded predictor X_2) then asks whether there is any difference between Drugs A and

	Group 1 *(A,T)*	*Group 2* *(B,T)*	*Group 3* *(P,T)*	*Group 4* *(A,C)*	*Group 5* *(B,C)*	*Group 6* *(P,C)*
	31	25	17	17	23	9
	31	25	18	15	18	10
	34	28	16	19	16	8
Mean	32	26	17	17	19	9

Contrast codes		*Group* *Predictor*	*1*	*2*	*3*	*4*	*5*	*6*
λ_1	Drugs vs. Placebo	X_1	1	1	–2	1	1	–2
λ_2	Drug A vs. Drug B	X_2	1	–1	0	1	–1	0
λ_3	Treatment vs. Control	X_3	1	1	1	–1	–1	–1
λ_4	Interaction: $\lambda 1 \times \lambda 3$	X_4	1	1	–2	–1	–1	2
λ_5	Interaction: $\lambda 2 \times \lambda 3$	X_5	1	–1	0	–1	1	0

Figure 10.5 Hypothetical data of Figure 10.1 coded for two-way ANOVA

B. We can verify that λ_1 and λ_2 are orthogonal by checking whether the sum of their crossproducts equals zero. That is

$$\sum_{k=1}^{6} \lambda_{1k}\lambda_{2k} = 1(1)+1(-1)-2(0)+1(1)+1(-1)-2(0) = 1-1+0+1-1+0 = 0$$

The orthogonality combined with the equal numbers of observations in each cell ensures that the coded predictors X_1 and X_2 are uncorrelated or not redundant. With three levels of the Drug categorical variable, we can only have two orthogonal codes so λ_1 and λ_2 are sufficient for the one-way analysis of that variable.

There are only two levels of the Psychotherapy categorical variable, so we need only one code. Hence, λ_3 codes the contrast between receiving psychotherapy treatment versus being in the control condition and provides the complete one-way ANOVA for that variable.

So far we have only three codes, λ_1, λ_2, and λ_3, for the two separate one-way analyses of the categorical variables, but five orthogonal contrast codes are required for the complete analysis of six groups. To generate the other two necessary codes, we simply multiply the contrast codes between the two categorical variables; that is, $\lambda_4 = \lambda_1 \times \lambda_3$ and $\lambda_5 = \lambda_2 \times \lambda_3$. These are the same kind of product terms we considered in Chapter 8 when we introduced the concept of interactions between variables. In the context of two-way ANOVA these product terms ask especially interesting questions. The code λ_4, the product of λ_1 and λ_3, asks whether the difference coded by λ_1 (Drugs versus Placebo) depends on the level of psychotherapy (Treatment versus Control). In other words, might there be one drug effect for those receiving psychotherapy treatment and a different drug effect for those in the control group. Similarly, λ_5, the product of λ_2 and λ_3, asks whether the comparison coded by λ_2 (Drug A versus Drug B) depends on the level of Psychotherapy (Treatment versus Control). The order of multiplication is arbitrary, so either statement could be reversed. That is, λ_4 asks whether the effectiveness of psychotherapy (the Treatment versus Control difference) depends on whether the patient also received a drug or the placebo. We consider the interpretation of interactions like these in greater detail later.

Note that we do not form a code by multiplying $\lambda_1 \times \lambda_2$ because that would yield another code for just the Drug categorical variable that could not be orthogonal to the other two codes for that variable. However, the two codes formed from products, λ_4 and λ_5, are orthogonal to each other and to the other codes. You may want to compute some of the crossproducts to verify this claim. This gives a total of five orthogonal contrast codes, precisely the number we need for the analysis of six groups.

Figure 10.6 shows the data matrix that we can analyze with a multiple regression program to model mood scores in terms of the X variables. That is, we can regress Y_i on X_1 to X_5. Note again that the mean for each predictor is zero because we are using contrast codes and there are an equal number of observations for each group.

Figure 10.7 presents the results of the regression analysis using the X contrast-coded predictors. Note that the sum of squares for the augmented model including all the predictors, and the F and PRE for the omnibus test of the complete model are *exactly* the same as when we used the Z predictors. This must be the case because in each analysis we used a complete set of codes so that, as in one-way ANOVA, the predicted value for each group is the mean Y for that group. If the predictions of the two models are the same, then the total error and the total error reduced must be the same. The only difference in the two analyses is how that total error reduction is divided into separate components. The X predictors ask different questions than the Z predictors, so the individual error reduction associated with each of the X predictors differs from that of the Z predictors, but the total error reduction must be the same.

Group		*Mood scores*	*Predictors*				
Drug	*Psychotherapy*	*Y*	X_1	X_2	X_3	X_4	X_5
A	T	31	1	1	1	1	1
A	T	31	1	1	1	1	1
A	T	34	1	1	1	1	1
B	T	25	1	−1	1	1	−1
B	T	25	1	−1	1	1	−1
B	T	28	1	−1	1	1	−1
P	T	17	−2	0	1	−2	0
P	T	18	−2	0	1	−2	0
P	T	16	−2	0	1	−2	0
A	C	17	1	1	−1	−1	−1
A	C	15	1	1	−1	−1	−1
A	C	19	1	1	−1	−1	−1
B	C	23	1	−1	−1	−1	1
B	C	18	1	−1	−1	−1	1
B	C	16	1	−1	−1	−1	1
P	C	9	−2	0	−1	2	0
P	C	10	−2	0	−1	2	0
P	C	8	−2	0	−1	2	0
Mean		20	0	0	0	0	0

Figure 10.6 Data matrix for analyzing mood scores using multiple regression and two-way coded predictors

Source	b_j	*SS*	*df*	*MS*	*F*	*p*	*PRE*
Between groups		960	5	192.0	46.1	.0001	.95
Drug		453	2	226.5	53.9	.0001	.90
X_1	3.5	441	1	441	105.8	.0001	.90
X_2	1.0	12	1	12	2.9	.11	.19
Psychotherapy		450	1	450	108.0	.0001	.90
X_3	5.0	450	1	450	108.0	.0001	.90
Drug x Psychotherapy		57	2	28.5	6.8	.01	.53
X_4	0.5	9	1	9	2.2	.16	.15
X_5	2.0	48	1	48	11.5	.005	.49
Within groups (MSE)		50.0	12	4.2			
Total		1010.0	17				

Figure 10.7 ANOVA table of the hypothetical mood scores using the model: $Y_i = \beta_0 + \beta_1 X_{1i} + \beta_2 X_{2i} + \beta_3 X_{3i} + \beta_4 X_{4i} + \beta_5 X_{5i} + \varepsilon_i$

There are nine PRE and F values in Figure 10.7. Each one corresponds to a comparison between a particular Model C and a Model A. To understand the meaning of each test, it is important to be precise about those models for each test. The omnibus test is reported in the row labeled "Between groups" and represents the comparison between the following two models:

MODEL A: $Y_i = \beta_0 + \beta_1 X_{1i} + \beta_2 X_{2i} + \beta_3 X_{3i} + \beta_4 X_{4i} + \beta_5 X_{5i} + \varepsilon_i$
MODEL C: $Y_i = \beta_0 + \varepsilon_i$
$H_0: \beta_1 = \beta_2 = \beta_3 = \beta_4 = \beta_5 = 0$

This corresponding null hypothesis, equivalent to assuming that all the group means are equal, is rejected by the large values of PRE and F.

The rows for each of the X_j compare a Model A that uses all the contrast-coded predictors to a Model C that includes all the predictors except X_j. For example, the PRE and F in the row labeled X_1 compare

$$\text{MODEL A: } Y_i = \beta_0 + \beta_1 X_{1i} + \beta_2 X_{2i} + \beta_3 X_{3i} + \beta_4 X_{4i} + \beta_5 X_{5i} + \varepsilon_i$$
$$\text{MODEL C: } Y_i = \beta_0 + \beta_2 X_{2i} + \beta_3 X_{3i} + \beta_4 X_{4i} + \beta_5 X_{5i} + \varepsilon_i$$
$$H_0: \beta_1 = 0$$

We generated the predictors X_1 and X_2 as one-way codes for the Drug categorical variable for the data matrix in Figure 10.1. The row labeled "Drug" in Figure 10.7 reports the results from testing a model including all the contrast codes to one that omits *both* the drug codes X_1 and X_2. That is, we are comparing

$$\text{MODEL A: } Y_i = \beta_0 + \beta_1 X_{1i} + \beta_2 X_{2i} + \beta_3 X_{3i} + \beta_4 X_{4i} + \beta_5 X_{5i} + \varepsilon_i$$
$$\text{MODEL C: } Y_i = \beta_0 + \beta_3 X_{3i} + \beta_4 X_{4i} + \beta_5 X_{5i} + \varepsilon_i$$
$$H_0: \beta_1 = \beta_2 = 0$$

In other words, we are asking whether the predictions would suffer if we were to ignore the Drug categorical variable. The large PRE and F indicate that we cannot ignore which drug a patient received. In the traditional language of ANOVA, this is known as the test of the *main effect* of the Drug categorical variable. In this case, we would conclude that there is a statistically significant main effect for the Drug categorical variable. Note that because X_1 and X_2 are uncorrelated in this instance, their sums of squares add to produce the sum of squares attributable to both of them.

We prefer the single-degree-of-freedom tests of the focused comparisons X_1 and X_2 to this global test of the drug effect because rejection of the hypothesis $\beta_1 = \beta_2 = 0$ is ambiguous. We do not know which part of this multiple-degree-of-freedom hypothesis is at fault. Maybe the global hypothesis is rejected because $\beta_1 \neq 0$ or because $\beta_2 \neq 0$ or because both β_1 and β_2 are not equal to zero. We present the omnibus test for the Drug variable in Figure 10.7 not as a recommended practice, but only to show what hypothesis is being tested by the "main effect" in the traditional approach to analysis of variance. Unfortunately, in the traditional approach usually *only* the omnibus tests in Figure 10.7 are presented in a source table. Doing so ignores readily available detailed information about the data revealed by the single-degree-of-freedom contrast codes.

We can similarly assess the main effect for the Psychotherapy variable. In this case there is only one code so that test is the same as the test of the null hypothesis that $\beta_3 = 0$. Finally, we can group together the two codes X_4 and X_5, which we constructed from products of other codes. Each of these two products involved one code for the Drug categorical variable and one code for the Psychotherapy categorical variables. So, in the traditional language of ANOVA, they are known together as the "Drug × Psychotherapy Interaction." Testing the Drug × Psychotherapy interaction is equivalent to comparing

$$\text{MODEL A: } Y_i = \beta_0 + \beta_1 X_{1i} + \beta_2 X_{2i} + \beta_3 X_{3i} + \beta_4 X_{4i} + \beta_5 X_{5i} + \varepsilon_i$$
$$\text{MODEL C: } Y_i = \beta_0 + \beta_1 X_{1i} + \beta_2 X_{2i} + \beta_3 X_{3i} + \varepsilon_i$$
$$H_0: \beta_4 = \beta_5 = 0$$

A traditional ANOVA source table similar to the one in Figure 10.7 would contain rows only for Drug, Psychotherapy, Drug × Psychotherapy, Within groups (or Error), and Total. Such a source table, because it aggregates individual contrasts into more global effects, omits information that is readily available in the individual contrasts. In other words, the traditional ANOVA table fails to analyze the variance as much as is possible using the focused comparisons of contrast codes.

Any standard multiple regression program can produce the analysis in Figure 10.7. Equivalently, one can use Equations 9.1 and 9.2 from Chapter 9 expressing the parameter estimates and their corresponding sums of squares as a function of the contrast codes and the cell means. As an illustration, we use the formulas to calculate the estimate b_1 and its associated SSR. The coefficient that estimates β_1 is

$$b_1 = \frac{\sum_k \lambda_{1k} \bar{Y}_k}{\sum_k \lambda_k^2} = \frac{1(32)+1(26)-2(17)+1(17)+1(19)-2(9)}{1^2+1^2+(-2)^2+1^2+1^2+(-2)^2} = \frac{42}{12} = 3.5$$

Then the sum of squared error reduced by including X_1 in the model along with all the other predictors is

$$\text{SSR}_{b1} = \frac{\left(\sum_k \lambda_{1k} \bar{Y}_k\right)^2}{\sum_k \lambda_{1k}^2 / n_k} = \frac{\left[1(32)+1(26)-2(17)+1(17)+1(19)-2(9)\right]^2}{\left[1^2+1^2+(-2)^2+1^2+1^2+(-2)^2\right]/3} = \frac{42^2}{12/3} = 441$$

Note that because two-way ANOVA is exactly the same as a one-way ANOVA with a clever set of codes, the formulas and all other details from Chapter 9 apply unaltered to two-way ANOVA. Thus, in terms of parameter estimates and statistical tests there is nothing new that needs to be learned to do two-way ANOVA—it is *exactly* the same as one-way ANOVA.

Interpretation of Coefficients

The one new difficulty introduced by the generalization of one-way ANOVA to two-way and higher ANOVA is the interpretation of the coefficients for the predictor variables that code the interactions—the products between one-way codes. But it is simply a challenge of interpretation, not a challenge that involves statistical procedures. And the strategies for interpreting coefficients in two-way and higher ANOVA are the same as the strategies we developed in Chapters 8 and 9. Newcomers to ANOVA, whether using our approach or more traditional approaches, often have initial difficulty understanding the concept of an interaction. We therefore devote considerable attention in this section to the interpretation of interactions. We do so by considering two different approaches to interpreting the coefficients in a model with two-way codes such as those in Figure 10.5. Those two approaches are (a) re-expression of the codes in terms of the cell means, as was done in Chapter 9, and (b) interpretation of product predictors as changes in simple slopes, as was done in Chapter 8. In each approach we also consider the interpretation of the coefficients for predictors not involving products because interactions are best understood in comparison.

Interpretation in Terms of Cell Means

The interpretation of the parameter estimates associated with the X predictors proceeds in the same manner as for any set of predictors in the one-way analyses of Chapter 8. The surprisingly large values, relative to Model C being correct, of $F_{1,12} = 105.8$ and PRE = .95 for X_1 lead us to reject the null hypothesis that $\beta_1 = 0$. To see exactly what the rejection of this null hypothesis means, we express β_1 in terms of the contrast code using Equation 9.1, which expresses the parameter estimate as a function of the codes and the cell means. That is, an equivalent statement of the null hypothesis $\beta_1 = 0$ is

$$\beta_1 = \frac{\sum_k \lambda_{1k}\mu_k}{\sum_k \lambda_{1k}^2} = \frac{\mu_{AT} + \mu_{BT} - 2\mu_{PT} + \mu_{AC} + \mu_{BC} - 2\mu_{PC}}{12} = 0$$

where μ_{AT} represents the true but unknown mean for the group that received Drug A and participated in Psychotherapy treatment, etc. Doing a little algebra (multiplying by 12 and moving the terms for the placebo conditions to the right side of the equation) yields the following equivalent statement of the hypothesis:

$$\mu_{AT} + \mu_{AC} + \mu_{BT} + \mu_{BC} = 2(\mu_{PT} + \mu_{PC})$$

Dividing each side of this equation by 2 gives

$$\frac{\mu_{AT} + \mu_{AC}}{2} + \frac{\mu_{BT} + \mu_{BC}}{2} = 2\left(\frac{\mu_{PT} + \mu_{PC}}{2}\right)$$

If we let $\mu_{A.} = (\mu_{AT} + \mu_{AC})/2$, which is the mean of the Drug A conditions averaged (indicated by the dot) across both the Treatment and Control levels of the Psychotherapy variable, $\mu_{B.} = (\mu_{BT} + \mu_{BC})/2$ and $\mu_{P.} = (\mu_{PT} + \mu_{PC})/2$, then the null hypothesis reduces to

$$H_0: \mu_{A.} + \mu_{B.} = 2\mu_{P.}$$

Dividing each side of this equation by 2 gives

$$\frac{\mu_{A.} + \mu_{B.}}{2} = \mu_{P.}$$

That is, concluding that $\beta_1 \neq 0$ is equivalent to concluding that the average of the drug conditions combined is not equal to the average for the placebo groups, ignoring whether the patients received psychotherapy treatment or not. Thus, unlike the ambiguous conclusions using the Z contrast codes, rejection of the null hypothesis for β_1 clearly implies that the mood scores for the drug groups, on average, differed from the mood scores of the placebo groups. With experience it is usually easy to go directly from the contrast codes to a statement of the null hypothesis without doing the formal derivation as earlier. For example, the (1, –1, 0) code for X_2 compares the Drug A groups with the Drug B groups so that the null hypothesis $\beta_2 = 0$ is equivalent to the following null hypothesis:

$$H_0: \mu_{A.} - \mu_{B.} = 0 \text{ or } H_0: \mu_{A.} = \mu_{B.}$$

The direction and magnitude of a difference revealed by rejecting the null hypothesis is given by b_j, the estimate of β_j. For X_1, this estimate is $b_1 = 3.5$. As with all regression coefficients, this means that our model prediction $\hat{Y}_i$ increases, on average, by 3.5 units for each unit increase in X_1. The change on X_1 from the placebo groups ($X_1 = -2$) to the drug groups ($X_1 = 1$) is a change of 3 units, so the predicted difference between the placebo and drug groups is $3 \times 3.5 = 10.5$. Indeed, the average of the four drug groups, $(32 + 26 + 17 + 19)/4 = 23.5$, exceeds by 10.5 the average of the two placebo groups, $(17 + 9)/2 = 13$. Figure 10.8(a) depicts graphically the means compared by the first contrast.

We cannot conclude in this analysis that the estimate $b_2 = 1.0$ for β_2 is significantly different from zero. Nevertheless, it is useful to interpret it to see which means are not significantly different from one another. On X_2 there is a two-unit difference between those taking Drug B and those taking Drug A, regardless of whether they were also receiving psychotherapy treatment or not. Thus, the mood scores of those taking Drug A are $2 \times 1 = 2$ points higher than the mood scores of those taking Drug B, but this difference is not statistically significant. Figure 10.8(b) depicts this comparison graphically.

The coefficient $b_3 = 5.0$ is significantly different from zero. The predictor X_3 codes the difference between the two psychotherapy groups, so we can conclude that, on average, there is a significant difference between the mood scores of those receiving treatment compared to those in the control group. That is, we can reject the null hypothesis

$$H_0\text{: } \mu_{T\cdot} = \mu_{C\cdot}$$

where the dots in the subscript indicate that we averaged across levels of drug. To be specific about the amount of difference in mood scores between receiving psychotherapy versus not, we need to interpret the meaning of the estimate $b_3 = 5.0$ as before. There is a two-unit difference on X_3 between receiving Treatment ($X_3 = 1$) and receiving the Control ($X_3 = -1$). Thus, the predicted average difference between the two groups is $2 \times 5 = 10$. Indeed, the average of the three groups receiving Treatment $(32 + 26 + 17)/3 = 25$, exceeds by 10 the average of the three Control groups $(17 + 19 + 9)/3 = 15$. Figure 10.8(c) depicts these means and their comparison.

Even though the coefficient for $b_4 = 0.5$ is not significantly different from zero, its interpretation provides a good introduction to understanding interactions: products of contrast codes. We begin by considering the full expression for β_4 in terms of the unknown means:

$$\beta_4 = \frac{\sum_k \lambda_{4k}\mu_k}{\sum_k \lambda_{4k}^2} = \frac{\mu_{AT} + \mu_{BT} - 2\mu_{PT} - \mu_{AC} - \mu_{BC} + 2\mu_{PC}}{12} = 0$$

Except for the signs on the last three terms, this is very similar to the expression for β_1 earlier. Putting the T and C terms on opposite sides of the equation gives

$$\frac{\mu_{AT} + \mu_{BT} - 2\mu_{PT}}{12} = \frac{\mu_{AC} + \mu_{BC} - 2\mu_{PC}}{12}$$

Dividing the numerators and denominators on both sides by 2 yields

$$\frac{1}{6}\left[\frac{\mu_{AT} + \mu_{BT}}{2} - \mu_{PT}\right] = \frac{1}{6}\left[\frac{\mu_{AC} + \mu_{BC}}{2} - \mu_{PC}\right]$$

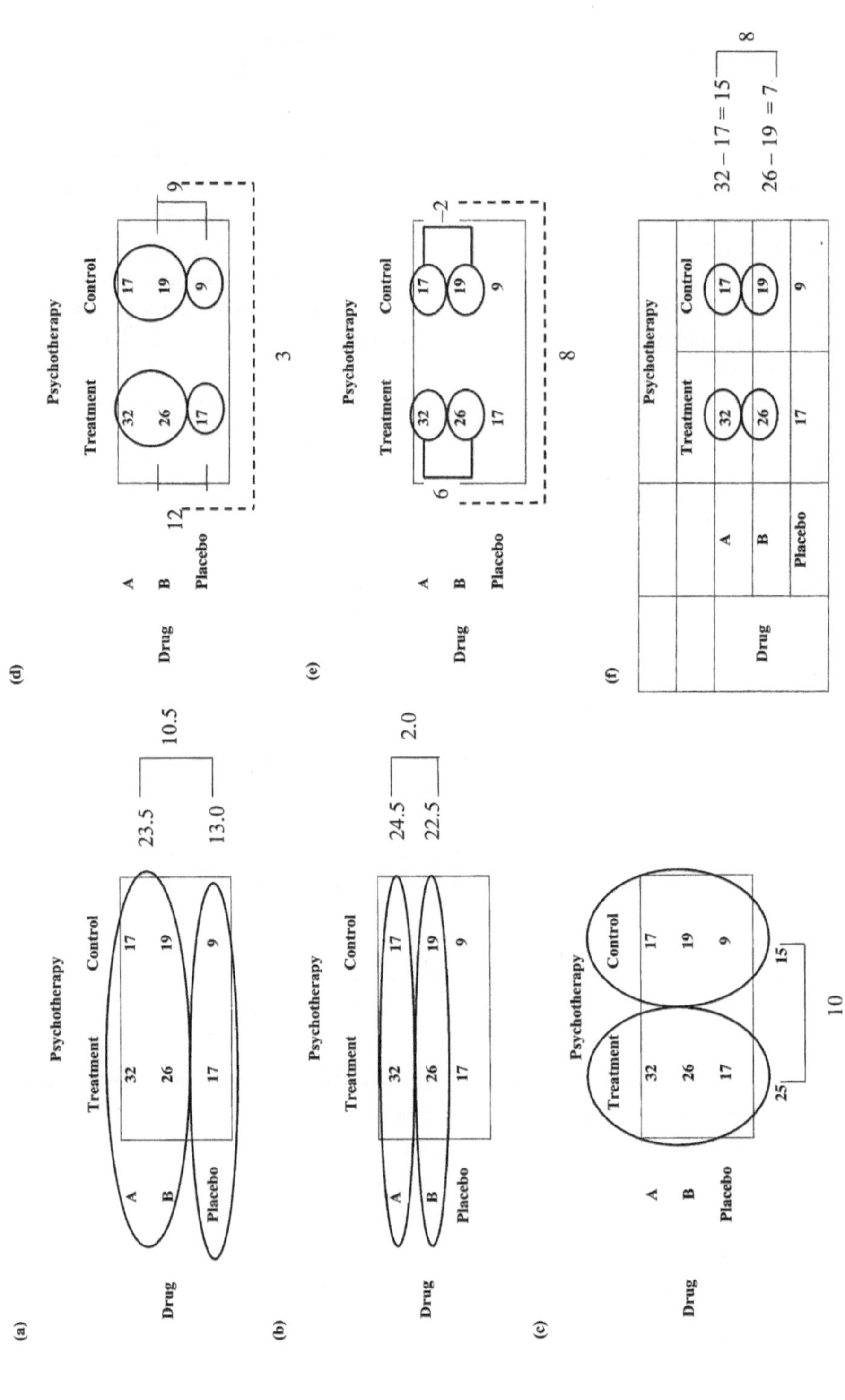

Figure 10.8 Means compared by contrast codes

Note that the terms in each of the two brackets are identical to the question asked by the first code: Is there a difference in mood scores between those taking either drug compared to those taking the placebo? The first coded predictor X_1 asked whether there was a difference regardless of whether the patient was also receiving psychotherapy treatment or not. The interaction code $X_4 = X_1 \times X_3$ asks whether that difference, estimated for those receiving psychotherapy treatment:

$$\frac{\mu_{AT} + \mu_{BT}}{2} - \mu_{PT}$$

is the same as the difference estimated for those in the control condition:

$$\frac{\mu_{AC} + \mu_{BC}}{2} - \mu_{PC}$$

In other words, the null hypothesis for this and all interactions is whether two differences are equal. Different differences imply an interaction. Figure 10.8(d) depicts the means and differences involved in this comparison. For the Treatment groups, the means for those receiving a drug (either A or B) are 32 and 26 (the top-left oval) and their mean of 29 differs from the mean of the Treatment group taking the placebo (17) by 12 points. For the Control group, the mean of those taking either Drug A or Drug B (the top-right oval) is (17 + 19)/2 = 18, which exceeds the mean of those taking the placebo (9) by 9 points. The statistical test for β_4 is therefore equivalent to asking whether these two differences—12 for the Treatment group and 9 for the Control group—are significantly different from each other. Or, equivalently, whether the difference of the differences (12–9 = 3) is significantly different from zero. In this case, because we could not reject the null hypothesis that $\beta_4 = 0$, we also cannot reject the null hypothesis that the Drug versus Placebo difference is the same for the Treatment group as it is for the Control group. Note that the aforementioned algebra shows that the estimated coefficient $b_4 = 0.5$ equals one-sixth of the difference of the differences. Indeed, $0.5 \times 6 = 3$, which is the difference of the differences depicted in Figure 10.8(d).

The algebra to determine which differences are being compared in an interaction may seem complicated at first, but again with experience it is usually easy to jump from the code to a specification of the differences being compared. For example, $X_5 = X_2 \times X_3$ so the null hypothesis

$$H_0: \beta_5 = 0$$

is equivalent to testing whether the difference coded by X_2—the difference between those taking Drug A versus Drug B—depends on the level coded by X_3, whether the patient is in the Treatment or Control groups. Thus, the equivalent null hypothesis is

$$H_0: \mu_{AT} - \mu_{BT} = \mu_{AC} - \mu_{BC}$$

In words, the null hypothesis is that the difference between Drugs A and B is the same for those receiving Psychotherapy treatment as it is for those in the Control group. Knowing that $b_5 = 2$ is significantly different from zero allows us to conclude that the drug effect is different for those receiving treatment than for those in the control group. Figure 10.8(e) depicts this difference and shows its magnitude. For those receiving Treatment, the difference in mood scores for those taking Drug A compared to those taking Drug B equals 32–26 = 6, whereas for those in the Control group the difference is in the other direction: 17–19 = –2. The difference of the differences is 8 and is significantly different from zero. Note that although it appears that Drug A is better for those receiving Treatment but Drug B is better for those in the Control group, we have no statistical test of these two individual differences. In other words, we do not know whether the drug difference of 6 for the Treatment group or the difference of –2 for the Control group differs significantly *from zero.* We simply know that 6 is significantly different from –2. We return later to address the specific question of whether there are significant differences between the drugs within each treatment condition.

In Chapter 8 we noted that we can always interpret an interaction with either variable in the product as the focal variable. The same is true for products of categorical predictors. Thus, we can focus on whether the Treatment versus Control difference is the same for those taking Drug A as it is for those taking Drug B. That is, the null hypothesis is

$$H_0: \mu_{AT} - \mu_{AC} = \mu_{BT} - \mu_{BC}$$

Figure 10.8(f) depicts this way of viewing the same interaction depicted in Figure 10.8(e) and displays the differences being compared. For those receiving Drug A, the Treatment–Control difference equals 32–17 = 15, but for those receiving Drug B the difference is only 26–19 = 7. The difference of the differences is 15 –7 = 8, the same as before, as it must be. In other words, Treatment is more effective if one takes Drug A instead of Drug B. We therefore have two equivalent interpretations for the interaction implied by rejecting the null hypothesis that $\beta_5 = 0$. Indeed, for any interaction there will be multiple ways to express it. It will often help to understand the interaction if all the possible interpretations are considered, but it is important to realize that they are all equivalent statements of the same effect.

Interpretation in Terms of Slopes

It is also useful to examine the coefficients for two-way contrast-coded predictors using the same procedures we developed in Chapter 8 for multiple regression models involving product terms. We begin with the estimated equation for Model A:

$$\hat{Y} = 20 + 3.5X_1 + X_2 + 5X_3 + 0.5X_4 + 2X_5$$

Then we substitute for the product definitions ($X_4 = X_1X_3$ and $X_5 = X_2X_3$) to get

$$\hat{Y} = 20 + 3.5X_1 + X_2 + 5X_3 + 0.5X_1X_3 + 2X_2X_3$$

Next we regroup the terms to express the "simple" linear relationship between $\hat{Y}$ and one of the three predictors X_1, X_2, or X_3. We will begin with X_3; regrouping terms yields

$$\hat{Y} = (20 + 3.5X_1 + X_2) + (5 + 0.5X_1 + 2X_2)\,X_3$$

which expresses the "simple" linear relationship between $\hat{Y}$ and X_3. The term in the first set of parentheses represents the intercept (given specific values of X_1 and X_2), and the term in the second set of parentheses represents the slope (i.e., as always, the change in $\hat{Y}$ given a unit change in X_3, given specific values *of* X_1 and X_2). Remember that X_3 codes whether the patient received psychotherapy treatment, so this linear relationship reflects the effectiveness of treatment versus control—the steeper the slope, the more effective the treatment relative to the control.

To understand the equation expressing $\hat{Y}$ as a linear function of X_3, we begin by examining Figure 10.9, which displays the relationship between $\hat{Y}$ and X_3 when $X_1 = 0$ and $X_2 = 0$.

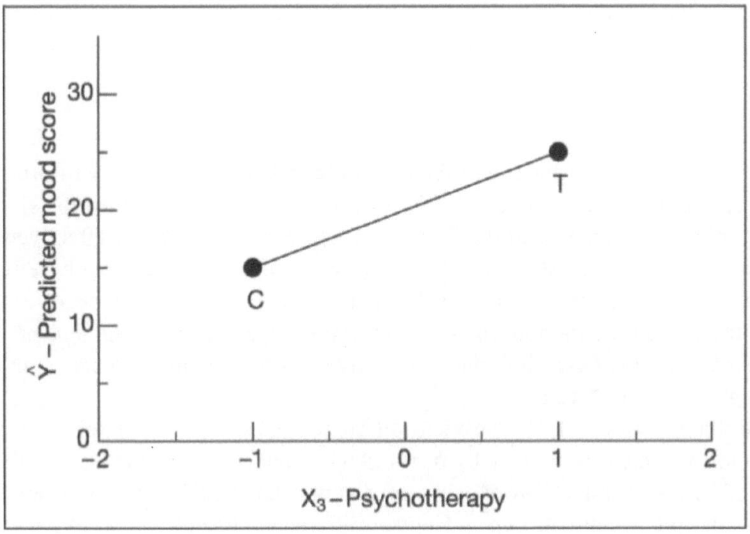

Figure 10.9 Simple relationship between $\hat{Y}$ and X_3 when $X_1 = 0$ and $X_2 = 0$

Both X_1 and X_2 are contrast-coded predictors with an equal number of observations in each cell, so average values of X_1 and X_2 are zero; hence, the line in Figure 10.9 depicts the average relationship between $\hat{Y}$ and X_3. For $X_1 = 0$ and $X_2 = 0$, the relationship between $\hat{Y}$ and X_3 reduces to

$$\hat{Y} = 20 + 5 \times 3$$

The intercept of 20 is, as always, the predicted value when $X_3 = 0$. Although X_3 is never zero (it is either +1 or −1), its average value is zero. Hence, we can interpret the intercept as the average value across levels of X_3. Indeed, 20 is the grand mean for these data. The slope for X_3 is 5, which implies that for every unit change in X_3 the predicted value $\hat{Y}$ changes by 5. The two-unit change from $X_3 = -1$ (Control) to $X_3 = +1$ (Treatment) predicts a change in $\hat{Y}$ (mood scores) of 2(5) = 10 points. This, as it must be, is the same interpretation we obtained for $b_3 = 5$ in terms of cell means. On average, ignoring the Drug condition, those receiving Treatment had mood scores 10 points higher than those in the Control condition.

What happens to the simple relationship between $\hat{Y}$ and X_3 if we do not focus on the average relationship, but instead allow X_1 and X_2 to be different from zero? The intercept term in the "simple" relationship is

$$(20 + 3.5X_1 + X_2)$$

so if $X_1 = 1$ (indicating Drug A or B) then the intercept will increase by 3.5 points and if $X_1 = -2$ (indicating Placebo) then the intercept will decrease by 7 points. These represent the changes in $\hat{Y}$ due to X_1 when X_3 equals zero (which it does equal on average). On average, then, $\hat{Y}$ for the four drug groups ought to be 3.5 + 7 = 10.5 points higher than $\hat{Y}$ for the two placebo groups. That is, on average, mood scores for those receiving Drug A or B are 10.5 points higher than for those in the Control group. This is necessarily the same interpretation we obtained in terms of cell means. Similarly, for $X_2 = +1$ (Drug A), $X_2 = 0$ (Placebo), and $X_2 = -1$ (Drug B), the intercept increases, stays the same, or decreases. These changes in $\hat{Y}$ (mood scores) are similarly interpreted.

Now we turn to the slope in the "simple" relationship between $\hat{Y}$ and X_3 and observe what happens when X_1 and X_2 are not equal to zero. The slope term is

$$(5 + 0.5X_1 + 2X_2)$$

so the slope relating $\hat{Y}$ and X_3 changes as a function of both X_1 and X_2. If $X_2 = +1$ (Drug A), then the slope for the prediction increases by $2 \times 1 = 2$ units. If $X_2 = -1$ (Drug B) then the slope decreases by $2 \times -1 = -2$ units. The steepness of the slope for X_3 represents the relative effectiveness of Treatment versus Control. Hence, the steeper slope for Drug A indicates that the relative effectiveness of Treatment versus Control is greater for Drug A than for Drug B. This is the essence of any ANOVA interaction. If there is an interaction, *the effect of one categorical variable depends on the level of another variable.* In this case, the relative effectiveness of Treatment depends on whether the patient is taking Drug A or Drug B.

Similarly, different values of X_1 also yield different slopes. If $X_1 = +1$ (either Drug A or B), then the slope for X_3 increases slightly by $b_4 = 0.5$ units, and if $X_1 = -2$ (Placebo) then the slope decreases by $0.5(-2) = -1$ units. However, $b_4 = 0.5$ is not statistically different from zero so there is no evidence that the Treatment versus Control difference depends on whether any Drug was administered ($X_1 = -1$) as compared to the Placebo.

Figure 10.10 depicts the changes in the intercept and the slope for X_3 as X_1 and X_2 change. Note that the different values of X_1 (Drug versus Placebo) yield different intercepts, but the changes in the slope for X_3 (Treatment versus Control) due to changes in X_1 are minimal. Although it may not appear to be the case, the slope for the bottom line ($X_1 = -2$, or Placebo) is slightly flatter than the average of the slopes of the top two lines ($X_1 = +1$, Drugs). In contrast, the intercept changes due to X_2 (Drug A versus Drug B) are noticeable but much smaller than those due to X_1 (any Drug versus Placebo). However, the slope changes due to X_2 are obvious. Indeed, the lines cross within the range of values used in this study. That is, mood scores are higher for Drug A ($X_2 = +1$) than for Drug B ($X_2 = -1$)

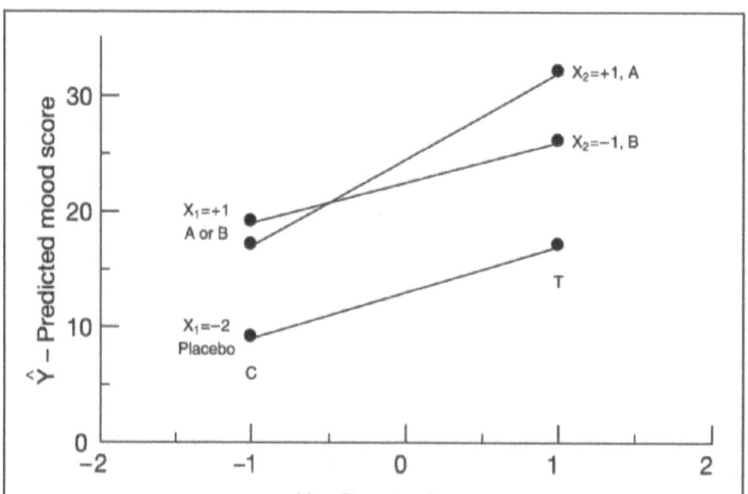

Figure 10.10 Simple relationship between $\hat{Y}$ and X_3 for different values of X_1 and X_2 (filled circles represent cell means)

for those receiving Treatment ($X_3 = +1$). For those in the Control condition ($X_3 = -1$) the reverse is true—mood scores for Drug B are higher than those for Drug A. These differences in slopes are the essence of an interaction.

Similar to the previous interpretation of the augmented model in terms of the simple linear relationship between $\hat{Y}$ and X_3, we could also do the interpretation in terms of $\hat{Y}$ and X_2. Although the interpretation of the interaction between X_2 and X_3 must necessarily be equivalent, the alternative interpretation in terms of the relationship between $\hat{Y}$ and X_2 may produce different insights. We do not provide that interpretation here, but it is recommended as an exercise for the reader to construct and interpret the relevant graph similar to Figure 10.10. Finally, we could do the interpretation in terms of the simple linear relationship between $\hat{Y}$ and X_1. However, that is unlikely to be useful for these data because the only interaction involving X_1 is $X_4 = X_1 \times X_3$ and the coefficient b_4 for that interaction was not significantly different from zero.

Summary of Interpretation

Although both interpretation strategies yield the same interpretation, each is useful for providing a slightly different perspective. For interpreting a given ANOVA, you should use as many of the strategies as necessary until a clear interpretation of the model emerges. Next we suggest a summary interpretation that one might include in a research report:

> On average, the two drug treatments produced higher mood scores ($M = 23.5$) than did the placebo ($M = 13$), PRE = .90, $F(1, 12) = 105.8$, $p < .0001$. The average mood scores produced by Drug A ($M = 24.5$) and Drug B ($M = 22.5$) did not significantly differ, PRE = .19, $F(1, 12) = 2.9$, $p = .11$. Averaged across the three drug conditions, patients receiving psychotherapy treatment had higher mood scores ($M = 25$) than did those in the control group ($M = 15$), PRE = .90, $F(1, 12) = 108$, $p < .0001$. However, a significant interaction between type of drug (A versus B) and psychotherapy treatment, PRE = .49, $F(1, 12) = 11.5$, $p < .005$, indicated that for patients who received psychotherapy, Drug A produced mood scores that were 6 points higher than Drug B, whereas for patients in the control group Drug A produced mood scores that were 2 points lower than Drug B.

Useful adjuncts to this journal summary are graphs of the cell means (which of course are the predicted values for the full model) as a function of the drug and psychotherapy variables. Figure 10.10 (which we developed in the section on interpreting slopes) and Figure 10.11 present two views of these data. In Figure 10.10, the differences between the lines for the drug treatments represent the "Drug" differences. In particular, the relatively large difference between the two lines for Drug A and Drug B, on the one hand, and the line for Placebo, on the other, corresponds to the large value for b_1, and the small difference between the Drug A and Drug B lines corresponds to the small value for b_2. That the three lines for the drug treatment groups are not parallel—the differences between lines are not constant—indicates the interaction between the drug and psychotherapy variables. In particular, the crossing of the Drug A and Drug B lines corresponds to the statistically significant value of b_5 ($X_5 = X_2 \times X_3$), which asks whether the difference between the Drug A and Drug B lines is constant or dependent on whether the patient was in the Treatment or Control groups. The relatively large difference between the two treatment lines in Figure 10.11 similarly corresponds to the large value of b_3 and the nonparallelism corresponds to the statistically significant value of b_5. Each of these graphs depicts the interaction, but from different perspectives, just as we were able to examine the interaction from different perspectives in each of the interpretation strategies presented earlier.

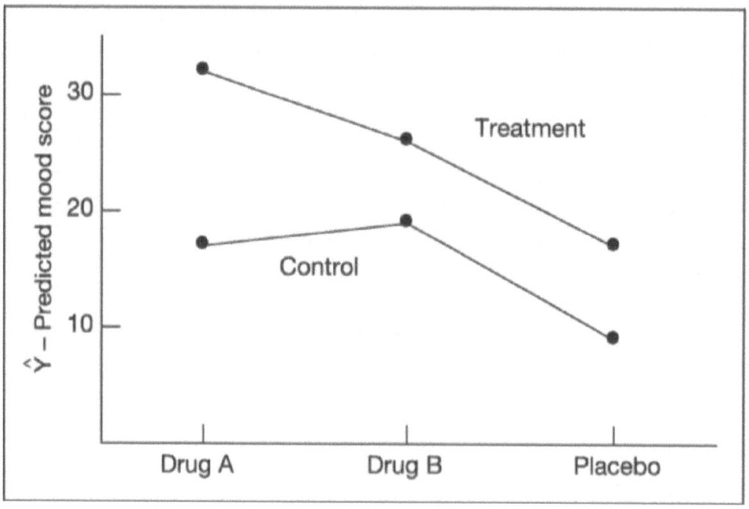

Figure 10.11 Cell means for mood scores by drug and psychotherapy treatment

Higher Order ANOVA

Our strategy for two-way ANOVA was to generate an appropriate set of contrast codes reflecting that two-way structure and then do the statistical analysis exactly as we did for one-way ANOVA. We can use that same strategy when we have three or more categorical predictor variables. Once we have an appropriate set of contrast codes for an experimental design with more than two categorical variables, the statistical analysis proceeds just as before. We therefore only need to learn a procedure for generating the appropriate codes. That procedure is a generalization of the same procedure we used for two-way ANOVA. For a factorial design of q categorical variables, the appropriate contrast codes are constructed according to the following rules:

1. For each of the q categorical variables, develop a set of orthogonal one-way contrast codes.
2. For each *pair* of categorical variables, construct additional contrast codes by multiplying all possible pairs of contrast codes, one from each categorical variable.
3. For each *triple* of categorical variables, construct additional contrast codes by multiplying all possible triads of contrast codes, one from each categorical variable.
4. Continue this procedure for each *quadruple*, *quintuple*, etc., until products are formed using codes from all q categorical variables simultaneously.

If $k_1, k_2, \ldots, k_q$ are the respective number of categories for each of the q categorical variables, then the aforementioned algorithm for generating contrast codes will produce the complete set of $k_1 \times k_2 \times \ldots \times k_{(q-1)}$ orthogonal contrast codes.

As an example, suppose we also wanted to use the patient's gender as a predictor variable in the study of mood scores. Figure 10.12 shows the design of such a study with 12 groups, for which we will need 11 orthogonal contrast codes. We can use the five contrast codes we have already specified for the Drug by Psychotherapy design. Let λ_6 be the code for gender, with –1 indicating men and +1 indicating women. Then we need to form the products between that code and each code from the other variables. For the Drug by Gender interactions,

$$\lambda_7 = \lambda_1 \times \lambda_6$$
$$\lambda_8 = \lambda_2 \times \lambda_6$$

	Gender			
	Male		*Female*	
	Psychotherapy		*Psychotherapy*	
Drug	*Control*	*Treatment*	*Control*	*Treatment*
A				
B				
Placebo				

Figure 10.12 A three-way factorial design: Drug (3) by Psychotherapy (2) by Gender (2)

and for the Psychotherapy by Gender interaction,

$$\lambda_9 = \lambda_3 \times \lambda_6$$

Finally, for the Drug by Psychotherapy by Gender three-way interactions, we form the products of each triple of categorical codes:

$$\lambda_{10} = \lambda_1 \times \lambda_3 \times \lambda_6$$
$$\lambda_{11} = \lambda_2 \times \lambda_3 \times \lambda_6$$

This provides the 11 orthogonal contrast codes required by the 12 groups of the three-way design. Figure 10.13 shows the codes for each group. Select some codes and verify that they sum to zero as required for contrasts. Select several pairs of codes and verify that their cross products sum to zero as required by orthogonality.

Interpretation of the one-way codes λ_1, λ_2, λ_3, and λ_6 and the two-way codes λ_4, λ_5, λ_7, λ_8, and λ_9 use the same strategy illustrated for interpreting codes in two-way ANOVA. The three-way codes λ_{10} and λ_{11} ask whether a given two-way interaction depends on the level of the third variable. For example, we can represent $\lambda_{10} = \lambda_1\lambda_3\lambda_6$ as $\lambda_1\lambda_3 \times \lambda_6$, asking whether the two-way interaction of λ_1 (Drug versus Placebo) by λ_3 (Treatment versus Control) is the same for each level of λ_6 (Women versus Men). Equivalently, the representation $\lambda_{10} = \lambda_1\lambda_6 \times \lambda_3$ shows that the three-way interaction equally asks whether the two-way interaction of λ_1 (Drug versus Placebo) by λ_6 (Women versus Men) depends on the level of λ_3 (Treatment versus Control). Finally, the representation $\lambda_{10} = \lambda_3\lambda_6 \times \lambda_1$ asks whether the two-way interaction of λ_3 (Treatment versus Control) by λ_6 (Women versus Men) depends on the level of λ_1 (Drug versus Placebo). At a fundamental level, these three questions are the same. However, depending on the focus of the study (is it about drug effects, psychotherapy effects, or gender differences?), one representation of each three-way interaction is likely to be most appropriate to report.

The global tests, as for one-way ANOVA, are not generally as useful as the focused questions represented by the one-degree-of-freedom contrasts. However, the testing of interactions is one instance for which global tests are sometimes useful. The proliferation of interaction contrasts in factorial designs makes testing and interpreting individual contrasts unwieldy. Interpretations of three-way and higher order interactions are often so theoretically ad hoc that some data analysts recommend examining only those interactions having a priori theoretical predictions. If the higher order interactions are not expected and not to be examined, then it might be better to eliminate them from the analysis, using only those contrasts that will be considered.

Group												
	A,C,M	*B,C,M*	*P,C,M*	*A,T,M*	*B,T,M*	*P,T,M*	*A,C,F*	*B,C,F*	*P,C,F*	*A,T,F*	*B,T,F*	*P,T,F*
λ_1	1	1	−2	1	1	−2	1	1	−2	1	1	−2
λ_2	1	−1	0	1	−1	0	1	−1	0	1	−1	0
λ_3	−1	−1	−1	1	1	1	−1	−1	−1	1	1	1
$\lambda_4 = \lambda_1\lambda_3$	−1	−1	2	1	1	−2	−1	−1	2	1	1	−2
$\lambda_5 = \lambda_2\lambda_3$	−1	1	0	1	−1	0	−1	1	0	1	−1	0
λ_6	−1	−1	−1	−1	−1	−1	1	1	1	1	1	1
$\lambda_7 = \lambda_1\lambda_6$	−1	−1	2	−1	−1	2	1	1	−2	1	1	−2
$\lambda_8 = \lambda_2\lambda_6$	−1	1	0	−1	1	0	1	−1	0	1	−1	0
$\lambda_9 = \lambda_3\lambda_6$	1	1	1	−1	−1	−1	−1	−1	−1	1	1	1
$\lambda_{10} = \lambda_1\lambda_3\lambda_6$	1	1	−2	−1	−1	2	−1	−1	2	1	1	−2
$\lambda_{11} = \lambda_2\lambda_3\lambda_6$	1	−1	0	−1	1	0	−1	1	0	1	−1	0

Figure 10.13 Orthogonal contrast codes for the three-way design of Figure 10.12

This effectively includes those eliminated interaction contrast codes in the "Error" term of the source table. Rather than eliminating the higher order interactions outright, a more conservative strategy that is frequently recommended is to do a global test of the higher order interactions and then pool those contrasts in the Error term only if the null hypothesis is not rejected for the global test. This strategy protects against missing something unusual in the data while greatly simplifying the data analysis and its presentation. An extra benefit is that the statistical power of the other contrasts will be increased if the F for the higher order interactions is < 1, because then more will be gained by adding degrees of freedom to the within-group (Error) term than will be lost by adding extra SS in that term.

When eliminating interaction terms one must remember the rule from Chapter 8 that products represent interactions only if all the components of the product are also included in the model. This implies, for example, that if the three-way interactions are retained, then none of the one-way contrasts can be eliminated. Less obviously, it also implies that none of the two-way contrasts can be eliminated because, as illustrated previously, any three-way interaction can be represented as the product of any of the two-way interactions and one of the one-way contrasts.

Other Details in Factorial Anova

In all other details factorial ANOVA with two or more categorical variables is the same as one-way ANOVA with one categorical variable. Specifically, confidence intervals, problems with nonorthogonal codes, handling unequal numbers of observations in each group, source tables, computational formulas for b_j and SSR_j for contrast-coded predictors, planned and post-hoc comparisons, and statistical power are exactly the same. However, we provide additional details on two issues—asking other questions and assessing statistical power—in order to highlight some issues that frequently arise in factorial ANOVA.

Asking Other Questions

It is sometimes not possible to generate a set of orthogonal contrast codes addressing all the theoretical questions that one might want to ask of a given set of data. This is often true for questions that span two or more categorical variables. The earlier strategy for generating codes for a standard factorial design allows choice of the one-way codes for each categorical variable but then determines the remaining interaction codes as products of those one-way codes. Those products may not ask the most relevant theoretical questions. For example, in a 2 × 2 design there is only one choice (+1, –1) for each one-way code, so there is no choice for the interaction code, which must be the product of the two one-way codes. Thus, the necessary codes for two variables A and B are

	A_1B_1	A_1B_2	A_2B_1	A_2B_2
λ_1	1	1	–1	–1
λ_2	1	–1	1	–1
λ_3	1	–1	–1	1

The interaction code λ_3 asks whether the average of the A_1B_1 and A_2B_2 groups equals the average of the other two groups. However, the theoretical question of interest might be whether the mean of one group, say A_2B_2, differs from the average of the other three groups. For example, the research hypothesis might be that both A_2 and B_2 must be present for an effect to exist. If so, the code of interest is [–1, –1, –1, 3]. This is not an interaction because it cannot be represented as the product of two contrast codes, one for each categorical variable. Nevertheless, if it is the question of interest, then it should be asked. We can compute its SSR with the usual formula and then compare it as a planned comparison (as in Chapter 9) to the mean within-group error (MSE) resulting from the analysis of variance using λ_1 through λ_3. Or, if that question is the focus of the study, one may simply ignore the factorial structure of the design and analyze the data using a set of one-way codes for four groups.

Other questions may also present themselves after the initial analysis. We can use the same post-hoc comparison procedures that we used in Chapter 9. For example, in the mood score data, whether Drug A was better than Drug B depended on whether the patient was also receiving psychotherapy. Examining the means, it appears that Drug A is more effective for those in Treatment but Drug B is more effective for those in the Control group. This suggests the recommendation that Drug A, even though it was superior on average, only be administered to those concurrently receiving psychotherapy. But before making such a recommendation we ought to test with post-hoc comparisons whether the difference between the means within levels of Psychotherapy are statistically significant. Comparisons such as this between categories (Drug A versus Drug B) of one variable within a single level (Treatment) are known as *simple effects*. We do the calculation for this comparison as an example. The codes for the two relevant simple effects are

	A,T	*B,T*	*P,T*	*A,C*	*B,C*	*P,C*
λ_6	1	–1	0	0	0	0
λ_7	0	0	0	1	–1	0

The code λ_6 asks whether there is a difference between Drug A and Drug B for those receiving Treatment and the code λ_7 asks the same drug difference question for those in the Control condition. Then

$$SSR_6 = \frac{\left(\sum_k \lambda_k \bar{Y}_k\right)^2}{\sum_k \lambda_k^2 / n_k} = \frac{(32-26)^2}{1^2/3+(-1)^2/3} = 54$$

The MSE (from Figure 10.4 or 10.7) is 4.2, and PA – PC for this focused test equals 1, so

$$F = \frac{\text{SSR}/(\text{PA}-\text{PC})}{\text{MSE}} = \frac{54}{4.2} = 12.86$$

If this were not a theoretically motivated, planned comparison, but only became a relevant question after examining the data, it would be advisable to conduct a Games-Howell test.

Although it is possible to calculate by hand estimates of t and F for Games-Howell tests, the formulas are burdensome, and not all statistical programs provide them. However, many provide adjusted p values, that is, the probability of finding values of F or *PRE* as large or larger than our calculated estimates, assuming that the null hypothesis is true, and adjusting for multiple comparisons. We will use these for our analyses. Recall that λ_6 tests whether people who took Drug A had significantly better moods than people who took Drug B among those in the treatment condition. The adjusted p-value in this case is .029. As such, there is evidence to suggest that among those who received psychotherapy, people who took Drug A had better moods than those who took Drug B. The adjusted p-value for λ_7 is .707. As such, people who received Drug A and B exhibited similar levels of mood in the control condition.

Compare these findings to the results of a Scheffé test of the same differences, which yields much more conservative F values. In this case the critical F value is easier to calculate and yields an adjusted critical value of

$$(m-1)\,F_{m-1;n-PA;\alpha} = 5F_{5,12;.05} = 5(3.11) = 15.55$$

Although we would reach the same conclusion for the test of λ_7, we would not reach the same conclusion for the test of λ_6. Our F value of 12.86 is below the adjusted critical value, so we would not be able to conclude that Drug A was significantly better than Drug B for those patients receiving Treatment. The Scheffé test also suggests that the effects of the drugs were different in the two conditions (the interaction coded by λ_5 was significant, so the Drug A versus Drug B difference of 6 for Treatment is significantly different from the difference of –2 for Control), but neither simple effect is significant (for Treatment, 6 is not statistically different from 0; and for Control, –2 is not statistically different from 0). The interaction and the simple effects tests ask different questions, so they certainly can have different answers. One remedy would be to conduct a replication study with more statistical power or with the simple effects as now legitimate planned comparisons. What this example illustrates is that decisions about post-hoc tests can alter the types of conclusions that we draw from our data, and researchers should consider the potential trade-offs of inflated Type I and Type II error rates given their data, goals, and the adjustment method used.

Power in Factorial ANOVA

The methods we have presented before for estimating power work equally well for estimating statistical power in factorial ANOVA. It is interesting to examine how power is affected for questions we ask about a given variable when additional categorical variables are added to the analysis. As an example, again consider the mood score data of Figure 10.1. If we were mainly interested in the questions about the effects of the drugs, we could have performed a one-way analysis of variance on these data, completely ignoring the psychotherapy variable. If we ignored whether patients were in the Treatment or Control groups, we would have three groups—Drug A, Drug

B, and Placebo—each with six observations. The respective means would be (32 + 17)/2 = 24.5, (26 + 19)/2 = 22.5, and (17 + 9)/2 = 13. The two contrast codes would be (1, 1, –2) and (1, –1, 0). Applying Equations 9.1 and 9.2, we obtain $b_1 = 3.5$ with SSR = 441 and $b_2 = 1$ with SSR = 12, the same values as before. However, PRE and F would be different because we are now testing the null hypothesis $\beta_1 = 0$ by comparing

MODEL A: $Y_i = \beta_0 + \beta_1 X_1 + \beta_2 X_2 + \varepsilon_i$
MODEL C: $Y_i = \beta_0 + \beta_2 X_2 + \varepsilon_i$

This is the same basic question asked when testing $\beta_1 = 0$ in the complete two-way analysis. However, the statistical power is not the same because the question is being tested in the context of a different Model A/C comparison.

To see the likely changes in power due to ignoring the other categorical variable, let us consider Figure 10.14, which presents the complete source table for the one-way ANOVA ignoring the psychotherapy variable. The parameter estimates and their respective sums of squares, degrees of freedom, and mean square errors reduced are exactly as they were in Figure 10.7. However, the error sum of squares and the error degrees of freedom have increased substantially. The reason is that the sums of squares that were reduced by X_3, X_4, and X_5 in the two-way analysis are now included in the error because those predictors are not used in this analysis. Given the same SSRs for X_1 and X_2 and increased SSE, then the F and PRE values must be considerably less than they were before. For example, in the complete two-way analysis of Figure 10.7, for X_1, $F_{1,12} = 105.8$ and PRE = .90, but in the one-way analysis only considering the Drug treatment variable, $F_{1,15} = 11.9$ and PRE = .44. The differences between the means for Drug A, Drug B, and Placebo have not changed, so clearly the two-way analysis provided considerably more statistical power for testing $\beta_1 = 0$ and $\beta_2 = 0$. This increase in power is generally the case in multi-way factorial ANOVA. Using one or more other categorical variables to reduce SSE allows a more powerful test of the hypothesis concerning the focal categorical variable. The only situation for which there would be a decrease in statistical power would be if the F values for the codes of the additional variables and the codes for the interactions were < 1.0. If $F < 1.0$ for a contrast, then the proportional reduction in error associated with that contrast is less than we would expect for a randomly chosen parameter. Hence, including that contrast-coded predictor would not be worth the loss of a degree of freedom due to the extra parameter, and so power goes down. Therefore, one should add other categorical variables for the purpose of increasing statistical power only if one expects those variables and their interactions to themselves reduce error in the criterion variable. Interactions proliferate rapidly as other categorical variables with many levels are added. One must be cautious that those many interactions do not thereby reduce statistical power.

Source	*bj*	*SS*	*df*	*MS*	*F*	*p*	*PRE*
Between groups (Drug)		453	2	226.5	6.1	.012	.45
X_1	3.5	441	1	441	11.9	.004	.44
X_2	1.0	12	1	12	0.3	.58	.02
Within groups (MSE)		557	15	37.1			
Total		1010.0	17				

Figure 10.14 One-way ANOVA of mood scores ignoring the psychotherapy variable

Summary

In a factorial design, there are two or more categorical predictor variables, and all levels of each categorical variable are combined with all levels of other categorical variables. We analyze data from a two-way or higher order factorial design by applying the one-way ANOVA techniques from Chapter 9 to an appropriate set of contrast codes. We generate this set of codes by developing separate sets of one-way contrast codes for each categorical variable and then forming appropriate products among these codes to represent the interaction between categorical variables. Interpreting the interaction contrasts can be difficult, but the same interpretative techniques from Chapters 8 and 9 apply.

Notes

1. We use Z here instead of the customary X because subsequently we will do an analysis of these same data using a different set of contrast codes. The Z values are the contrast-coded predictor variables for the first analysis, and the X values will be the contrast-coded predictor variables for the second analysis.
2. Note that the Z values are independent of each other because they are based on orthogonal contrast codes and there are equal numbers of observations in each group. Hence, the sums of squares in Figure 10.4 for the individual contrast-coded predictors sum to the overall sum of squares for the complete model.

11 ANCOVA

Models With Continuous and Categorical Predictors

The material covered in this chapter represents an integration of the material covered in Chapters 7 through 10. In Chapters 7 and 8, we considered models with continuous predictors; Chapters 9 and 10 were concerned with models with categorical predictors. We will now consider models that include both sorts of predictors. Such models are more traditionally covered under the heading of analysis of covariance (ANCOVA). We feel that this is a confusing label for these models because, as we will demonstrate shortly, these models are estimated no differently from models involving only continuous or only categorical predictors. We thus prefer to think of them simply as models involving both categorical and continuous predictor variables.

In Chapter 8, we saw that an extension of models with continuous predictors included products of those predictors, or interactions among them, as predictors as well. Similarly, in the last chapter we considered interactions among categorical predictor variables. In this chapter, we consider models that include not only categorical and continuous predictors but also interactions between the two kinds of predictor variables.

Interest in these sorts of models originally developed within the tradition of methods for the analysis of experimental and quasi-experimental research designs. In this tradition, researchers were primarily interested in the effects of experimental factors, or the categorical predictor variables, and wished to estimate those effects while controlling for some continuously measured concomitant variable, called a *covariate*. As is detailed later, this control might be desired because of the increase in power that such control might produce or because the concomitant variable was redundant with the categorical ones. Models involving both categorical and continuous predictor variables have, however, a wide range of application outside of experimental research designs. For instance, one might be interested in the effects of both a categorical variable (e.g., political party affiliation) and a continuous one (e.g., personal income) on some dependent variable, and might wish to estimate each of these effects when controlling for the other. In other words, unlike in the experimental design domain, we might be interested in the effects of the continuous predictor variable and how those effects change when we control for a categorical one as well as the effects of the categorical variable and what happens to those effects when a continuous predictor variable is controlled

Given the traditional use of these models in experimental design, we start the chapter by considering their use in an experimental design, where the primary interest is in the effects of the categorical variables. We use this context to illustrate the reasons we might want to control for a continuously measured concomitant variable. In later sections of the chapter, we illustrate the use of these models in a context where the primary interest is in the effect of the continuous predictor when a categorical one is controlled.

DOI: 10.4324/9781003438335-16

Controlling for an Orthogonal Continuous Variable in a Factorial Design

Suppose we are evaluating a new teaching method. Students are randomly assigned to either the new or the old method. In addition, students are randomly assigned to one of two tutors (the tutors were trained in both methods, so they teach each student using the appropriate randomly assigned method). Thus, the experimental design is a two-factor crossed design, with two levels of both the method and tutor factors. Ten students have been randomly assigned to each of the resulting four conditions. The dependent variable is the student's score on an achievement test given at the end of the tutoring sessions. In addition, we have a pretest measure from each student, indicating his or her achievement in the domain in question prior to being randomly assigned to conditions. The hypothetical raw data are given in Figure 11.1. The Z_i variable represents each student's pretest achievement score. The X_{1i} variable is a contrast-coded variable that codes method (–1 if Old Method; 1 if New Method); X_{2i} is a contrast-coded variable that codes tutor (–1 if Tutor A; 1 if Tutor B); and Y_i represents each student's post-test score.

The means for both Z_i and Y_i for each of the four cells of the design are given in Figure 11.2. Notice that the data have been constructed so that all four pretest means equal 50. Because students

Y_i	X_{1i}	X_{2i}	Z_i	Y_i	X_{1i}	X_{2i}	Z_i
58	1	−1	50	57	1	−1	49
63	1	−1	49	61	1	−1	52
65	1	−1	53	57	1	−1	50
56	1	−1	47	67	1	−1	51
60	1	−1	53	56	1	−1	46
50	−1	−1	49	62	−1	−1	54
58	−1	−1	51	55	−1	−1	48
52	−1	−1	50	63	−1	−1	52
55	−1	−1	47	50	−1	−1	46
57	−1	−1	52	58	−1	−1	51
61	1	1	47	59	1	1	49
71	1	1	53	65	1	1	54
68	1	1	52	60	1	1	46
58	1	1	48	65	1	1	51
68	1	1	51	65	1	1	49
47	−1	1	46	62	−1	1	51
56	−1	1	51	51	−1	1	49
63	−1	1	53	54	−1	1	51
53	−1	1	48	58	−1	1	50
52	−1	1	47	54	−1	1	54

Figure 11.1 Hypothetical experimental data

Condition	$\bar{Z}_k$	$\bar{Y}_k$
Old Method, Tutor A	50	56
Old Method, Tutor B	50	55
New Method, Tutor A	50	60
New Method, Tutor B	50	64

Figure 11.2 Pretest and post-test means by condition

were randomly assigned to the four treatment conditions after they took the pretest, we would expect their mean pretest scores to be similar, although the chances are that they would not all be identical. (They have been constructed to be identical here for didactic purposes.)

The reason for constructing these data with equal pretest means is that as a result the pretest is not correlated with condition. To see this, suppose we regressed the pretest Z_i on X_{1i}, X_{2i}, and their interaction. With all three contrast codes as predictors, the predicted values of Z_i would equal the four cell means. Each of these cell means also equals the grand mean. Hence, a model with all three contrast-coded predictors generates the same predicted values for every case as the simple model that predicts the grand mean for every case. Thus, condition is unrelated to pretest scores.

ANOVA of the Data

Let us conduct a straightforward two-way ANOVA on the post-test data, ignoring the pretest variable for the moment. We regress Y_i on X_{1i}, X_{2i}, and X_{3i}, where X_{3i} is the interaction contrast code, formed by multiplying X_{1i} and X_{2i} together. The resulting model is

$$\hat{Y}_i = 58.75 + 3.25X_{1i} + 0.75X_{2i} + 1.25X_{3i} \qquad \text{SSE} = 710$$

The value of the regression coefficient for X_{1i}, 3.25, equals half the difference between the average Y_i under the new method (collapsing across tutors) and the average Y_i under the old method (collapsing across tutors). The regression coefficient for X_{2i}, 0.75, equals half the difference between the average under Tutor B (collapsing across methods) and the average under Tutor A (collapsing across methods). The regression coefficient for the interaction contrast code, 1.25, equals half the difference between the average of the Old Method/Tutor A and New Method/Tutor B conditions and the average of the Old Method/Tutor B and New Method/Tutor A conditions. Conceptually it tells us that the method difference is larger under Tutor B than under Tutor A.

We can calculate the sum of squares due to each contrast-coded predictor by computing a series of compact models that omit each contrast code in turn. Alternatively, we can use Equation 9.2, the formula for the SSR associated with a contrast-coded predictor that we have used before:

$$\text{SSR} = \frac{\left(\sum_k \lambda_k \bar{Y}_k\right)^2}{\sum_k \left(\lambda_k^2 / n_k\right)}$$

The sum of squares explained by X_1 equals 422.5, that explained by X_2 equals 22.5, and that explained by X_3 equals 62.5. We have the same number of observations, 10, in each of the four conditions. Thus, the three predictors are nonredundant and these sums of squares add up to the between condition sum of squares. The ANOVA source table for this analysis is presented in Figure 11.3.

Including an Orthogonal Covariate

In this analysis, we have made no use of the pretest Z_i. We might decide to add it as a predictor to the model for several reasons. For instance, we might be interested in asking how method and tutor affect post-test performance when we control for the pretest or when we look within levels of the pretest. In other words, we might be interested in these effects over and above differences in performance that existed at the time of the pretest. A seemingly different reason for including it as a predictor in the model is that we might expect it to be highly correlated with the post-test, as presumably it is only an earlier version of the post-test, measuring the same domain of achievement. If

Source	*b*	*SS*	*df*	*MS*	*F*	*PRE*
Between or model		507.5	3	169.17	8.58	.42
Method	3.25	422.50	1	422.50	21.42	.37
Tutor	0.75	22.50	1	22.50	1.14	.03
Method x Tutor	1.25	62.50	1	62.50	3.17	.08
Error		710.00	36	19.72		
Total		1217.50	39			

Figure 11.3 Two-way ANOVA source table

so, then, as we shall see, it might make our tests of method and tutor effects, and their interaction, considerably more powerful.

In these data, the pretest scores are highly related to the post-test scores. The sum of squared errors from a simple regression model in which Y_i is regressed on Z_i equals 855.57. From the last or "Total" row of the source table of Figure 11.3, we see that the sum of squares of Y_i for the simplest single-parameter model equals 1217.5. Accordingly, a comparison between the two-parameter simple regression model using Z_i to predict Y_i and a one-parameter model yields a PRE of .297 and $F(1, 38) = 16.05$. Thus, the relationship between the pretest and the post-test is substantial and significant.

As we have seen, the pretest Z_i is uncorrelated with condition because the pretest conditions means are identical. Accordingly, the sum of squares of Y_i that could be explained by Z_i will not overlap with the sums of squares of Y_i explained by the three condition contrast codes. Because the sum of squares explained by these three contrast-coded predictors equals 507.5, the sum of squares explained by Z_i equals 361.93, and Z_i is nonredundant with the three contrast-coded predictors, the sum of squares explained by the pretest, Z_i, plus the three contrast-coded predictors ought to equal 507.5 + 361.93 or 869.43. Accordingly, the sum of squared errors for a model in which Y_i is regressed on Z_i, X_{1i}, X_{2i}, and X_{3i} should equal 1217.5–869.43 or 348.07.

In the ANOVA source table of Figure 11.3, the denominator for each of the F statistics testing the significance of the condition differences equals the mean square error within or the mean square error from the final augmented model that includes all predictor variables. If we include the pretest Z_i as an additional predictor, this mean square error ought to be reduced substantially because the sum of squares potentially explainable by Z_i has been controlled for or removed from the sum of squared errors. Thus, by including Z_i as an additional predictor in the model, we would expect our tests of condition differences in Y_i to be more powerful, yielding larger F values.

The resulting augmented model, with both the pretest and the three contrast-coded predictors, is estimated as

$$\hat{Y}_i = -4.52 + 1.27Z_i + 3.25X_{1i} + 0.75X_{2i} + 1.25X_{3i} \qquad \text{SSE} = 348.07$$

Notice that the coefficients for the three contrast-coded predictors have not changed from the model where Z_i was not included as a predictor. Because Z_i is nonredundant with all of the contrast-coded predictors, its inclusion has no effect on the value of their coefficients or on their interpretations. The coefficient for X_{1i} continues to tell us about the magnitude of the difference in Y_i due to method. The coefficient for X_{2i} continues to tell us about the magnitude of the mean tutor difference. And the coefficient for X_{3i} continues to inform us about the degree to which the method difference is larger under Tutor B than under Tutor A.

We could test whether the coefficient for Z_i significantly differs from zero by comparing this augmented model with the model presented earlier that included only the contrast-coded predictors. The value of PRE that results from this comparison is

$$\text{PRE} = \frac{710.00 - 348.07}{710.00} = .510$$

with an associated F statistic of

$$F_{1,35} = \frac{.510/1}{(1-.510)/35} = 36.39$$

Accordingly, we can conclude that, independent of condition, or on average within condition, pre-test scores significantly relate to post-test scores.

We can test whether the condition means differ by testing the regression coefficients in this model for each of the three contrast-coded predictors. Let us start with the omnibus test of whether there are any differences in the condition means of Y_i. To do this test, we want to compare the augmented model that includes Z_i and the three contrast-coded predictors with a compact one that includes only Z_i. We have already said that the sum of squared errors from this compact simple regression model equals 855.57. Accordingly, the value of PRE for the omnibus test is

$$\text{PRE} = \frac{855.57 - 348.07}{855.57} = .593$$

And the omnibus F statistic, having 3 and 35 degrees of freedom, is

$$F_{3,35} = \frac{.593/3}{(1-.593)/35} = 17.01$$

An equivalent expression for the F statistic is given in terms of the sums of squares

$$F_{3,35} = \frac{507.50/3}{348.07/35} = 17.01$$

where 507.50 is the reduction in the sum of squared errors as we move from the compact to the augmented model, and 348.07 is the sum of squared errors of the augmented model.

Notice that this F statistic is nearly twice as large as the F statistic for the omnibus test based on the augmented model that did not include the pretest as a predictor (given in Figure 11.3). The reason for this difference is that the sum of squared errors from the augmented model is now considerably less than it was without the pretest included. It is also true that the degrees of freedom for error have been reduced by 1 because of the additional parameter for the pretest estimated in the augmented model. In combination, however, the substantially smaller sum of squared errors and the slightly smaller degrees of freedom for error result in a considerably smaller mean square error.

In other words, the denominator of the *F* ratio for the omnibus test of condition differences when the pretest is included equals 348.07/35 or 9.94. In the model that did not include the pretest, the denominator of the *F* ratio for the omnibus test of condition differences equals 710.0/36 or 19.72. The net result of including the pretest as a predictor, then, has been to reduce the mean square error and increase the power of tests of the condition effects. Such an increase in power happens whenever there is a significant relationship between the continuously measured predictor variable, in this case the pretest, and the dependent variable in the augmented model that includes the categorical predictor variables. The full source table for the model that includes the pretest and all three contrast-coded predictors is given in Figure 11.4. Let us compare this ANCOVA source table with the ANOVA source table of Figure 11.3.

First, notice that the test of the overall model—comparing this model that includes four predictor variables with the single-parameter model that includes no predictors—yields substantially higher values of both PRE and *F* than the test of the overall model that did not include the pretest. These larger values are entirely attributable to the fact that the pretest is highly related to the dependent variable. Accordingly, the sum of squares attributable to the model, 869.43, has dramatically increased from the ANOVA table, while the error sum of squares has dramatically decreased. The decrease in the sum of squared errors for the model, and equivalently the increase in the sum of squares explained by the model, is equal to the sum of squares associated with the pretest, that is, 361.93.

Second, notice that the rows for the three contrast-coded variables, representing method, tutor, and their interaction, have the same sums of squares, degrees of freedom, and mean squares as they did in the ANOVA source table. Because the pretest is unrelated to method, tutor, and their interaction, neither the regression coefficients for these contrast-coded predictor variables nor their sums of squares are affected by including the pretest in the model. Including the pretest, however, has a major effect on the row in the source table referring to error in the model, as just described. As a result, all the *F* statistics used to test the tutor effect, method effect, and the interaction effect on the post-test dependent variable are substantially larger than they were in the ANOVA source table of Figure 11.3. Whereas the Method X Tutor interaction was not significant in the ANOVA source table, it now is. The positive coefficient associated with this interaction contrast code tells us that the new-old method difference, while significant on average across tutors, is significantly larger for Tutor B than it is for Tutor A.

This analysis has illustrated one of the major reasons for including a continuously measured predictor variable (also called a covariate) in randomized experimental research designs. If that predictor variable is measured prior to randomization of participants to conditions, on average it will be unrelated to the contrast-coded variables that code experimental condition. The purpose of including such a variable is to increase the power of the analysis that tests for condition differences

Source	*b*	*SS*	*df*	*MS*	*F*	*PRE*
Model		869.43	4	217.36	21.87	.71
Pretest	1.27	361.93	1	361.93	36.39	.51
Between conditions		507.50	3	169.17	17.01	.59
Method	3.25	422.50	1	422.50	42.48	.55
Tutor	0.75	22.50	1	22.50	2.26	.06
Method x Tutor	1.25	62.50	1	62.50	6.28	.15
Error		348.07	35	9.94		
Total		1217.50	39			

Figure 11.4 ANCOVA source table

in the dependent variable. If the covariate is in fact unrelated to condition, then neither the regression coefficients for the various contrast-coded predictors that code condition differences nor their sums of squares will be affected by its inclusion in the model. The null hypothesis associated with the test of a given contrast-coded predictor will also not change. Regardless of the inclusion of the covariate, we will still be testing for differences among the condition means on the dependent variable, as coded by the λ values. Tests for condition differences will be more powerful as a result of including a covariate whenever the test of whether the covariate's regression coefficient in the full augmented model significantly differs from zero. When the covariate is unrelated to the dependent variable, then the decrease in the sum of squared errors resulting from including the covariate may not offset the decrease in the degrees of freedom for error in the model. The ideal covariate, therefore, in this situation, is one that is as highly associated as possible with the dependent variable controlling for the categorical variables or within levels of the categorical variables.

Even with random assignment of participants to condition after measuring the covariate, it will almost never be the case that the covariate will be entirely nonredundant with the condition contrast codes. In other words, it will be a very rare event for all of the pretest or covariate means in the various experimental conditions to be exactly equal. Our example is designed simply to illustrate what happens in the pure case, when the covariate is completely independent of condition. In any given study, there will likely be some nonsignificant relationships between the covariate and the contrast codes that represent condition. Nevertheless, including a covariate increases the statistical power of tests of condition differences, given a covariate that is significantly related to the dependent variable within levels of the categorical variables.

As we said in the introduction to this chapter, in experimental designs the usual interest in including a continuously measured predictor variable with a set of categorical ones is to examine what happens to tests of condition differences when we control for the continuously measured covariate. As we have seen, with a covariate measured prior to random assignment of participants to condition, the result will generally be an increase in statistical power for tests of condition differences. There is no necessary reason, however, for confining our interpretations of the model that includes both kinds of predictors to this typical interest. In other words, there is nothing to prevent us from turning the interpretation of this model around—concentrating not on the tests of mean differences while controlling for the covariate, but on a test of the pretest–post-test relationship while controlling for condition differences on the post-test. If we simply regress the post-test on the pretest, the pretest's regression coefficient equals 1.27. The sum of squared errors for this simple regression model equals 855.57. A test of the simple pretest–post-test relationship yields a PRE of .297 and an F of 16.05 with 1 and 38 degrees of freedom. When we examine the pretest–post-test relationship controlling for the categorical variables, as given in the ANCOVA source table of Figure 11.4, the pretest's coefficient is still 1.27, but a test of whether it is significantly related to the post-test within condition levels yields a PRE of .51 and an F of 36.39 with 1 and 35 degrees of freedom. Thus, we might also look at this analysis as a way of increasing the power of tests of the pretest–post-test relationship by controlling for experimental conditions.

Analysis of Post-Test–Pretest Difference Scores

In giving a rationale for the analysis that includes pretest as a predictor, we suggested that it might make sense to examine the effects of method and tutor on the post-test when controlling for pretest differences. An equivalent way of doing this analysis would be to examine condition differences in improvement from the pretest to the post-test. To do such an analysis, we might compute a new dependent variable equal to $Y_i - Z_i$, assuming they were both measures of the same thing—the pretest taken before the experiment and the post-test at its conclusion. This *difference score* tells us about each individual's improvement in achievement during the course of the study. We would then be interested in condition differences in improvement (the mean $Y_i - Z_i$ difference). Because the Z_i condition means are identical, the mean differences among conditions on the $Y_i - Z_i$ difference scores will be equivalent to the mean differences among conditions on Y_i.

To examine condition effects on this improvement difference score, let us regress it on the three contrast-coded predictors that define condition. The following estimated parameters result:

$$Y_i - Z_i = 8.75 + 3.25X_{1i} + .075X_{2i} + 1.25X_{3i} \qquad \text{SSE} = 364.00$$

The intercept in this model equals the average of the condition means on the difference score or, equivalently, the difference between the means of the condition means of Y_i and of the condition means of Z_i. Somewhat surprisingly, perhaps, the regression coefficients for the three contrast-coded predictors have not changed as a result of changing the dependent variable to the $Y_i - Z_i$ difference score. Both in this difference score analysis and in the analysis where the post-test was the dependent variable and the pretest was included as a predictor, the regression coefficients continue to equal what they did in the simple analysis of variance with Y_i as the dependent variable and no pretest. Once again, this invariance occurs because the pretest is uncorrelated with condition.

To illustrate algebraically why these regression coefficients have not changed, let us examine the algebraic expression for the regression coefficients associated with contrast-coded predictors in this difference score analysis. Using our standard formula, but substituting the difference score means for the usual $\bar{Y}$ values, we get

$$b = \frac{\sum_k \lambda_k \left(\overline{Y_k - Z_k}\right)}{\sum_k \lambda_k^2} = \frac{\sum_k \lambda_k \left(\bar{Y}_k - \bar{Z}_k\right)}{\sum_k \lambda_k^2} = \frac{\sum_k \lambda_k \bar{Y}_k - \sum_k \lambda_k \bar{Z}_k}{\sum_k \lambda_k^2}$$

Because all $\bar{Z}_k$ are identical, the expression $\Sigma\lambda_k\bar{Z}_k$ equals zero and this expression for the regression coefficient for each of the predictors reduces to what it is when simply Y_i is the dependent variable in the model. In sum, these regression coefficients equal the coded condition differences in the mean difference scores ($\bar{Y}_k - \bar{Z}_k$), which, given equal $\bar{Z}_k$, are equivalent to the coded differences in $\bar{Y}_k$.

This difference score analysis gives us the difference score ANOVA source table of Figure 11.5. Notice that just as the regression coefficients for the contrast-coded predictors in this difference score analysis equal what they were in the analysis that included pretest as a predictor variable, so too are their sums of squares equal to what they have been all along. Once again, this equivalence occurs because Z_i is uncorrelated with condition.

This analysis, however, is different from both the ANOVA and the ANCOVA in terms of the sum of squares for error and the sum of squares total. The total sum of squares is now equal to the sum of squared variation in the $Y_i - Z_i$ difference scores, which in this case is less than the total sum of squares in Y_i. Because this total sum of squares has been reduced and because the sums of squares explained by the three contrast-coded predictors are unchanged, the error sum of squares in this analysis must be less than the sum of squared errors in the earlier analysis of variance with Y_i as the dependent variable. As a result, the *F* and PRE statistics for the omnibus test of any condition differences and for the three single-degree-of-freedom tests are all larger than they were in the original ANOVA source table of Figure 11.3.

Notice, however, that the sum of squared errors in this difference score source table is larger than the sum of squared errors in the analysis that included the pretest as a predictor variable (ANCOVA source table of Figure 11.4). As a result, the *F* and PRE statistics testing condition differences in this difference score analysis are all slightly smaller than they were in the source table of

Source	b	SS	df	MS	F	PRE
Between or model		507.50	3	169.17	16.73	.58
Method	3.25	422.50	1	422.50	41.79	.54
Tutor	0.75	22.50	1	22.50	2.22	.06
Method x Tutor	1.25	62.50	1	62.50	6.18	.15
Error		364.00	36	10.11		
Total		871.50	39			

Figure 11.5 Difference score ANOVA source table

Figure 11.4. In sum, although this difference score analysis is more powerful in this case than the simple ANOVA of Y_i, it is not as powerful as the ANCOVA in which the dependent variable was Y_i and Z_i was included as a predictor variable.

To understand why this is so, let us examine the difference score model:

$$Y_i - Z_i = \beta_0 + \beta_1 X_{1i} + \beta_2 X_{2i} + \beta_3 X_{3i} + \varepsilon_i$$

We can re-express this model by adding Z_i to both sides of the equation:

$$Y_i = \beta_0 + Z_i + \beta_1 X_{1i} + \beta_2 X_{2i} + \beta_3 X_{3i} + \varepsilon_i$$

This difference score model now looks very similar to the ANCOVA model in which the pretest was used as a predictor variable. There is, however, one major difference. Instead of estimating a parameter for the pretest variable as we did in the ANCOVA model, we have set the parameter value equal to 1.0. By doing the difference score analysis, we have in effect assumed that the parameter value for the pretest equals 1.0, rather than letting it be a free parameter and deriving its least-squares estimate.

By definition, the least-squares estimates are those that minimize the sum of squared errors. Accordingly, the sum of squared errors in a model where the coefficient for the covariate is fixed at 1.0 cannot be less than the sum of squared errors in the ANCOVA model because in the latter model the coefficient for the covariate is the least-squares estimate. Hence, the difference score analysis will generally be less powerful than the ANCOVA. Frequently, it will be substantially less powerful and may, in fact, be even less powerful than the simple ANOVA model.

Once we realize that this difference score model is identical to the ANCOVA model, except that we have fixed the coefficient for the covariate at 1.0 instead of estimating it, we can rewrite the difference score analysis source table as we have in Figure 11.6. In this revised table, the sum of squares total refers to the total sum of squares of Y_i rather than the total sum of squares in the $Y_i - Z_i$ difference score.

Notice in this source table that the degrees of freedom for the pretest, Z_i, equal zero since in this model its coefficient has been set at 1.0 rather than estimated from the data. The sum of squares associated with Z_i equals the difference between the sum of squares total for Y_i and the sum of squares total for the $Y_i - Z_i$ difference score.

In these data, the difference score analysis is only slightly less powerful than the ANCOVA. This near equivalence results from the fact that the estimated parameter for the covariate in the ANCOVA model, 1.27, is rather close to the value of 1.0 at which it is fixed in the difference score analysis. We could test whether the ANCOVA model results in significantly smaller errors of prediction than the difference score model. This is equivalent to testing whether the parameter associated with the

Source	*b*	*SS*	*df*	*MS*	*F*	*PRE*
Model		853.50	3	284.5	28.14	.70
Pretest	1.00	346.0	0			
Between conditions		507.50	3	169.17	16.73	.58
Method	3.25	422.50	1	422.50	41.79	.54
Tutor	0.75	22.50	1	22.50	2.22	.06
Method x Tutor	1.25	62.50	1	62.50	6.18	.15
Error		364.00	36	10.11		
Total		1217.50	39			

Figure 11.6 Revised difference score ANOVA source table

covariate in the augmented ANCOVA model significantly differs from 1.0. For this test, the ANCOVA model, in which the covariate's parameter is estimated, is the augmented one:

$$\text{MODEL A: } Y_i = \beta_0 + \beta_1 Z_i + \beta_2 X_{1i} + \beta_3 X_{2i} + \beta_4 X_{3i} + \varepsilon_i$$

The compact model is the difference score model in which the pretest's parameter is fixed at 1.0:

$$\text{MODEL C: } Y_i = \beta_0 + Z_i + \beta_1 X_{1i} + \beta_2 X_{2i} + \beta_3 X_{3i} + \varepsilon_i$$

Comparing the sums of squared errors for these two models, we get

$$\text{PRE} = \frac{364.00 - 348.07}{364.00} = .044$$

which converts to an F of 1.61 with 1 and 35 degrees of freedom. Hence, in these data, we cannot conclude that the parameter for the covariate is significantly different from 1.00. This conclusion means that the ANCOVA model for these data is not significantly more powerful than the difference score analysis.

Having said that the ANCOVA model will generally be more powerful than the difference score analysis, we should note that the ANCOVA uses up a degree of freedom in estimating the covariate's parameter whereas the difference score analysis does not. It is thus possible, if the estimated slope for the covariate is close to 1.00, for the mean square error from the difference score model to be smaller than that from the ANCOVA model. The difference score analysis may also be easier than the ANCOVA to describe to lay audiences.

The Case of a Partially Redundant Continuous Predictor

So far, we have illustrated one major reason for including a continuously measured concomitant variable in an analysis of condition differences. By including a covariate that is measured prior to randomization of participants to conditions, the covariate will generally be unrelated to condition and its inclusion in the model will increase the power of the tests of mean condition differences, so long as the covariate is highly related to the dependent variable of interest.

But there are other important reasons for including a continuously measured covariate in an analysis of mean condition differences. These reasons arise precisely *because* the covariate is correlated with condition and we wish to examine the condition differences over and above, or controlling for, covariate differences. There are two primary occasions when this is of interest. Analytically, these occasions are identical—we simply will be estimating an ANCOVA model with a covariate that is partially redundant with the categorical predictor(s)—but they differ in important theoretical ways that give rise to rather different interpretations.

The first occasion on which controlling for a partially redundant covariate is of interest occurs when, for whatever reason, there are pre-existing differences among participants in the various conditions of the design, and the researcher wishes to attempt to examine the effects of the categorical variable(s) free from those differences. Suppose, for instance, in the example we have been using, that students had not been randomly assigned to the four cells defined by the tutor and method factors. If the decision about which students got which tutor–method condition was not based on a random decision rule but was instead based on some unknown assignment rule (e.g., student and tutor schedules), then we might expect differences among the various conditions on the pretest long before the students had been exposed to the tutor or to the method. We would then quite reasonably want to control statistically for these pretest differences in looking at condition differences on the post-test. In other words, we might then like to control for the pretest in our analysis so that we could look at differences among conditions on the post-test free from the pre-existing differences on the pretest.

This use of ANCOVA to control for or adjust for known pre-existing differences among participants is particularly common in what have come to be known as quasi-experimental research designs where participants are not randomly assigned to condition (Judd & Kenny, 1981a; Shadish et al., 2002). Although this adjustment function of ANCOVA (attempting to equate participants on pre-existing covariate differences) is widely used in such quasi-experimental designs, it is important to recognize that this analytic approach is not a general solution to the problem of causal inference in nonexperimental research. The ability to reach causal conclusions about the effects of independent variables depends not on the statistical analysis that one undertakes but on the research design.

The second occasion on which it becomes important to conduct an ANCOVA with a covariate that is partially redundant with condition is when one is attempting to test a mediational model (Baron & Kenny, 1986; Judd & Kenny, 1981b). Suppose one has conducted an experiment in which the effect of some treatment difference on a dependent variable is to be estimated. If there is an effect of the experimental conditions (as estimated through an analysis of variance), then there certainly must be an underlying mediating mechanism or process that is responsible for that effect. Often one has an idea about the nature of that mechanism, and one may attempt to measure one or more variables that intervene in a causal chain between the experimental manipulation and the ultimate dependent variable that it affects. These intervening variables are known as mediating variables. If, in fact, the hypothesized mediating mechanism is partially responsible for the effect of the experimental treatment on the ultimate dependent variable, then that effect should no longer be as apparent once one conducts an ANCOVA, controlling for the mediating variable.

Although the analytic models in these two cases are essentially identical—to determine if there are condition differences once one controls for another variable that is known to be correlated with condition—they represent different theoretical models about the underlying causal process linking the categorical independent variable(s) with the dependent variable. This difference is perhaps most clearly shown by representing the two cases in terms of hypothesized causal models, which are shown in Figure 11.7. In these causal models, single-headed arrows represent hypothesized causal effects of one variable upon another, whereas curved double-headed arrows represent a correlation between two variables with no causal claims about the process responsible for that correlation. As previously, we use X to represent a categorical independent variable (here with only two levels, thus needing only one contrast-coded predictor), Z represents the continuously measured covariate, and Y represents the dependent variable.

The first causal model represents the quasi-experimental case, in which participants have not been randomly assigned to the two levels of the categorical independent variable and therefore the two groups differ on some continuously measured variable Z. (X and Z are thus partially redundant with each other.) Here one wishes to estimate the effect of X on Y free from the potential confounding

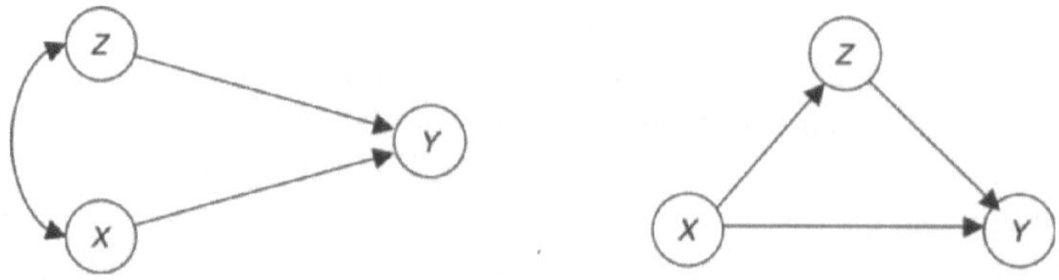

Figure 11.7 Theoretical causal models for two sorts of partially redundant covariates

influence of Z on Y. The goal then is to somehow come up with a more accurate estimate of the X effect, controlling for or eliminating any confounding due to the partially redundant covariate.

The second case is one in which participants have been randomly assigned to levels of the categorical independent variable, and one measures both the ultimate dependent variable, Y, as well as another variable, Z, that is thought to be affected by X. As the model indicates, here X may exert an effect on Y in two ways: The first is through the mediating variable, Z, and the second is over and above that mediating variable. In this case, one controls for the mediating variable, Z, in an ANCOVA because one is interested is examining whether there is any residual direct effect of X on Y over and above the hypothesized mediating process through Z.

As we have said, the analytic approach in these two cases is identical—estimating effects while controlling for a continuously measured covariate that is partially redundant with the categorical independent variable. Nevertheless, to illustrate both and the differences in interpretations that result from the different causal models underlying them, we provide an example of each one.

Examining Partial Effects in a Quasi-Experimental Situation

In Figure 11.8, we present data that have been somewhat modified from those used in the example earlier in this chapter. The values of Y_i are identical to what we had earlier in Figure 11.1, as are the definitions of the contrast-coded predictor variables, with X_{1i} coding method and X_{2i} coding tutor. What we have modified slightly are the values of the pretest variable, Z_i, so that now there are differences among the four cells of the design (resulting from the crossing of the two factors) on the pretest measure. The four cell means for both the pretest, Z_i, and the post-test, Y_i, are given in Figure 11.9.

There is now a definite relationship in these data between condition and Z_i, the pretest, since its condition means are no longer equal. If we regressed Z_i on the three contrast-coded predictors (X_{3i} being defined as earlier to capture the Method X Tutor interaction), we get the following parameter estimates:

$$\hat{Z}_i = 50.25 + 1.75X_{1i} + 0.25X_{2i} + 0.75X_{3i}$$

with a sum of squared errors of 225.94. The estimated slopes in this model tell us about the differences in the mean pretest values, according to Equation 9.1, the expression for the slopes of contrast-coded predictor variables:

$$b = \frac{\sum_k \lambda_k \bar{Z}_k}{\sum_k \lambda_k^2}$$

To test whether these pretest means are significantly different from each other we could compare this as Model A with a Model C that simply predicted the grand mean of Z_i for every observation. This Model C has a sum of squared errors of 373.5, resulting in a PRE of .40 and an F of 7.84 with 3 and 36 degrees of freedom. Thus, there is significant redundancy between the pretest and the categorical variables that code condition.

Y_i	X_{1i}	X_{2i}	Z_i	Y_i	X_{1i}	X_{2i}	Z_i
58	1	−1	51	57	1	−1	50
63	1	−1	50	61	1	−1	53
65	1	−1	54	57	1	−1	51
56	1	−1	48	67	1	−1	52
60	1	−1	54	56	1	−1	47
50	−1	−1	48	62	−1	−1	53
58	−1	−1	50	55	−1	−1	47
52	−1	−1	49	63	−1	−1	51
55	−1	−1	46	50	−1	−1	45
57	−1	−1	51	58	−1	−1	50
61	1	1	50	59	1	1	52
71	1	1	56	65	1	1	57
68	1	1	55	60	1	1	49
58	1	1	51	65	1	1	54
68	1	1	54	65	1	1	52
47	−1	1	44	62	−1	1	49
56	−1	1	49	51	−1	1	47
63	−1	1	51	54	−1	1	49
53	−1	1	46	58	−1	1	48
52	−1	1	45	54	−1	1	52

Figure 11.8 Modified pretest–post-test data

Condition	$\bar{Z}_k$	$\bar{Y}_k$
Old Method, Tutor A	49	56
Old Method, Tutor B	48	55
New Method, Tutor A	51	60
New Method, Tutor B	53	64

Figure 11.9 Modified pretest and post-test means by condition

Turning to the analysis of post-test scores, if we conducted that analysis while ignoring the pretest, the exact same model would result as before:

$$\hat{Y}_i = 58.75 + 3.25X_{1i} + 0.75X_{2i} + 1.25X_{3i}$$

and the ANOVA source table would be identical to the source table in Figure 11.3 because we have done nothing to alter the Y_i variable in this modified dataset.

However, the ANCOVA model, controlling for the partially redundant pretest variable, looks very different from what it was previously:

$$\hat{Y}_i = -4.84 + 1.26Z_i + 1.04X_{1i} + 0.43X_{2i} + 0.30X_{3i} \qquad \text{SSE} = 348.07$$

Note that the regression coefficients for the contrast-coded predictors have been dramatically affected by including the pretest variable Z_i. No longer does the coefficient for X_{1i} equal half the difference between the average Y_i under the new and old curricula. Similarly, the coefficients for X_{2i} and X_{3i} can no longer be interpreted as they were previously in terms of differences among various $\bar{Y}_k$. In other words, when a covariate that is partially redundant with the contrast-coded predictors is included in the model, the partial regression coefficients for the contrast-coded predictors can no longer be derived from Equation 9.1:

$$b = \frac{\sum_k \lambda_k \bar{Y}_k}{\sum_k \lambda_k^2}$$

and therefore a test of whether the parameters for these contrast-coded predictors equal zero is no longer a simple test of a comparison among the Y_i means in the various conditions.

With the inclusion of a covariate, the regression coefficient for a contrast-coded predictor variable is given by

$$b = \frac{\sum_k \lambda_k \bar{Y}_k}{\sum_k \lambda_k^2} - b_z \frac{\sum_k \lambda_k \bar{Z}_k}{\sum_k \lambda_k^2}$$

where b_z is the regression coefficient associated with the covariate in the full model that includes both the covariate and the set of contrast-coded predictors.

To illustrate this expression, let us calculate the value of the regression coefficient for X_{1i} in the model that includes Z_i. Notice that the first half of this expression equals the regression coefficient for X_{1i} in the model that did not include the covariate, i.e., 3.25. The second half of the expression equals the parallel difference coded by the contrast weights among the *covariate or pretest* condition means, weighted by the regression coefficient for the pretest. Numerically, for the coefficient for X_{1i} the second half of this expression equals

$$b_z \frac{\sum_k \lambda_k \bar{Z}_k}{\sum_k \lambda_k^2} = 1.26\left(\frac{53+51-48-49}{4}\right) = 1.26(1.75) = 2.21$$

In sum, then, according to this expression the regression coefficient for the X_{1i} contrast-coded predictor equals

$$b_1 = 3.25 - 1.26(1.75) = 1.04$$

This new expression for the regression coefficient for a contrast-coded predictor in the presence of a covariate is equal to the magnitude of the difference among the $\bar{Y}_k$ coded by the contrast weights, adjusting for or subtracting off the magnitude of the same difference among the covariate condition means $\bar{Z}_k$. The degree to which the coded comparison among the $\bar{Y}_k$ is adjusted by the same comparison among the $\bar{Z}_k$ depends on the magnitude of the covariate's regression coefficient, b_z. In sum, the regression coefficients for contrast-coded predictors in the presence of a covariate tell us about the differences among the $\bar{Y}_k$ coded by the contrast weights, adjusting that difference for the parallel difference that exists in the covariate condition means $\bar{Z}_k$. The degree to which this adjustment is performed depends on the magnitude of the within-condition relationship between the covariate and the dependent variable, i.e., the partial regression coefficient for the covariate.

In the case of the regression coefficient for X_{1i}, we know that half the difference between the mean Y_i score under the new and old methods equals 3.25, i.e., the regression coefficient for X_{1i} not controlling for the pretest. The regression coefficient for X_{1i} controlling for the pretest equals 1.04. This equals half the difference in Y_i associated with the difference in method over and above any pretest differences associated with method.

An equivalent but slightly different way to think about the coefficient for a contrast-coded predictor in the presence of a covariate is that it tells us about the magnitude of differences among adjusted values of $\bar{Y}_k$, adjusting those condition means to get rid of differences in the covariate condition means. More precisely, we can compute for each condition the adjusted mean:

$$\bar{Y}_k' = \bar{Y}_k - b_z\left(\bar{Z}_k - \bar{Z}\right)$$

where $\bar{Z}$ is the mean of the condition means for the covariate. We can then use these adjusted means to derive the regression coefficient for a contrast-coded predictor in the model that includes the covariate, using the old formula for the regression coefficient for a contrast-coded predictor. In other words, once we have the adjusted means, the regression coefficient for a contrast-coded predictor in the model that includes the partially redundant covariate equals

$$b = \frac{\sum_k \lambda_k \bar{Y}'_k}{\sum_k \lambda_k^2}$$

where $\bar{Y}_k$ are the adjusted cell means as just defined.

To illustrate, in Figure 11.10 the values of these adjusted cell means, $\bar{Y}'_k$, are given for the four Method × Tutor conditions of our design. These were derived using the formula for the previous adjusted cell means. For example, the value of the adjusted mean for the Old Method × Tutor A condition is given by

$$\bar{Y}'_k = 56 - 1.26(49 - 50.25) = 57.58$$

where 56 is the value of $\bar{Y}_k$ for this condition, 1.26 is the regression coefficient for the pretest in the model that includes both the pretest and the condition contrast-coded predictors, 49 equals the pretest mean for this condition, $\bar{Z}_k$, and 50.25 equals the mean of the four pretest condition means.

We can now use these adjusted $\bar{Y}_k$ to compute the regression coefficients for the contrast-coded predictors. For instance, the coefficient for X_{1i} in the model that includes the covariate equals

$$b_1 = \frac{\sum_k \lambda_k \bar{Y}'_k}{\sum_k \lambda_k^2} = \frac{(-1)57.58 + (-1)57.85 + (+1)59.05 + (+1)60.52}{4} = 1.04$$

Conceptually, then, the coefficients for contrast-coded predictors in the presence of a covariate tell us about the magnitude of the coded differences among adjusted condition means, adjusting those dependent variable means by the extent to which the covariate means depart from each other. If all $\bar{Z}_k$ are equal to each other there will be no adjustment, as we saw in the case of an orthogonal

Condition	$\bar{Y}'_k$
Old Method, Tutor A	57.58
Old Method, Tutor B	57.85
New Method, Tutor A	59.05
New Method, Tutor B	60.52

Figure 11.10 Post-test condition means adjusted for the pretest($\bar{Y}'_k$)

covariate. But with a partially redundant covariate, the differences among the $\bar{Z}_k$ will result in some adjustment among the dependent variable means that are compared when examining the regression coefficient for a contrast-coded predictor.

The source table for the ANCOVA model with these modified data is given in Figure 11.11. Notice that the sums of squares and, as a result, the F statistics for the omnibus test of condition differences and the individual contrast tests are all considerably smaller than they were in the ANOVA source table. This is so because they are testing different null hypotheses than they were in the ANOVA model. In the ANOVA model the omnibus test was testing whether there were any differences among the condition means $\bar{Y}_k$. The tests of the contrasts were testing specific comparisons among these condition means. With the inclusion of the pretest in the model, the omnibus test is now testing for the presence of differences among the adjusted condition means. Similarly, the contrast tests are now testing specific comparisons among these adjusted condition means. In other words, the tests are now examining condition differences in $\bar{Y}_k$ having adjusted for condition differences on the pretest.

Because the pretest is now partially redundant with the contrast codes that code condition, the sums of squares in this source table are not additive as they were in the earlier ANOVA table. Therefore, some of them must be derived through the estimation of compact models. To get the sum of squares for the omnibus condition test, we must estimate a model with only the pretest used as a predictor since this omnibus sum of squares is no longer equal to the sum of the three sums of squares explained by the three contrast-coded predictors. Even though the contrast codes are still orthogonal and even though there are still an equal number of observations in each condition, the three contrast-coded predictors differ in how redundant or correlated they are with the covariate. Thus, the sum of their three individual sums of squares is not equal to the difference in the sum of squares explained if they are all omitted from the model.

Models with two or more partially redundant covariates are simple extensions to the single covariate case that we have just examined. The formula for the adjusted means adjusts for or subtracts off mean differences on each covariate, each weighted by its regression coefficient. For instance, if Z_{1i} and Z_{2i} are two covariates, then the regression coefficient for contrast-coded predictors in the model that included these two covariates would examine comparisons among the following (doubly) adjusted cell means:

$$\bar{Y}'_k = \bar{Y}_k - b_{z1}(\bar{Z}_{1k} - \bar{Z}_1) - b_{z2}(\bar{Z}_{2k} - \bar{Z}_2)$$

As a final comment, reiterating what we said before, although ANCOVA permits comparisons among condition means adjusting for one or more partially redundant covariates, it is not a general solution to problems of internal validity in research designs where random assignment of

Source	*b*	*SS*	*df*	*MS*	*F*	*PRE*
Model		869.43	4	217.36	21.87	.71
Pretest	1.26	361.93	1	361.93	36.39	.51
Between conditions		34.30	3	11.43	1.15	.09
Method	1.04	27.81	1	27.81	2.80	.07
Tutor	0.43	7.44	1	7.44	0.75	.02
Method x Tutor	0.30	3.29	1	3.29	0.33	.01
Error		348.07	35	9.94		
Total		1217.50	39			

Figure 11.11 ANCOVA source table

observations to conditions has not been used. Adjustment for redundant covariates is a powerful procedure, but it does not solve the problems that limit causal inferences in nonexperimental research designs.

Examining Partial Effects: The Case of Mediation

All of what we have just said about the analysis of categorical predictor variables when we control for a partially redundant covariate applies as well to the second occasion when such models are of interest, namely, when we wish to examine the process that mediates the effects of some experimental categorical variable on a dependent variable. Let us consider a new example to illustrate the interpretations that ensue from such an analysis in this case. Suppose that participants were randomly assigned to a family counseling intervention designed to improve outcomes for adolescents who suffer from bipolar disorder. Either they receive the family counseling intervention along with the usual pharmacological care or they receive only the usual care. And the question is whether this intervention affects the manifestation of bipolar symptoms eight weeks later. The dependent variable is assessed by a clinical psychologist who is blind to experimental treatment, using a 10-point rating scale (higher scores equal more symptoms).

If the family counseling intervention is effective in reducing symptoms, it is thought that it must operate by affecting the amount of criticism the parents direct at the bipolar adolescent during the interim period. Hence, at week 7 of the study, another clinical psychologist, again blind to experimental treatment, interviews the parents and rates the degree to which the parents spontaneously criticize the adolescent, this time on a five-point scale (higher scores equal more criticism). Hence, for each of 20 families there are three variables: whether they received the experimental treatment or not, assessed parental criticism at 7 weeks, and adolescent symptoms at 8 weeks. Hypothetical scores on these variables for the families are given in Figure 11.12.

First, one assesses whether the treatment had an effect on the ultimate outcome variable: symptoms at 8 weeks. If it did, then the theoretical expectation is that those effects are mediated through the parental criticism variable, measured at 7 weeks. In other words, the thinking is that the treatment effect ought to work, if it does, by affecting the manifestation of parental criticism of the adolescent and this in turn is responsible for lower symptom levels.

	Treatment		*Control*	
	Criticism (C_i)	*Symptoms (S_i)*	*Criticism (C_i)*	*Symptoms (S_i)*
	3	4	4	7
	2	3	4	8
	4	4	3	5
	2	3	2	5
	2	4	4	7
	3	5	4	6
	1	3	5	6
	2	5	3	4
	3	7	4	5
	3	5	4	6
Mean	2.5	4.3	3.7	5.9

Figure 11.12 Hypothetical data for 20 families

These expectations imply the following:

1 There will be an overall treatment effect on symptoms at 8 weeks such that treatment will reduce the number of symptoms. (Note, however, that under some conditions, mediation may be present even when this condition is not met [Kenny & Judd, 2014; MacKinnon et al., 2002; O'Rourke & MacKinnon, 2018]).
2 The treatment will lead to lower parental criticism.
3 Parental criticism will predict fewer symptoms at 8 weeks holding constant treatment condition.
4 The effect of the treatment on symptoms will be reduced when parental criticism is controlled.

The first two of these conditions can be assessed by a separate two-group ANOVA model for each of the dependent variables, first symptoms at 8 weeks and then parental criticism at 7 weeks. The third and fourth conditions can be assessed by a single ANCOVA model, using both treatment and parental criticism as predictors of symptoms. Here, we expect that criticism will predict fewer symptoms and that the effect of treatment on symptoms will be reduced compared to its effect in the simple ANOVA model. These are the classic conditions for establishing mediation, as identified by Judd and Kenny (1981b) and Baron and Kenny (1986).

We first contrast code the treatment variable ($X_i = +.5$ if treatment; $-.5$ if control) and then examine its effects on both symptoms and criticism with two two-group ANOVA models, examining the first two of the aforementioned four expectations. The overall treatment effect on symptoms is estimated as

$$\hat{S}_i = 5.10 - 1.60X_i \qquad \text{SSE} = 27.00$$

This model predicts the two treatment means, and the estimated slope for X_i equals the difference between those means, given the codes we have used. A test of whether the slope differs from zero, and thus whether the two group means on symptoms differ, yields a PRE of .32 and an F of 8.53 with 1 and 18 degrees of freedom. Hence, there is an overall treatment effect; the treatment resulted in fewer symptoms measured 8 weeks later.

The effect of treatment on criticism at 7 weeks is examined in the following two-group ANOVA model:

$$\hat{C}_i = 3.10 - 1.20X_i \qquad \text{SSE} = 12.60$$

Again, the slope for the treatment variable equals the mean difference between the treatment and control groups in criticism and this difference is significant: PRE = .36, $F_{1,18} = 10.29$.

To assess the third and fourth mediation expectations, the ANCOVA model is estimated as

$$\hat{S}_i = 2.84 - 0.72X_i + 0.73C_i \qquad \text{SSE} = 20.28$$

In the previous paragraph we showed that the means for criticism differ significantly between the two conditions, demonstrating that the two predictors in this ANCOVA model are redundant with each other. Accordingly, the coefficient for treatment no longer estimates the difference between the two condition means; rather, it estimates the difference between the two conditions when adjusting or controlling for differences between them on the criticism variable:

$$b_{SX.C} = \frac{\sum_k \lambda_k \bar{S}_k}{\sum_k \lambda_k^2} - b_{SC.X} \frac{\sum_k \lambda_k \bar{C}_k}{\sum_k \lambda_k^2} = -1.60 - (0.73)(-1.20) = -0.72$$

Notice in this expression that we are indicating each parameter estimate in this model with multiple subscripts, the first letter indicating the dependent variable, the second the predictor involved, and then, following the dot, any other variables controlled for in the model. The reason for this complete notation will become apparent shortly.

In this model, the test of the slope associated with the covariate, criticism, is significant (PRE = .25, $F_{1,17} = 5.63$), while that for the treatment variable no longer is (PRE = .08, $F_{1,17} = 1.40$). Accordingly, it seems that the four conditions for establishing mediation outlined earlier have been met. Namely, the treatment has an overall effect on the ultimate outcome variable as well as on the mediator. The mediator significantly affects the outcome when controlling for treatment, whereas when controlling for the mediator the effect of treatment on symptoms is no longer significant.

However, such a demonstration does not necessarily mean that there has been a significant reduction in the treatment effect once criticism is controlled. In other words, although the treatment effect no longer differs from zero, it might not significantly differ from its effect in the initial ANOVA model. For this and other reasons, tests of whether the partial regression coefficient for the treatment in the ANCOVA model is significantly smaller than its overall or total effect have been developed. Rather than develop those tests here, we refer the interested reader to MacKinnon et al. (2002), MacKinnon (2008), and Judd et al. (2014). Importantly, however, they rest on the following equality:

$$b_{SX} - b_{SX.C} = b_{SC.X} b_{CX}$$

which means that the reduction of the overall effect of the treatment when the mediator is controlled equals the product of the treatment's effect on the mediator and the mediator's partial effect on the outcome controlling for the treatment.

This equality is none other than our formula for the partial slope associated with a contrast-coded predictor when a redundant covariate is controlled, as just given:

$$b_{SX.C} = \frac{\sum_k \lambda_k \bar{S}_k}{\sum_k \lambda_k^2} - b_{SC.X} \frac{\sum_k \lambda_k \bar{C}_k}{\sum_k \lambda_k^2} = b_{SX} - b_{SC.X} b_{CX}$$

A small amount of algebraic manipulation gives the desired equality: $b_{SX} - b_{SX.C} = b_{SC.X} b_{CX}$. In other words, the tests that have been developed to determine whether the treatment effect is significantly reduced when the mediator is controlled are tests of the partial regression slope associated with the treatment variable. If such tests are significant (i.e., the regression slope is significantly smaller in the ANCOVA model), the treatment variable is said to have an indirect effect on the outcome variable via the mediator.

This mediation example is rather simple, and our treatment does not do justice to the extensive literature on the subject. The point of this section is to illustrate that everything we have said about ANCOVA with a partially redundant covariate applies in the case of mediation assessment as well as in the case of a confounded treatment variable. These two differ not in the analyses conducted but in their underlying theoretical model.

The Homogeneity of Regression Assumption in ANCOVA

Most classic treatments of ANCOVA specify that an assumption, referred to as the homogeneity of regression assumption, is crucial to the use and interpretation of ANCOVA results. To clarify this assumption, why it is potentially important, and what may be done if it is violated, let us return to the data example used earlier, with two crossed independent variables, Method (new versus old) and Tutor (A versus B) and a confounded covariate, the pretest. The data we will use were given in Figure 11.8 and the cell means, for both the post-test and the pretest, were given in Figure 11.9. We use the same notation we did earlier: the post-test is Y_i, the pretest is Z_i, X_{1i} is the contrast-coded predictor representing Method (+1 if new; –1 if old), X_{2i} is the contrast-coded predictor representing Tutor (–1 if A; +1 if B), and X_{3i} is their product, representing the Method × Tutor interaction.

The ANCOVA model that we presented earlier for these data was estimated as

$$\hat{Y}_i = -4.84 + 1.26Z_i + 1.04X_{1i} + 0.43X_{2i} + 0.30X_{3i} \qquad \text{SSE} = 348.07$$

And we saw that the slopes associated with the contrast-coded predictors could be interpreted either as differences among Y_i cell means, adjusting for parallel differences among the Z_i cell means:

$$b = \frac{\sum_k \lambda_k \bar{Y}_k}{\sum_k \lambda_k^2} - b_z \frac{\sum_k \lambda_k \bar{Z}_k}{\sum_k \lambda_k^2}$$

or as differences among the adjusted Y_i cell means, where those adjusted means are given by the expression

$$\bar{Y}'_k = \bar{Y}_k - b_z(\bar{Z}_k - \bar{Z})$$

The assumption of homogeneity of regression that underlies this analysis, and the interpretation of its parameter estimates, is that the relationship between the pretest and the post-test is invariant across the four conditions defined by the contrast-coded predictors. To understand this, note that in these expressions for the interpretation of the slopes associated with the contrast-coded predictors we use one value for the slope of the pretest, b_Z, rather than different values for the four different conditions. That is, when calculating the adjusted cell means that are compared, a single value of b_Z is used regardless of the value of k. If the relationship between the dependent variable, Y_i, and the covariate, Z_i, differs substantially in magnitude across the various conditions, then we should not be assuming a single adjustment weight b_Z, but instead should be allowing for different adjustment weights for the various conditions. Accordingly, in controlling for a covariate and interpreting the resulting parameter estimates for the contrast-coded predictors, it makes sense to examine whether the relationship between Y_i and Z_i is invariant or homogeneous across conditions.

To suggest that the relationship between the pretest and the post-test depends on condition is to suggest that condition and the pretest interact in affecting the post-test. Therefore, to test the homogeneity of the relationship between Y_i and Z_i across conditions, we need to test whether the interactions between Z_i and the condition-defining contrast-coded predictors are significant. To examine these interactions, we follow the standard procedure of computing products of the variables whose interactions we wish to test and then entering those product variables as separate predictor variables into a model that includes the variables that are components of the products. We then test whether the augmented model that includes the product terms (i.e., the interactions) generates significantly better predictions than a compact model that omits them.

Let us illustrate this using the data contained in Figure 11.8. Because we have three contrast codes that define condition, we need three interaction or product terms to examine the condition by pretest interaction: $X_{1i}Z_i$, $X_{2i}Z_i$, and $X_{3i}Z_i$. When these are included in the model, the slope associated with the first product predictor will estimate the extent to which the pretest–post-test relationship differs across the two methods (i.e., the extent to which the homogeneity assumption is violated). The second slope will estimate the extent to which the pretest–post-test relationship differs across the two tutors. The third will examine the triple interaction, i.e., whether the pretest–post-test relationship differs across levels of the Method × Tutor interaction.

The model that includes all three of these pretest × condition interactions is estimated as

$$\hat{Y}_i = -4.40 + 1.26Z_i + 6.11X_{1i} - 1.32X_{2i} - 3.93X_{3i} - 0.10X_{1i}Z_i + 0.03X_{2i}Z_i + 0.08X_{3i}Z_i \qquad \text{SSE} = 344.23$$

To demonstrate that this model in fact does allow the slope for the covariate to vary between the four cells of our design, let us examine the expression for the "simple" effect of the covariate in the various cells. The model can be rewritten as

$$\hat{Y}_i = (-4.40 + 6.11X_{1i} - 1.32X_{2i} - 3.93X_{3i}) + (1.26 - 0.10X_{1i} + 0.03X_{2i} + 0.08X_{3i})\, Z_i$$

And from this, we can derive the "simple" Y_i:Z_i regression models in each of the four cells of the design, substituting for the various values of X_{1i}, X_{2i}, and X_{3i}. For instance, for the New Method, Tutor B cell, we get

$$\hat{Y}_i = (-4.40 + 6.11(+1) - 1.32(+1) - 3.93(+1)) + (1.26 - 0.10(+1) + 0.03(+1) + 0.08(+1))\, Z_i$$

which reduces to

$$\hat{Y}_i = -3.54 + 1.27Z_i$$

Parallel expressions for the other three cells of the design give

New Method, Tutor A: $\hat{Y}_i = 6.96 + 1.05Z_i$

Old Method, Tutor B: $\hat{Y}_i = -7.90 + 1.31Z_i$

Old Method, Tutor A: $\hat{Y}_i = -13.12 + 1.41Z_i$

Clearly this model allows different "simple" slopes for Z_i in the four cells of the design.

Because we wish to examine whether there are any differences among conditions in these "simple" slopes, and because we have no specific expectations that predict such differences, we can conduct an omnibus three-degree-of-freedom test to examine the homogeneity of regression assumption, testing whether the set of three interaction terms leads to a significant improvement in the fit of the model. Thus, this interactive model becomes Model A and the earlier ANCOVA model is the Model C with which we want to compare it. This comparison yields the following PRE and F values:

$$\text{PRE} = \frac{348.07 - 344.23}{348.07} = .011$$

$$F_{3,32} = \frac{.011/3}{(1-.011)/32} = \frac{3.845/3}{344.226/32} = 0.119$$

There is no reason to prefer Model A over Model C. Thus, there is no evidence in these data that the homogeneity of regression assumption of ANCOVA has been violated. Equivalently, we have found no evidence to suggest that the relationship between the pretest and the post-test differs depending on which of the four cells of the research design we are looking at. Using a single slope, and thus a single adjustment weight, to compute the four adjusted means suffices.

As a check on the homogeneity of regression assumption, this omnibus model comparison seems sufficient. However, there may be times when one has prior expectations that one or more of the interactions between the pretest and the contrast-coded predictors would be found. For instance, it is possible that one might have expected the pretest–post-test relationship to be stronger under the old method than under the new one. In such a case, it seems appropriate to examine the slope associated with that particular interaction term, $X_{1i}Z_i$, comparing the augmented model that includes all three interaction terms to a compact one that only leaves out this one hypothesized interaction.

Should it be found that the homogeneity of regression assumption has been violated, either through the omnibus test of the set of interaction terms, or the more focused examination of single interaction terms in the case of prior expectations about their possible existence, then one is left with the model that includes one or more covariate × condition interactions and one is forced to interpret them. Importantly, then, the homogeneity of regression assumption is an assumption whose violation implies simply that a model more complicated than the ANCOVA model is required, and one must interpret the obtained significant interaction term(s). Thus, this assumption is not like the assumptions about the distribution of residuals that we made very early on in this book (i.e., that they are normally distributed, have a single variance, and are independent). Violations of those assumptions mean that our inferential tests are biased. In this case, violations of the homogeneity of regression assumption simply mean that life is a bit more complicated, and one must interpret the data in light of the resulting covariate × condition interactions.

Suppose, for instance, in the present case, that the interaction between Z_i and X_{1i} had been significant. One would then be compelled to provide interpretations in the context of the more complicated interactive model that included at least this significant interaction. And in this case, the resulting interaction could be interpreted by focusing either on the extent to which the pretest-post-test relationship depends on the new versus old method or on the extent to which the magnitude of the new—old method difference depends on the value of the pretest. Although these two interpretations are fundamentally equivalent, given our focus on the condition differences, the second is likely to be the preferred interpretation of the interaction coefficient. We will proceed to illustrate the interpretation with the present data, even though our omnibus test clearly failed to show any evidence of the interactions, and the more focused single-degree-of-freedom test of just $X_{1i}Z_i$ was also not significant in these data.

As we saw in Chapter 8, parameter estimates in models with interactions can be difficult to interpret. This difficulty emerges because slopes of components of product predictors estimate the effects of those variables only when the value of the other component variable equals zero. In the case of the current Model A, the estimated slope for X_{1i} (i.e., 6.11) informs us about half the expected method difference for someone scoring zero on the pretest variable, Z_i, but there are no pretest values in our dataset that are close to zero. Parameter estimates are more meaningful if one centers or mean-deviates the covariate in models that include products of the covariate and the contrast-coded predictors. Recall that, in addition to centering the covariate, as a predictor in its own right, the product predictors must also be recomputed using the centered variable. The resulting Model A with a centered covariate is

$$\hat{Y}_i = 58.86 + 1.26Z_{Ci} + 1.00X_{1i} + 0.36X_{2i} + 0.27X_{3i} - 0.10X_{1i}Z_{Ci} + 0.03X_{2i}Z_{Ci} + 0.08X_{3i}Z_{Ci}$$

where Z_{Ci} is the centered version of the pretest (i.e., $Z_{Ci} = Z_i - \bar{Z}$). In a deep sense, of course, this centered model is identical to the model prior to centering, in that it makes the same predictions and has the same sum of squared errors (i.e., 344.23). But now the slope estimates associated with the three contrast-coded predictors are considerably more interpretable: They estimate the "simple" condition differences at the mean value of the pretest. For instance, the slope associated with X_{1i} in this model (i.e., 1.00) estimates half the predicted difference in post-test scores between the old and new methods when the pretest equals its mean and when the magnitude of the old—new method difference is allowed to depend on the value of the pretest. The degree to which that difference depends on the value of the pretest is indicated by the parameter estimate associated with the $X_{1i}Z_{Ci}$ product predictor. As the value of the pretest increases by one unit, the estimated "simple" difference between the old and new method decreases by 0.10 units. Had this value been significant, we would have concluded that method effects are larger for students who performed worse on the pretest.

An equivalent interpretation for this parameter estimate can be given by focusing on the "simple" slopes of the pretest in the two different method conditions derived earlier. That "simple" slope, on average, across the two conditions is 1.26. The parameter estimate associated with the $X_{1i}Z_{Ci}$ product predictor tells us that the "simple" slope decreases by 0.10 units in the new method condition and increases by 0.10 units in the old method condition. In other words, the pretest–post-test relationship is stronger in the old method than in the new one.

Our interpretation of covariate × condition interactions in this example has been a bit forced because the data suggest that the homogeneity of regression assumption is not violated. Our purpose has been simply to illustrate how one might proceed if it happened that the assumption *was* violated. One would simply live with the more complex model and interpret the significant covariate × condition interactions. But there are other conditions where such interactions are the primary focus of research, where one is interested in whether the magnitude of the relationship between some continuous predictor variable (heretofore called a covariate) and the dependent variable depends on the levels of one or more categorical variables. And it is to such an example that we now turn.

Models With Continuous and Categorical Predictors Outside of Experimental Contexts

Our hypothetical data consist of the academic records of 2740 first-year students, either enrolled in the College of Arts and Sciences or in the College of Engineering at a large university. We have four variables available on these students: their university aptitude test score, taken during their last year in secondary school, on a scale of 20–80; their cumulative grade point average (GPA) at the end of the first year of university; the college in which they were enrolled; and whether they were first-generation (i.e., neither parent attended college) or continuing generation (i.e., one or more parents attended college). We are interested in knowing whether applicants' aptitude test scores predict their first-year GPA at university and whether that relationship depends on college and/or first-generation status.

The mean aptitude scores and first-year GPAs for each of the four groups, defined by college and first-generation status, are given in Figure 11.13.

We start by examining mean differences as a function of first-generation status and college in each of these variables, defining X_{1i} as –1 if First-Generation, +1 if Continuing Generation, X_{2i} as +1 if Arts and Sciences, –1 if Engineering, and X_{3i} as their product. The two ANOVA models, one for aptitude and one for GPA, are given here:

$$\widehat{\text{Aptitude}_i} = 59.95 - 0.99X_{1i} - 2.44X_{2i} + 0.11X_{3i} \qquad SSE = 115139.16$$

	Aptitude		*GPA*	
	First-Generation	*Continuing-Generation*	*First-Generation*	*Continuing-Generation*
Engineering				
Mean	63.50	61.30	2.88	3.03
SD	(6.02)	(5.60)	(0.70)	(0.63)
n	329	91	329	91
Arts and Sciences				
Mean	58.39	56.62	2.68	2.93
SD	(6.64)	(6.54)	(0.76)	(0.67)
n	968	1352	968	1352

Figure 11.13 Aptitude scores and freshman GPAs by college and first-generation status

$$\widehat{GPA}_i = 2.880 + 0.100X_{1i} - 0.077X_{2i} + 0.025X_{3i} \qquad SSE=1355.40$$

For both variables, the first-generation status difference is significant, although in opposite directions. First-generation students have lower aptitude scores on average than do continuing-generation students ($F_{1,2736} = 124.71$, PRE = .04) but they end up with higher first-year GPAs ($F_{1,2736} = 58.56$, PRE = .02). There is also a significant college difference for both variables, with students in Engineering having both higher aptitude scores ($F_{1,2736} = 196.53$, PRE = .07) and higher first-year GPAs ($F_{1,2736} = 21.39$, PRE = .01) than Arts and Sciences students.

A simple regression model in which aptitude scores are used to predict first-year GPAs yields the following parameter estimates:

$$\widehat{GPA}_i = 1.018 + 0.013\ \text{Aptitude}_i$$

As might be expected, students with higher aptitude scores when they enter university end up with better first-year GPAs ($F_{1,2738} = 270.97$, PRE = .09).

We have already seen that aptitude scores are significantly related to the two categorical variables of college and first-generation status. Therefore, when we estimate an ANCOVA model, with aptitude and X_{1i} –X_{3i} as predictors of GPA, these predictors will be partially redundant. The resulting model is

$$\widehat{GPA}_i = 0.38 + 0.136X_{1i} - 0.010X_{2i} + 0.021X_{3i} + 0.036\ \text{Aptitude}_i \qquad \text{SSE} = 1208.49$$

And the source table in Figure 11.14 summarizes the statistical results.

Clearly, aptitude remains a significant predictor of first-year GPA when controlling for first-generation status, college, and their interaction. Additionally, of the categorical predictors, only first-generation remains significant once aptitude is controlled. The effect of college, which was significant in the ANOVA model, no longer is. It seems that once students are equated in terms of aptitude scores, there is no longer a college difference in GPAs. As with any ANCOVA model, the slopes for the categorical predictors inform us about the magnitude of differences among the adjusted GPA cell means. These are computed as

$$\overline{GPA}'_k = \overline{GPA_k} - b_{Aptitude}\left(\overline{Aptitude_k} - \overline{Aptitude}\right)$$

Source	SS	df	MS	F	PRE
Model	187.16	4	46.79	105.89	.13
First-Generation Status (X_1)	18.57	1	18.57	41.99	.02
College (X_2)	0.10	1	0.10	0.22	.00
First-Generation Status x College (X_3)	0.46	1	0.46	1.04	.00
Aptitude	146.91	1	146.91	332.33	.11
Error	1208.49	2735	0.44		
Total	1395.65	2739			

Figure 11.14 ANCOVA source table

and their values are:

For continuing-generation Engineering students:	2.69
For first-generation Engineering students:	2.92
For continuing-generation Arts and Sciences students:	2.68
For first-generation Arts and Sciences students:	2.99

As we have already discussed, this model assumes that the "simple" slope of aptitude does not vary across the four groups. That is, this model makes the homogeneity of regression assumption that we discussed earlier. It is of central interest in these data to ask whether the relationship between aptitude and first-year GPA is the same across all four student groups. Perhaps aptitude scores are more predictive of GPA in one college or the other. If so, then the greater diagnosticity of the test among the students in one college than the other could, perhaps, be taken into account at the time of admissions, weighting more heavily the aptitude scores in making admissions decisions for some students than for others. Similarly, a first-generation status difference in the diagnosticity of the test would indicate predictive bias, that is, that the test makes better predictions for one group than the other, which would likely need to be addressed in other ways, for example by revising the test itself. In any case, if aptitude scores are more diagnostic among some student groups than others, then it implies that the homogeneity of regression assumption would be violated in these data: We would need different aptitude slopes for students who differ in their first-generation status, their course of study, or the interaction of these two factors.

To test whether aptitude is differentially related to GPA among the four different groups, we estimate a model that includes as predictors all products of aptitude with the contrast-coded predictors that code the four categories of students: $Aptitude_i X_{1i}$, $Aptitude_i X_{2i}$, and $Aptitude_i X_{3i}$. But this time, because we suspect that aptitude scores vary in their diagnosticity across the four groups, we will examine the individual contributions of each of these product predictors rather than simply conducting the omnibus test of whether they, as a set, increase the explanatory power of the model. The estimated Model A is

$$\widehat{GPA}_i = 0.200 - 0.26X_{1i} + 0.676X_{2i} + 0.289X_{3i} + 0.044\ \text{Aptitudei} +$$
$$0.007\ \text{Aptitudei}X_{1i} - 0.011\ \text{Aptitudei}X_{2i} - 0.004\text{Aptitudei}X_{3i}$$

$$SSE = 1204.18$$

Tests of the individual product predictors are summarized in the partial source table of Figure 11.15. Although the effects sizes are not large here, there is evidence to suggest that the relationship

Source	*SS*	*df*	*MS*	*F*	*PRE*
Aptitude × First-Generation Status ($AptitudeX_1$)	1.45	1	1.45	3.28	.001
Aptitude × College ($AptitudeX_2$)	3.92	1	3.92	8.88	.003
Aptitude × First-Generation Status × College ($AptitudeX_3$)	0.64	1	0.64	1.46	.001
Error	1204.18	2732	0.44		

Figure 11.15 Portion of source table testing interactions between continuous and categorical predictors

between aptitude and first-year GPA depends on whether one is an Engineering student or an Arts and Sciences student.

To interpret these differences, let us re-express the model in terms of the "simple" effect of aptitude for each of the four groups:

$$\widehat{GPA}_i = (0.200 - 0.260X_{1i} + 0.676X_{2i} + 0.289X_{3i}) + (0.044 + 0.007X_{1i} - 0.011X_{2i} - 0.004X_{3i})\,\text{Aptitude}_i$$

This re-expression can be used to generate the "simple" GPA: Aptitude prediction functions for each of the four cells by substituting the appropriate values for the contrast-coded predictors.

For continuing-generation Engineering students

$$\widehat{GPA}_i = (0.200 - 0.260(+1) + 0.676(-1) + 0.289(-1) + (0.044 + 0.007(+1) - 0.011(-1) - 0.004(-1))\,\text{Aptitude}_i = -1.025 + 0.066\,\text{Aptitude}_i$$

For first-generation Engineering students:

$$\widehat{GPA}_i = (0.200 - 0.260(-1) + 0.676(-1) + 0.289(+1)) + (0.044 + 0.007(-1) - 0.011(-1) - 0.004(+1))\,\text{Aptitude}_i = 0.073 + 0.044\,\text{Aptitude}_i$$

For continuing-generation Arts and Sciences students:

$$\widehat{GPA}_i = (0.200 - 0.260(+1) + 0.676(+1) + 0.289(+1)) + (0.044 + 0.007(+1) - 0.011(+1) - 0.004(+1))\,\text{Aptitude}_i = 0.906 + 0.036\,\text{Aptitude}_i$$

For first-generation Arts and Sciences students:

$$\widehat{GPA}_i = (0.200 - 0.260(-1) + 0.676(+1) + 0.289(-1)) + (0.044 + 0.007(-1) - 0.011(+1) - 0.004(-1))\,\text{Aptitude}_i = 0.847 + 0.031\,\text{Aptitude}_i$$

In Figure 11.16, we have graphed these four "simple" relationships. As the differences in these graphed slopes indicate, the two Engineering groups have steeper slopes than do the two Arts and Sciences groups. This is the implication of the significant Aptitude × College interaction, with its coefficient, –0.011, equaling half the difference between the average of the two "simple" slopes for the Arts and Sciences student groups and the average of the two "simple" slopes for the Engineering student groups. In other words, when averaging across first- and continuing-generation groups, aptitude is more diagnostic of first-year GPA among Engineering students than it is among Arts and Sciences students.

It is perhaps somewhat surprising that, if we were to conduct four simple regressions, one for each of the four student groups and regressing first-year GPA on aptitude scores for just the students in those four groups, we would get exactly the parameter estimates that we calculated above.

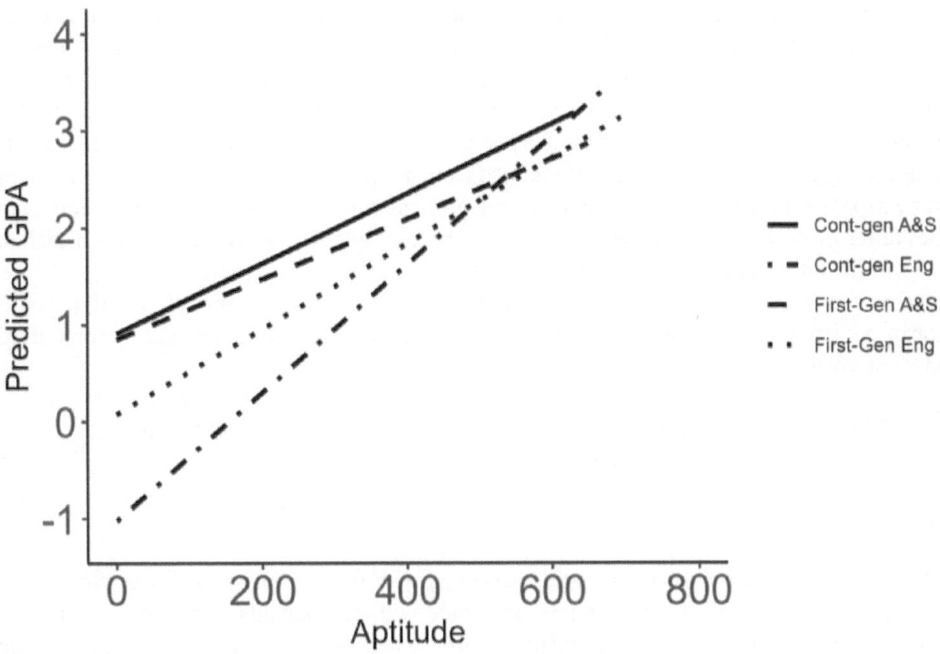

Figure 11.16 Four-group "simple" relationships between Aptitude and GPA

That is, we might split our sample into the four groups, and in each group separately regress GPA on Aptitude, with the following results:

For continuing-generation Engineering students:

$$\widehat{GPA}_i = -1.025 + 0.066\ \text{Aptitude}_i \quad SSE = 23.77$$

For first-generation Engineering students:

$$\widehat{GPA}_i = 0.073 + 0.044\ \text{Aptitude}_i \quad SSE = 135.75$$

For continuing-generation Arts and Sciences students:

$$\widehat{GPA}_i = 0.906 + 0.036\ \text{Aptitude}_i \quad SSE = 524.41$$

For first-generation Arts and Sciences students:

$$\widehat{GPA}_i = 0.847 + 0.031\ \text{Aptitude}_i \quad SSE = 520.25$$

In a deep sense, then, the full model that includes all three contrast-coded predictors and their interactions with aptitude scores is equivalent to four simple regression models, one from each of the four student groups, regressing GPA on aptitude. Accordingly, if we add up the four sums of squared errors from these four within-group simple regression models, we get the sum of squared errors from the overall interactive model (i.e., 135.75 + 23.77 + 520.25 + 524.41 = 1204.18).

Each of the within-group simple regressions provides us with a test of whether there is a significant association between aptitude and GPA in that group (and all four are significant in this case), but it is only by testing the various interaction terms in the full model, using all the data, that we are able to examine statistically whether the various simple slopes differ from each other across the four groups. And in this case, although all four simple slopes are significant, the slopes for the Engineering students are significantly larger than those for the Arts and Sciences students.

In general, then, to examine whether two variables are related more strongly to each other in some groups than in others, we encourage a test of whether the group × continuous predictor interactions are significant in a model using the data from all groups together. This might be accompanied by separate model estimates in each of the groups, to test whether the group-specific simple slopes differ from zero. But to determine whether these simple slopes are different from each other, the test of the interactions in the full model, with all the data, is necessary.

Before concluding this example, we re-estimate the full interactive model, this time centering or mean-deviating aptitude (represented as $Aptitude_{Ci}$), both as a predictor and in computing the product predictors:

$$\widehat{GPA}_i = 2.784 + 0.124X_{1i} + 0.045X_{2i} + 0.033X_{3i} + 0.044\ \text{Aptitude}_{Ci} + 0.007\text{Aptitude}_{Ci}X_{1i} - 0.011\ \text{Aptitude}_{Ci}X_{2i} - 0.004\ \text{Aptitude}_{Ci}X_{3i}$$

$$\text{SSE} = 1204.18$$

Centering makes the parameter estimates associated with the contrast-coded predictors (X_{1i} –X_{3i}) more interpretable. Recall that the slopes associated with components of product predictors represent "simple" slopes of that component variable when and only when the other component equals zero. In the model with Aptitude uncentered, the parameter estimates associated with the contrast-coded predictors thus inform us about "simple" differences among the students in the four cells of the design (defined by first-generation status × college) when and only when Aptitude = 0. Although an aptitude score of 0 may be possible in some cases (tests may vary from country to country), it seems unlikely. To render these coefficients more interpretable, we therefore center the aptitude variable. Thus, in this re-estimated model, the slope associated with X_{1i} estimates the first- (versus continuing-generation) status difference in first-year GPA for students whose aptitude scores are at the average for the sample. In a deep sense, of course, this model is identical to the one estimated prior to centering. Note, additionally, that we do not need to center the contrast-coded predictors. Although their means are not equal to zero (given the very unequal n values of the cells), the zero values do represent the mean of the first-generation status and college categories.

Summary

In many ways, this chapter marks the final point in the process of developing more complex models for our data to ask more complex, and perhaps more interesting, questions of those data. Throughout the development of these more complex models we have kept the same basic machinery to determine whether the increase in complexity is worthwhile as more parameters are added to the model. This machinery depends on comparisons of augmented and compact models regardless of the complexity of the models we wish to test. While this machinery has remained constant across the chapters, our models have developed from the simplest involving a single parameter (Chapter 4) to models making predictions conditional on a single continuous variable (Chapters 6) to models involving multiple continuous predictors, including product terms (Chapters 7 and 8), to models with categorical variables including products of those categorical variables (Chapters 9 and 10), and finally to models involving continuous and categorical predictor variables and their products (this chapter). This is as far as we wish to extend the complexity of the models we have considered. We believe that nearly every interesting substantive question that social science researchers might like to ask of their data can be answered using the range of models that we have explored. This is not to say that the limits of model complexity have been reached. Rather, this is to suggest that the models we have considered are those that are most likely to be of use to the data analyst. Further, when other more complicated models are appropriate, we hope that the reader by now is equipped to adapt the model comparison approach to new, more complicated situations.

So what remains to be done? The remaining chapters of the book are devoted to problems that are frequently encountered in data that violate various assumptions underlying the model comparison approach we have developed. In the next two chapters, we consider how our models and data analyses need to be modified when dealing with data for which we cannot assume independence of errors. In the final chapters, we focus on violations of the assumptions that residuals are normally distributed and come from a single population, having a single variance.

Section B Recap

This marks the conclusion of Section B. This section presented a series of models. It began with a model that estimates only an intercept and one slope parameter, which reflects the relationship between the outcome and a single continuous predictor. The section ended with models that simultaneously estimate parameters reflecting slopes of continuous and categorical predictors as well as nonlinear and interactive effects. These chapters provide an extensive suite of tools to analyze data, allowing the researcher to test a number of different questions.

It is relatively easy to focus on what seems to be the primary lesson of any given chapter. For example, in light of our treatment of a single continuous linear predictor, a reader might focus on the idea that the chapter explains how to test the relationship between the predictor and outcome variables. In light of our discussion of multiple continuous predictors, a reader might focus on the possibility of testing the predictors as a set, assessing their combined capacity to account for variation in the outcome variable. In light of our discussion of interactions, a reader might focus on the idea that the chapter explains how to test whether the slope of one predictor depends on another. Though these chapters certainly demonstrate those particular lessons, we encourage readers to consider the full range of model comparisons allowed by any particular Model A. Researchers often fail to fully understand the flexibility that these models afford. Accordingly, they sometimes fail to examine the subtle questions that most directly test their hypotheses.

In each chapter, we explain a wide variety of potential model comparisons. Given a Model A that estimates an intercept and one slope parameter, there are in fact infinite possible specifications for Model C. Certainly, one can use that model to test whether there is a relationship between the predictor and the outcome (does the slope differ from zero?). But model comparisons can also test the intercept, and they can test the combination of the intercept and the slope. Further, they can test each parameter (or combination of parameters) against a range of potential a priori values. Does the slope differ from zero? Does it differ from one? Does it differ from ten?

Similarly, a Model A that allows for an interaction between X_1 and X_2 can obviously test the interaction. But importantly, model comparisons also allow us to test the simple effects of X_1 at particular values of X_2, etc.

Moreover, by carefully choosing where to center predictor(s), we can test whether the predicted value of Y (represented by the intercept) differs from some a priori value for participants with whatever X values we specify. In parallel fashion, interactive models allow for elegantly focused simple effects tests, examining the effect of X_1 when X_2 equals 0 or 1 or 10, etc. By considering the full range of model comparisons, a researcher can gain precise control of the analysis, ensuring that the statistical test accurately and powerfully examines the relevant theoretical question.

DOI: 10.4324/9781003438335-17

Section C

Violations of Assumptions About Error

Introduction to Section C

Violations of Assumptions About Error

Section B introduced a series of increasing complex model comparisons. Let us again consider two very general expressions:

DATA = MODEL + ERROR

$Y_i = \beta_0 + \beta_1 X_1 + \beta_2 X_2 + \ldots + \beta_p X_p + \varepsilon$

The chapters in Section B gradually expanded the right-hand side of the equation, namely, the MODEL portion of the equation. allowing us to estimate more and more parameters that reflect the influence of more and more predictor variables (including interactions between predictors). The ultimate models, introduced in Chapter 11, provide great flexibility in accounting for variation in the outcome variable, *Y*. But notice that our treatment of the outcome variable, represented on the left-hand side of the equation, which constitutes the DATA, has not changed at all. From Chapter 1 to Chapter 11, we have restricted our DATA to a single continuous variable with one observation per participant. We have also required that the errors conform to our assumptions, namely that they have constant variance and are unbiased, independent, and normally distributed. What if our outcome violates these assumptions? In many cases, researchers must be able to handle complex or problematic data.

For example, in repeated-measures designs, researchers collect multiple outcome variables. Marketing researchers might record evaluations of one advertisement that includes photographs of people and another advertisement that does not include photographs of people. In a more extreme example, a two week-long study using "ecological momentary assessment" might send ten notifications per day to a participant's phone with a short survey to assess mood. This study could involve more than one hundred outcome measures for each participant. How can a researcher accommodate multiple outcome measures for each participant?

Further, even when data involve only one outcome variable, the errors may not be unbiased, independent, and normally distributed with constant variance. A researcher analyzing counts of aggressive behaviors in preschool children might find that the outcome is badly skewed (violating the normality assumption). A researcher analyzing memory might find that, as memory improves, error variance diminishes (violating the assumption of constant variance). And in any study, a researcher might find quirky observations, called outliers, that violate a number of assumptions.

The value of the tools developed in Sections A and B would be greatly enhanced if we could extend them to data that are not well-behaved. This is the goal of our final section. Section C develops a suite of tools to handle various violations of our assumptions. The general approach will be to cleverly modify the outcome measure, *Y*, in such a way that the transformed variable satisfies our assumptions. Once this transformation has been achieved, we can apply all of the tools introduced earlier in the book.

DOI: 10.4324/9781003438335-19

Chapter 12 introduces the analysis of repeated measures. Given multiple outcome measures, the chapter explains how to transform the data, computing an independent score for each participant. These transformed measures can answer questions about differences between participants as well as the way participants respond differently to the conditions under which the outcome measures are collected.

Chapter 13 expands Chapter 12's discussion of strategies for dealing with repeated measures. This chapter briefly introduces linear mixed models, which offer a powerful and elegant solution when the data involve multiple outcome measures for a given observation.

Chapter 14 examines idiosyncratic cases that violate our assumptions about error (i.e., outliers) as well as more systematic violations of those assumptions. We offer tools for diagnosing violations and strategies that address those problems.

Chapter 15 considers the special case of an outcome variable that is categorical, rather than continuous. This chapter introduces logistic regression, with an extended discussion of binary outcomes (e.g., Did the participant vote or not?, Was the patient depressed or not?).

12 Repeated-Measures ANOVA

Models With Nonindependent Errors

The regression model that has been used in all of the preceding chapters is

$$Y_i = \beta_0 + \beta_1 X_{1i} + \beta_2 X_{2i} + \ldots + \beta_{p-1} X_{1p-1} + \varepsilon$$

In this model, we have made no assumptions about the distributions of the predictor variables, the X_i. They can be continuously measured, contrast-coded predictors used to represent categorical variables, or products of variables. We have, however, made assumptions all along about the residuals, ε_i, in this model. Specifically, we have assumed three things about ε_i throughout the preceding chapters:

1 They are normally distributed.
2 They have constant variance.
3 They are independent of each other.

In this and subsequent chapters we consider necessary modifications to the regression model when these assumptions are violated. Specifically, in this chapter and the next we consider violations of the third assumption concerning the independence of ε_i, or, more colloquially, the independence of observations. Subsequent chapters consider violations of the normality and constant variance assumptions.

Defining Nonindependence

Independence of observations is formally defined using conditional probabilities (Kenny & Judd, 1986). Two observations are nonindependent if the conditional probability of one of them having a particular value, given knowledge of the other one, is different from the unconditional probability. All this means is that if two observations are nonindependent, then knowing the value of one of them gives us information about the value of the other, allowing us to make a reasonable guess about the other's value. To illustrate, suppose we collected data from heterosexual married couples, asking both spouses to rate, on a scale of 1–7, how satisfied they are with their marriage. Presumably, there would be some agreement between spouses on the quality of their marriage. The resulting data might look as follows:

Couple	*Male spouse*	*Female spouse*
1	1	1
2	4	3
3	6	7
4	5	6

DOI: 10.4324/9781003438335-20

The relative agreement between spouses in these data results in nonindependent observations. Observations within couples on average are more similar than observations between couples. Or, in terms of conditional probabilities, if we know that the score of one spouse is below the mean, then that gives us information about whether the score of the other spouse is above or below the mean. Because couples partially agree about the quality of their marriage in these data, if one spouse's score is below the mean, the other's is likely to be as well. Similarly, if one spouse's score is above the mean, the other's is likely to be as well. Observations that are linked, in the sense that they come from the same couple, are more similar on average than observations that are not linked, i.e., that come from different couples. This illustrates what we call *positive nonindependence*: greater similarity between linked observations than between nonlinked observations.

Negative nonindependence occurs when linked observations are more dissimilar on average than nonlinked observations. Consider hypothetical ratings from these same four couples about the percentage of housework each does:

Couple	*Male spouse*	*Female spouse*
1	30%	80%
2	40%	75%
3	70%	40%
4	20%	100%

If there were perfect agreement between spouses in these data, the sum of each couple's percentage estimates should be 100%. That is, as one spouse estimates that he or she does a higher percentage of housework, the other estimates a lower percentage. While there is not perfect agreement in these data, there is considerable nonindependence of observations within couples. If one spouse's estimated percentage is below the mean, the other's is quite likely to be above it. Observations that come from the same couple are on average more dissimilar than are observations from different couples. They are negatively nonindependent.

Nonindependence in data is likely to be found in two classic situations: grouped data and sequential data. The couples data that we have just used to define nonindependence illustrate nonindependence due to grouped data. Observations are grouped together in the sense that pairs of them come from the same couples. Another common grouping of observations occurs when each individual provides multiple observations: Each person is measured more than once or provides more than a single response. It is then likely that observations taken from the same individual will be more or less similar on average than observations taken from different individuals. Thus, the observations are nonindependent because they are grouped within individuals. A further example of grouped data is typically encountered in social psychological research on small group interaction. If different individuals within interacting groups are measured, nonindependence is very likely a problem.

Like grouped data, sequential data are also commonly encountered. Whenever lots of data are collected over time from one or more individuals, nonindependence due to sequence is likely: Observations that are close together in the sequence or in time are likely to be more or less similar on average than observations that are further away in time. As a simple illustration, suppose we measured the number of pages read by a college student every day for 50 days. Because of exams, varying academic pressure, and other factors, we would expect cycles in the data: groups of days right before exams would be heavy reading periods; groups of days right after exams might show fewer pages read. In other words, there is likely to be more similarity in the number of pages read on adjacent days, on average, than on days that are nonadjacent. Thus, the observations would be nonindependent due to time sequence.

Because nonindependence is likely to be present in both grouped and sequential data, the usual regression model that we have employed needs to be modified to deal with the fact that one of its underlying assumptions has been violated. For both kinds of data, the corrective strategy consists of removing the nonindependence from the data through a transformation of those data. How this is done, however, typically depends on the origin of the nonindependence. Problems of non-independence in sequential data are typically handled with somewhat different transformations from those employed in grouped data. We focus in this chapter on nonindependence problems in grouped data. Included here are data in which a few observations are gathered from each of a set of participants over time. When many observations are gathered from each of the same set of participants over time, then procedures known as multilevel analyses for longitudinal data are used (e.g., Hoffman, 2015). An introduction to multilevel models, which have several applications, is provided in Chapter 13. When many observations are gathered from a single subject over time, then procedures, known as *time-series analysis*, are recommended for dealing with nonindependence due to sequence.

Let us return to our married couple example to make one further distinction relevant to the nonindependence of grouped data. Suppose in our example we are interested in knowing whether male spouses give significantly different ratings of marital satisfaction than do female spouses. If we worried not at all about the violation of the independence assumption in these data, we would construct a contrast code to represent gender of spouse and regress our dependent variable (Y_i, each individual's rating) on this contrast-coded predictor. In this case, the groups that give rise to the nonindependence problem are said to be *crossed* with the independent variable of interest (i.e., gender). That is, each couple or group provides one and only one rating under each level of that independent variable: From within each couple there is one observation coded –1 on the contrast-code predictor and one observation coded +1. Another way of saying this is that the independent variable of interest varies *within* each couple, that is, gender varies *within* the groups that give rise to the nonindependence (i.e., couples).

Now suppose that the independent variable we are interested in is not gender of spouse but, rather, some characteristic of couples. Say, for instance, that we have two sets of couples, some newly married and others who have been married more than 10 years, and we wanted to know whether the two sets of couples give significantly different ratings of marital satisfaction. Again, if we did not worry about nonindependence, we might code duration of marriage as short or long using a contrast code, and then regress ratings, one for each individual, on the resulting contrast-coded predictor. This time, however, the groups or couples that give rise to nonindependence are *nested* under the levels of the independent variable of interest. That is, each couple or group provides two ratings of marital satisfaction, both of which receive the same code on the predictor variable. If one member of the couple is coded as –1, meaning that he or she comes from a relatively recent marriage, the other necessarily has the same code on the predictor variable. The independent variable of interest in this case is said to vary *between* couples.

In general, if groups of nonindependent observations are nested under the independent variable of interest, then observations from within a group all have the same value on that variable and the variable varies *between* groups. If groups of observations are crossed with the independent variable, then the observations within a group are under different levels of the independent variable and that variable varies *within* groups.

To clarify the *between–within* distinction further, consider a second example. Suppose that a researcher was interested in exploring differences in short-term memory as a function of stress levels. She intends to manipulate stress levels by testing participants' memories in the presence of loud white noise and in a control condition without any noise. She further decides that she will measure each participant's memory twice. These two observations from any given participant can either be crossed with the experimental treatment variable of interest (i.e., the presence or absence of white noise) or nested under it. In the case where the treatment is crossed with participants, each participant's memory would be measured once with the white noise present and once with it not present. The treatment variable in this case varies *within subjects*. In the case where the two observations from each participant are nested under the treatment, each participant's memory would be measured twice, but now it would be measured both times either with the noise present or with it not present. In this case the treatment variable varies only *between subjects*.

For educational purposes we will start by treating these two different cases separately. That is, we will first consider how nonindependence problems are overcome when groups of nonindependent observations are nested under the independent variable or factor whose effect is to be estimated. We will then consider solutions to nonindependence when groups are crossed with a factor of interest. In both cases, we will assume in this chapter that these independent variables, whether they vary between or within groups, are categorical and that each group has the same number of observations. With these restrictions we will thus be considering what have come to be known in the experimental design literature as repeated-measures ANOVA models. Our consideration will also include multiple independent variables that vary within or between groups.

In the next chapter, we will expand our consideration of nonindependence to include predictor variables that vary continuously rather than categorically and that may vary both within and between groups of nonindependent observations simultaneously. To illustrate such variables, imagine in our couples example that we were interested in the effects of age, rather than years of marriage, on marital satisfaction. The mean ages of couples are likely to vary between couples. But there is also likely to be some variation in age within couples—spouses are unlikely to be exactly the same age. This then would be an independent variable that varies both between and within couples. We will leave the consideration of such variables to the next chapter.

We remind the reader again that nonindependence can arise whenever observations are grouped, regardless of whether they come from the same person, the same small group, the same family, the same classroom, or whatever. While the ANOVA literature on repeated measures often assumes that observations are nonindependent because they come from the same person or research participant (thus, independent variables are known as between-*subject* or within-*subject* variables), nonindependence may also arise from other sorts of groupings.

Nonindependence in Nested Designs

Consider the following study. A researcher is interested in whether competition in small task groups affects satisfaction with the group product. Groups of three participants come into the laboratory. They are given some task to complete as a group. Half the groups are told that their performance will be compared with the performance of other groups and that they will be rewarded accordingly. The other half of the groups are not told this. Thus, half of the groups think they are competing with other groups, while the other half do not. After completing the task, all participants are asked to rate their satisfaction with the group product on a 1–9 scale, with higher numbers indicating greater satisfaction. The resulting data, with three groups (of three participants each) in each of the two experimental conditions, are given in Figure 12.1. Unfortunately, our notation must become a bit more complicated, with each observation designated as Y_{hi}. The first subscript, h, refers to the individual observation (in this case person) within each group; it varies between 1 and s, that is, the number of participants in the group. The second subscript, i, refers to the group from which the

	Instructions					
	Competition			*No competition*		
	Group 1	*Group 2*	*Group 3*	*Group 4*	*Group 5*	*Group 6*
Y_{1i}	8	6	7	6	7	5
Y_{2i}	9	7	8	6	6	5
Y_{3i}	7	6	8	5	7	4
$\hat{Y}_k$		7.33			5.67	

Figure 12.1 Data for nested design

observation comes, with a total of n groups (in this case six). Additionally, as before, we will use k to refer to condition or the level of the categorical independent variable under which groups are nested. There is a total of m such levels, which in this case equals 2.

Two issues are worth discussing about this notation. First, as will become clear shortly, it is no accident that we are now using i to refer to group rather than to the individual observations within each group. Second, the value of h that refers to the individual observation within each group is completely arbitrary because the participants within each group are interchangeable. That is, there are no factors that distinguish one person in a group from another.

We want to know whether the mean satisfaction level under the competition instructions is significantly different from the mean level under the no-competition instructions (i.e., whether the two mean values of 7.33 and 5.67 differ). We will first proceed to analyze these data using the methods of Chapter 9, treating all 18 observations as independent. In fact, however, it is highly unlikely that the observations within the groups are independent of each other; there are probably some groups in which interactions proceed more smoothly and thus the mean satisfaction levels are higher than other groups. Once we have analyzed these data ignoring this possible nonindependence, we will conduct the appropriate analysis that corrects the possible assumption violation.

To examine the effect of the categorical independent variable (i.e., instructions), we code it using a contrast-coded predictor, X_{hi}, having values of +1 for competition and −1 for no competition. The inappropriate analysis, ignoring the potential nonindependence due to groups in these data, regresses Y_{hi} on X_{hi} across all 18 observations in the dataset. The resulting model is

$$\hat{Y}_{hi} = 6.50 + .83X_{hi} \qquad \text{SSE} = 16.00$$

As explained in earlier chapters, the intercept in this model equals the mean of the two condition means and the coefficient for the contrast-coded predictor equals half the difference between the mean of the competition condition and that of the no-competition condition. To examine whether these two means differ from each other, we want to compare this augmented model with a compact one in which the mean of all 18 observations is the predicted model, setting the parameter associated with the contrast-coded predictor equal to zero:

$$\hat{Y}_{hi} = 6.50 \qquad \text{SSE} = 28.50$$

This comparison generates the source table of Figure 12.2. According to this analysis that inappropriately treats all 18 observations as independent, there is a significant difference between the two condition means, with higher satisfaction in the competition condition.

Source	*b*	*SS*	*df*	*MS*	*F*	*PRE*
Condition	.83	12.50	1	12.50	12.50	.44
Error		16.00	16	1.00		
Total		28.50	17			

Figure 12.2 Source table for analysis that ignores nonindependence

	Instructions					
	Competition			*No competition*		
	Group 1	*Group 2*	*Group 3*	*Group 4*	*Group 5*	*Group 6*
e_{1i}	0.67	−1.33	−0.33	0.33	1.33	−0.67
e_{2i}	1.67	−0.33	0.67	0.33	0.33	−0.67
e_{3i}	−0.33	−1.33	0.67	−0.67	1.33	−1.67

Figure 12.3 Residuals from analysis of nested data

To demonstrate the nonindependence that the aforementioned analysis ignores, let us examine the residuals from Model A for these data. Recall that the independence assumption focuses on these residuals. In Figure 12.3, we give the residuals from this Model A for all 18 observations. These, of course, equal each observation minus the appropriate condition mean.

Notice that the residuals within each group tend to have the same sign. In Groups 2 and 6 all the residuals are negative, while in Group 5 they are all positive. In other words, if one observation within a group is lower than the respective condition mean, then the others in that same group tend to be lower as well. And the same tends to be true for observations that are above their respective condition mean. This is characteristic of positive dependence due to groups.

How do we handle the assumption violation? The general strategy is to transform the data so that the nonindependence is removed. Suppose, for instance, that instead of including three scores from each group in the analysis we only analyzed one randomly selected observation from each group. In such an analysis nonindependence would cease to be a problem. Rather than having three linked observations from each group, there would only be a single observation, unlinked to every other observation in the analysis.

But clearly, we would not want to throw out a randomly selected two-thirds of the data. Nevertheless, the general goal of having a single score, rather than three linked scores, to represent each group seems a viable way to handle the nonindependence problem. One solution, to use all the information in the data and yet include only one score from each group in the analysis, is to compute the mean value for each group and to analyze the six resulting group means. In essence, this is what we will do, except that rather than computing the mean of the three scores from each group we will calculate a composite score that we call W_i, defined as

$$W_i = \frac{\sum_h \delta_h Y_{hi}}{\sqrt{\sum_h \delta_h^2}} \tag{12.1}$$

More specifically, in the present case, we will calculate W_{0i} for which all values of δ_h equal 1. To do so, we add up the three scores from any given group and divide this sum by the square root of three, rather than dividing by three (which would give the simple average of the three scores):

$$W_{0i} = \frac{Y_{1i} + Y_{2i} + Y_{3i}}{\sqrt{1^2 + 1^2 + 1^2}}$$

We compute (and analyze) this W_{0i} composite score rather than the simple average so that we can compare the sums of squares from the appropriate analysis that corrects the nonindependence problem to the inappropriate analysis that we have just conducted. If we analyzed the simple average, we would get the appropriate test statistics (F and PRE), but the sums of squares would differ. We square the values of δ_h because at a later point we will compute and analyze other W_i scores where the values of δ_h will be contrast codes, with some of them negative and some positive. The simple sum of these δ_h values would thus always equal zero. And, again, we want to keep the sums of squares in the appropriate metric to help us understand the different sources of variation in our data. Thus, we will compute W_{0i} using the above formula with all values of δ_h set to 1.

For the first group, then, we compute W_{0i} as

$$W_{0i} = \frac{8 + 9 + 7}{\sqrt{1^2 + 1^2 + 1^2}} = \frac{24}{\sqrt{3}} = 13.856$$

Values on this composite variable for all groups are given in Figure 12.4. With these W_{0i} scores, we have now combined all three scores from each group into a single score, and we can now ask whether the two experimental conditions, competition versus no competition, have different mean values on this composite W_{0i} score. Those means values are given in the last row of Figure 12.4.

We can now use our regular procedures to answer this question, because we have eliminated the nonindependence problem in these data by combining the three nonindependent observations from each group into a single composite one. Regressing W_{0i} on X_i (with values of +1 for competition and of −1 for no competition), we get the following model:

$$W_{0i} = 11.258 + 1.443X_i \qquad \text{SSE} = 10.667$$

	Instructions					
	Competition			*No competition*		
	Group 1	*Group 2*	*Group 3*	*Group 4*	*Group 5*	*Group 6*
Y_{1i}	8	6	7	6	7	5
Y_{2i}	9	7	8	6	6	5
Y_{3i}	7	6	8	5	7	4
W_{0i}	13.856	10.970	13.279	9.815	11.547	8.083
$\bar{W}_{0k}$		12.702			9.815	

Figure 12.4 Data for nested example with W_{0i} computed

Because each group only contributes a single value of W_{0i} to this analysis, we have omitted the subscript h from this model. In other words, the group has become our unit of analysis. There are now six observations contributing to this analysis rather than the 18 in the earlier analysis that ignored the potential nonindependence problem.

The coefficients in this model can be interpreted just as all coefficients for models with contrast-coded predictors. The only complication is that they tell us about values of W_{0i} rather than about values of Y_{hi}. In other words, the value of b_0 in this model, 11.258, equals the mean of the two condition means of W_{0i}, and the value of b_1, 1.443, equals half the difference between those two mean values. Accordingly, Equation 9.1 for the slope of a contrast-coded predictor (in terms of the condition means) that we gave in Chapter 9 continues to apply in the present case, except that it makes reference to the condition means of W_{0i} rather than the condition means of Y_{hi}:

$$b_1 = \frac{\sum_k \lambda_k \bar{W}_{0k}}{\sum_k \lambda_k^2}$$

Because the metric of W_{0i} is not the same as the metric of Y_{hi}, it makes sense to convert these parameter estimates back into the Y_{hi} metric. If we were to divide each W_{0i} by its denominator, $\sqrt{\sum_h \delta_h^2}$, which in this case is $\sqrt{3}$, we get back to the Y_{hi} metric. We can do the same to the parameter estimates. Thus, if we take the value of b_0 from this W_{0i} model, i.e., 11.258, and divide it by $\sqrt{3}$, we get 6.50, which equals the mean of the two condition means in the original Y_{hi} metric. And if we take the value of b_1, 1.443, and divide it by $\sqrt{3}$, the resulting value (.833) equals half the difference between the two $\bar{Y}_k$ values.

Comparing this augmented model with a compact single-parameter model provides a test of whether X_i is a significant predictor of W_{0i}. This is equivalent to a test of whether the two values of $\bar{W}_{0k}$ are significantly different from each other. And because these mean values are perfect functions of the mean values of $\bar{Y}_k$ (dividing each by $\sqrt{3}$), this test is equivalent to asking whether the two values of $\bar{Y}_k$ differ significantly. The resulting source table is presented in Figure 12.5. Notice that the value of the regression coefficient for X_i that we have included in this table is given in the metric of Y_{hi} rather than in the less meaningful metric of W_{0i}. We will do this throughout this chapter when we report regression coefficients in source tables. Notice also that, compared with the incorrect analysis that failed to remove the nonindependence in the data (the source table of Figure 12.2), the error sum of squares and degrees of freedom have been substantially affected by correcting the problem, while the sum of squares and degrees of freedom due to the independent variable have not been affected.[1]

In essence, what we have done here is remove from the analysis, and from the sum of squares and degrees of freedom for error, the within-group variability. We are now only examining variability between groups rather than variability that is partly between and partly within groups. In other

Source	*b*	*SS*	*df*	*MS*	*F*	*PRE*
Condition	.83	12.50	1	12.50	4.68	.54
Error		10.67	4	2.67		
Total		23.17	5			

Figure 12.5 Analysis of nested data solving the nonindependence problem

words, the original Y_{hi} scores varied within groups (across the individual observations) as well as between them. By collapsing observations within groups into a single W_{0i} score, we have eliminated any variability within groups from this analysis. All that is left is variability between groups.

Recognizing that we have divided up the total variation in the Y_{hi} scores into within- and between-group variation and then analyzed the effect of the independent variable, which varies between groups, we can now write out the full source table, integrating both between and within sources of variation in Y_{hi}. This source table is presented in Figure 12.6. Notice that in these data there is relatively little variability within groups. The sum of squares within groups equals 5.33, which is the difference between the total sum of squares in the original Y_{hi} scores (i.e., 28.50) and the total sum of squares in the W_{0i} scores that ignored within-group variability (i.e., 23.17).

We can derive this value for the within-group sum of squares by going back to the original Y_{hi} values and calculating the sum of the squared deviations of each individual Y_{hi} score from the mean of the group from which it was taken, $\bar{Y}_i$. Thus, the three observations for the first group are 8, 9, and 7, with a mean of 8. The sum of the squared deviations of these three scores around their mean is 2 [i.e., $(7 - 8)^2 + (9 - 8)^2 + (8 - 8)^2$]. If we do this calculation for each of the six groups and then add up the resulting values, we get the total sum of squares within groups of 5.33. We do not have to calculate the sum of squares within groups in this manner because we can also get it by subtracting the total sum of squares in W_{0i} from the total sum of squares in Y_{hi}. Nevertheless, it is useful to know that the within-group sum of squares calculated in this way is equal to the sum of the squared deviations of all individual observations from their group means.

Comparing this analysis to the earlier incorrect analysis that failed to remove the within-group variability, the sum of squares for error has been reduced slightly, by the magnitude of the within-group variability (i.e., 5.33). The degrees of freedom for error, however, have been reduced substantially, from 16 to 4, because 12 of the original degrees of freedom in Y_{hi} were within groups rather than between. Since the error sums of squares has been reduced proportionally less than the error degrees of freedom, our correct analysis that removes the nonindependence from the data results in a substantially smaller value of F for testing the treatment effect. With positive nonindependence (i.e., linked observations more similar on average than nonlinked observations) and an independent variable that varies between groups (i.e., groups of linked observations are nested under its levels), it will always be the case that if one conducts an analysis that inappropriately ignores the nonindependence in the data the value of F will be too large, resulting in too many Type I errors. Should the nonindependence be negative, with relatively less similarity of observations within groups than between, then the F that results from the analysis that inappropriately ignores the nonindependence will always be too small, resulting in too many Type II errors.

If one examines the mean square error between groups and the mean square within groups, it is clear why nonindependence of observations within groups is a problem in these data. If the observations were independent of each other, then variation between observations within the six groups would be approximately the same as variation of the observations between groups. In such a case,

Source	*b*	*SS*	*df*	*MS*	*F*	*PRE*
Condition	.83	12.50	1	12.50	4.68	.54
Error between groups		10.67	4	2.67		
Total between groups		23.17	5			
Total within groups		5.33	12	0.44		
Total		28.50	17			

Figure 12.6 Full source table for nested example

these two sources of variation might be combined or pooled in a common error term, as is done in the analysis that ignores the nonindependence issue. The fact that the mean square error within groups is substantially less than the mean square between groups indicates substantial positive dependence due to groups: On average, observations from the same group are more similar than observations from different groups.

A statistic known as the *intraclass correlation* provides an index of the degree of nonindependence due to grouped observations, based on the previous rationale. That statistic is computed as follows:

$$\text{ICC} = \frac{MS_b - MS_w}{MS_b + MS_w(s-1)}$$

where MS_b and MS_w are the mean squares between and within groups and s is the total number of observations within a group. This intraclass correlation varies between +1 and $-1/(s-1)$, with positive values indicating that observations within groups are on average more similar than between them and negative values indicating the reverse. For the present data, the intraclass correlation equals[2]

$$\text{ICC} = \frac{2.67 - .44}{2.67 + .44(2)} = .63$$

Thus, these data show evidence of considerable positive nonindependence due to groups: Observations are more similar within groups than between, and the test of the between-group condition effect is substantially biased if this nonindependence is ignored (i.e., the incorrect F with 1 and 16 degrees of freedom equals 12.50, while the correct F with 1 and 4 degrees of freedom equals only 4.68). As this example illustrates, violations of the independence assumption can lead to substantial biases in the computed test statistics.[3]

We have illustrated this analysis with a very simple example. The same general strategy holds, however, whenever the groups that give rise to nonindependence are nested under the independent variable(s) of interest, regardless of the number of such variables and regardless of the numbers of their levels. Suppose, for instance, a study were conducted to examine impression formation between dyads under various conditions. Eighteen pairs of previously unacquainted participants are brought into the laboratory. They each fill out a questionnaire about themselves and then engage in a structured interaction. Each dyad is randomly assigned to one of six conditions, defined by two independent variables—one with two levels and one with three levels—that are crossed with each other. The two-level variable is whether they see each other's completed questionnaire prior to the interaction. The three-level variable is whether the interaction is competitive, cooperative, or no instructions. At the end of the interaction, they rate their impression of the other person in the dyad, thus yielding 36 observations. The design is diagrammed in Figure 12.7, with three dyads randomly assigned to each of the six cells, and two observations from each dyad.

	Other's questionnaire seen?																	
	Yes									*No*								
	Competitive			*Cooperative*			*None*			*Competitive*			*Cooperative*			*None*		
Dyad	1	2	3	4	5	6	7	8	9	10	11	12	13	14	15	16	17	18

Figure 12.7 Design with two independent variables under which dyads are nested

Source	*df*
Between dyads	
Questionnaire seen	1
Instructions	2
Questionnaire × Instructions	2
Error between	12
Total between	17
Total within	18
Total	35

Figure 12.8 Rows of the source table for the analysis of Figure 12.7 design

In this case, the dyad is the grouping that likely induces nonindependence in the observations, and both independent variables vary between dyads. From the dependent variable, measured at the level of the individual observations, we would compute a W_{0i} score for each dyad, collapsing across the two scores from each. These W_{0i} scores would then be regressed on five contrast-coded predictors (one coding the questionnaire factor, two coding the instructions factor, and two coding the interaction between questionnaire and instructions). Given that three dyads were randomly assigned to each of the six cells of this design, the rows of the resulting source table (including the appropriate degrees of freedom) would be as indicated in Figure 12.8 (with only the omnibus rows for the instructions factor and the interaction).

Nonindependence in Crossed Designs

In the introduction to this chapter, we said that nonindependence of observations can occur whenever observations are grouped in some way, be they from the same individual, the same family, the same small group, or the same classrooms. In the two examples that we used to illustrate the analysis for nested designs, the grouping factor was first a small group of three participants and then a dyad. In many designs, individual subjects provide multiple observations in different experimental conditions over time. In this case, the individual subjects give rise to nonindependent observations because subjects are repeatedly measured in different conditions. And it is from this sort of design that "repeated-measures" ANOVA takes its name. Although we will use data examples that illustrate crossed designs with repeated measures on the same subjects, we again remind the reader that the procedures that we explicate would apply to any sort of grouping factor that induces nonindependence, when that grouping factor is crossed with one or more independent variables of interest.

Suppose we were interested in response facilitation effects, showing that the presence of another enhances performance. Participants are brought into the laboratory and asked to complete two sets of math puzzles. For one set, the experimenter is present during the task; for the other set, the experimenter is absent. Performance is measured on both puzzle sets, with higher numbers indicating better performance.[4] It is hypothesized that performance should be better when the experimenter is present.

The hypothetical data are given in Figure 12.9. Here we have 16 individual observations, Y_{hi}, taken from eight different subjects in the two experimental conditions. The subscript i refers to the subject from whom the observation is taken; it varies from 1 to 8 (n). The subscript h continues to refer to which observation within a set of grouped observations (i.e., within a subject) we are looking at. Now, however, it also refers to the levels of the independent variable of interest since that variable now varies within subjects. Thus, in these data h either equals 1 or 2, referring to the two levels of the independent variable.

As we did before, we will first test the effect of the experimental treatment variable while ignoring the likely nonindependence in these data that derives from repeated observations from the same subjects. We will then compare these results with those of the appropriate analyses.

Treating all 16 observations as if they were independent, we create a contrast-coded predictor, X_{hi}, with values of –1 if the experimenter is absent and +1 if present. The dependent variable, Y_{hi}, is then regressed on this contrast-coded predictor across all 16 observations:

$$\hat{Y}_{hi} = 6.75 + 0.375X_{hi} \qquad \text{SSE} = 20.75$$

As always with a simple ANOVA using a contrast-coded predictor, the intercept equals the mean of the two condition means, and the regression coefficient for the contrast-coded predictor equals half their mean difference.

A comparison of this model with a compact single-parameter model gives the source table of Figure 12.10. We conclude from this erroneous analysis that the treatment effect is not significant because the obtained F and PRE do not exceed their critical values at $\alpha = .05$.

The residuals from this analysis for all of the observations are presented in Figure 12.11. Notice the pattern in these residuals. With two exceptions, if subjects have a negative residual in one treatment, they do in the other as well. Similarly, positive residuals are found in the same subjects across treatments. Positive nonindependence due to the grouping of observations within subjects is thus illustrated by these residuals.

To eliminate this nonindependence problem, we need to do an analysis on a single summary score from each group of observations (i.e., each subject), just as we did in the case where groups of nonindependent observations were nested under treatment. Here, subjects are crossed with treatment, so rather than using a single summary score that is something like the average or sum of the within-subject observations, as we did in the nested case (i.e., W_{0i}), we want to compute something like a within-subject difference score, using what amount to within-subject contrast weights. In

Subject	*Experimenter absent*	*Experimenter present*
1	7	8
2	5	5
3	6	6
4	7	9
5	8	8
6	7	7
7	5	6
8	6	8
$\bar{Y}_h$	6.375	7.125

Figure 12.9 Data for crossed design

Source	*b*	*SS*	*df*	*MS*	*F*	*PRE*
X_1	.375	2.25	1	2.25	1.52	.10
Error		20.75	14	1.48		
Total		23.00	15			

Figure 12.10 Source table for analysis that ignores nonindependence

Subject	*Experimenter absent*	*Experimenter present*
1	0.625	0.875
2	−1.375	−2.125
3	−0.375	−1.125
4	0.625	1.875
5	1.625	0.875
6	0.625	−0.125
7	−1.375	−1.125
8	−0.375	0.875

Figure 12.11 Residuals from analysis of crossed data

other words, if we want to ask whether the difference between the two treatment means is significant, we might compute for each subject a composite score that is essentially the difference between the subject's performances in the two treatments and then ask whether this within-subject difference score on average differs from zero.

We want to compare the results we obtain in this way with those that we obtained in the inappropriate analysis that ignored the nonindependence problem. So rather than analyzing a simple difference score for each subject, we will compute a W score for each subject, using Equation 12.1. However, this time we will set the values of δ_h equal to +1 if the observation is taken with the experimenter present and to −1 if the observation is taken with the experimenter absent. These are the same contrast values used to construct the contrast-coded predictor in the analysis just described that ignored the nonindependence in these data. But now the comparison is within each subject and our composite score reflects the difference within each subject. To indicate that this within-subject composite score is a contrast or difference composite, we will refer to it as W_{1i}. We use W_{0i} to represent a composite score that is like an average of all the within-subject (or within-group) observations, with all values of δ_h set to 1.

Setting the values of δ_h to −1 if the experimenter is absent and to +1 if the experimenter is present, the computation of W_{1i} for each subject is accomplished as follows:

$$W_{1i} = \frac{\sum_h \delta_h Y_{hi}}{\sqrt{\sum_h \delta_h^2}} = \frac{(-1)Y_{1i} + (+1)Y_{2i}}{\sqrt{(-1)^2 + (+1)^2}}$$

where Y_{1i} is the ith subject's score when the experimenter is absent and Y_{2i} is the observation when the experimenter is present.

In Figure 12.12 we have calculated W_{1i} for each subject. In the column labeled $\sum_h \delta_h Y_{hi}$, we have calculated the numerator of W_{1i}. This numerator amounts to the simple difference between Y_{2i} and Y_{1i} for each subject. Because the denominator of W_{1i} is a constant for all subjects, W_{1i} will correlate perfectly with the simple within-subject difference scores. It can also equal zero only when the simple difference score equals zero.

If the treatment difference is significant, then the average of these W_{1i} scores across subjects should be significantly different from zero, because the average of subjects' difference scores equals the difference between the treatment averages. In other words, the mean value of W_{1i} equals $(\bar{Y}_2 - \bar{Y}_1)/\sqrt{2}$, and hence $(\bar{Y}_2 - \bar{Y}_1)$ will equal zero only when the mean value of W_{1i} equals zero.

Subject	Y_{1i}	Y_{2i}	$\Sigma_h \delta_h Y_{hi}$	W_{1i}
1	7	8	1	0.707
2	5	5	0	0.0
3	6	6	0	0.0
4	7	9	2	1.414
5	8	8	0	0.0
6	7	7	0	0.0
7	5	6	1	0.707
8	6	8	2	1.414
Mean	6.375	7.125	0.75	0.530

Figure 12.12 Values of W_{1i}: crossed design

We will use the simplest single-parameter model of Chapter 4 to test whether the mean of the W_{1i} scores differs from zero. The augmented and compact models are

MODEL A: $W_{1i} = \beta_0 + \varepsilon_i$
MODEL C: $W_{1i} = 0 + \varepsilon_i$

A comparison of these two models is a test of whether the mean W_{1i} score differs from zero. Equivalently, this provides a test of whether $\bar{Y}_h$ with the experimenter present differs from $\bar{Y}_h$ with the experimenter absent. Note that because each subject appears only once in this analysis, the dependence problem (due to repeated observations taken from the same subject) is no longer problematic.

Compact and augmented models are estimated as

MODEL A: $\hat{W}_{1i} = 0.53$ SSE = 2.75
MODEL C: $\hat{W}_{1i} = 0$ SSE = 5.00

Because we have a total of eight W_{1i} scores, one from each subject, the degrees of freedom for error equal 7 for the augmented model and 8 for the compact one. Comparing these two models gives PRE and F values of

$$\text{PRE} = \frac{5.00 - 2.75}{5.00} = \frac{2.25}{5.00} = .45$$

$$F_{1,7} = \frac{.45}{(1-.45)/7} = \frac{2.25}{2.75/7} = 5.73$$

The sum of squares reduced, 2.25, can be derived from the expression: $\sum_i \left(\hat{W}_{1iA} - \hat{W}_{1iC}\right)^2$ where $\hat{W}_{1iA}$ and $\hat{W}_{1iC}$ are the predicted values from the augmented and compact models. Since each of these models makes the same prediction for all subjects, the sum of squares reduced can be equivalently expressed as

$$n(\hat{W}_{1iA} - \hat{W}_{1iC})^2$$

And since $\hat{W}_{1iA}$ equals $\bar{W}_1$ and $\hat{W}_{1iC}$ equals zero, this reduces to

$$n(\hat{W}_{1iA})^2 = n(\bar{W}_{1i})^2 = 8(0.53)^2 = 2.25$$

Before presenting all of this in a source table, let us once again interpret the value of the estimated parameter in the augmented model. As we saw in the nested design, if we take the estimated parameters and divide them by the denominator of W_i, $\sqrt{\sum_h \delta_h^2}$, we get a parameter estimate in the same metric as Y_{hi} rather than in the W_i metric. In this case, the denominator of W_{1i} equals $\sqrt{2}$. If we divide the single-parameter estimate in the augmented model (0.530, which is also $\bar{W}_1$) by $\sqrt{2}$, we get 0.375. This value is the same as the value of b_1 in the model where the nonindependence in the data was inappropriately ignored. Again, it equals half the difference between the mean Y_{hi} when the experimenter is present and the mean Y_{hi} when the experimenter is absent.

The source table that summarizes our test is presented in Figure 12.13. Notice that the sum of squares for the treatment difference in this source table is identical to what it was before in the source table of Figure 12.10, which resulted when we ignored the nonindependence in the data.[5]

As in the nested design, eliminating the nonindependence in the data has dramatically affected the sum of squared errors and the degrees of freedom for error. Instead of having a sum of squared errors of 20.75, as we had in the inappropriate analysis of Figure 12.10, we now have a sum of squared errors of 2.75. The degrees of freedom for error have been reduced from 14 to 7. Rather than counting each value of Y_{hi} as an observation, we now have only eight observations, one from each subject; hence the degrees of freedom have been cut in half.

Conceptually, the sum of squared errors in this source table can be thought of as the interaction of treatment with subjects. If all of the between-treatment differences were the same for all subjects, then there would be no variation in the values of W_{1i} around their mean, and the augmented model that predicts the mean value of W_{1i} for each case would fit the data without error. That is, the treatment effect would not vary from subject to subject. The sum of squared errors from the augmented model derives from the fact that the magnitude of the difference between the treatment conditions varies from subject to subject. Hence, the error term in this source table can be thought of as the sum of squares associated with the treatment × subject interaction.

There is still considerable variation in the Y_{hi} scores that is not represented in this source table. In the source table of Figure 12.10 where nonindependence was ignored, we saw that the total sum of squares of Y_{hi} equals 23.00. The difference between this value and the total sum of squares for

Source	*b*	*SS*	*df*	*MS*	*F*	*PRE*
Treatment	.375	2.25	1	2.25	5.73	.45
Error		2.75	7	0.39		
Total		5.00	8			

Figure 12.13 Preliminary crossed source table

W_{1i} in the source table that we have just presented, 5.00, equals the total sum of squares between subjects. In other words, we have once again divided up the sum of squares of Y_{hi} into two portions: one between subjects and one within subjects. Since the treatment variable in this crossed example varies within subjects rather than between them, we have looked only at the within-subject variation in the data to estimate and test the treatment effect. We have entirely omitted the between-subject variation from this analysis. In the earlier section of this chapter, we considered the case where groups of nonindependent observations were nested under the treatment variable or, equivalently, the treatment variable varied between groups. There, we conducted our analysis by only examining between-group variation in the data. Here, we have conducted our analysis only by examining the within-subject variation (i.e., within linked pairs of observations) in the data since the treatment variable varies within participants. What we now need to do, as we did in the earlier section, is to report both the between- and within-subject variation in the values of Y_{hi} in a single source table.

To calculate the between-subject sum of squares in these data, let us use the same procedure we did earlier when considering a treatment variable that varied between groups of linked observations. There, to capture only the between-group variation, we computed values of W_{0i} for each group, assigning δ_h weights of 1 to all observations within groups. Let us do the same thing here, where our groups of observations come from the same subject. The resulting W_{0i} values are given in Figure 12.14. The total sum of squares of these W_{0i} values around their mean equals 18, the difference between the total sum of squares of Y_{hi} and the total within-subject sum of squares. This sum of squares total for W_{0i} equals the total between-subject sum of squares, just as it did in the earlier design where groups of linked observations were nested under treatment conditions.

We can now write out the full source table for these data, incorporating variation both within and between subjects, and including the sum of squares due to treatment as a component of the within-subject variation. This source table is presented in Figure 12.15.

Subject	Y_{1i}	Y_{2i}	$\Sigma_h \delta_h Y_{hi}$	W_{0i}
1	7	8	15	10.61
2	5	5	10	7.07
3	6	6	12	8.49
4	7	9	16	11.31
5	8	8	16	11.31
6	7	7	14	9.90
7	5	6	11	7.78
8	6	8	14	9.90
Mean	6.375	7.125	13.50	9.55

Figure 12.14 Values of W_{0i}

Source	*b*	*SS*	*df*	*MS*	*F*	*PRE*
Total between subjects		18.00	7	2.57		
Treatment	.375	2.25	1	2.25	5.73	.45
Error within		2.75	7	0.39		
Total within subjects		5.00	8	0.63		
Total		23.00	15			

Figure 12.15 Full crossed source table

Observations in this example are linked because they come from the same participant. This induces positive dependencies in the data: Observations within participants are more similar on average than are observations between participants. This is revealed by a comparison of the mean squares within and between subjects from this full source table, using the formula given earlier for the intraclass correlation:

$$\text{ICC} = \frac{MS_b - MS_w}{MS_b + MS_W(s-1)} = \frac{2.57 - 0.39}{2.57 + 0.39(1)} = .74$$

The positive nonindependence in these data, due to observations being linked within subjects, means that the error term was too large when the nonindependence was ignored. That is, the F was considerably smaller under the incorrect analysis of Figure 12.10 than in the analysis of Figure 12.15, which more correctly uses only the within-subject variation as the error term. If there were negative dependencies among linked observations, the within-group error term would be larger than the between-group error term and the situation would be reversed.

Summary of Analyses

Before dealing with more complicated designs, with multiple independent variables, let us summarize what we have discussed so far in this chapter. First, we have defined nonindependence as links between observations that are grouped in some manner, which makes them either more similar (positive dependence) or less similar (negative dependence) than nonlinked observations. The groupings that give rise to nonindependence can derive from many sources: Observations may come from the same subject, the same family, the same small group, the same class, and so forth. What matters is simply that they are grouped and as a result, the variation between observations is not homogeneous. Observations are either less variable (i.e., more similar) within the groups than between them (positive dependence) or more variable (i.e., less similar) within the groups than between them (negative dependence).

Second, the independent variables that we have considered to this point vary either within sets of grouped or linked observations or between them. In the former case, the groups that give rise to the nonindependence are said to be crossed with the independent variable of interest. In the latter case, the groups are said to be nested under the independent variable of interest.

When the independent variable of interest varies between groups of linked observations (i.e., observations are nested under the levels of the independent variable), one should examine its effects by examining only the between-group variation in the data, regressing W_{0i} (where i indicates group) on a contrast-coded predictor(s) that codes the between-group independent variable. When the independent variable of interest varies within groups of linked observations (i.e., observations are crossed with levels of the independent variable), one should examine its effects by examining only the within-group variation in the data. Here one computes a within-group contrast or difference score (i.e., $W_{1i,}$ $W_{2i,}$ and so on) and asks whether the mean difference score differs from zero across groups.

If dependencies in grouped data are ignored in the analysis, the bias in the resulting F statistic is a function both of whether the nonindependence is positive or negative and whether groups are crossed with the independent variable or nested under it. The table in Figure 12.16 summarizes the resulting bias. As we have seen, the potential biases can be far from trivial. One should attend conscientiously to violations of the independence assumption.

	Nonindependence	
	Positive	*Negative*
Groups nested under independent variable	F too large	F too small
Groups crossed with independent variable	F too small	F too large

Figure 12.16 Direction of bias in F if nonindependence is ignored

Multiple Crossed Independent Variables

In the example discussed earlier (nonindependence in crossed designs) there was only a single independent variable that varied within the groups of nonindependent observations (i.e., within subjects). The procedure that we discussed there for handling nonindependence (testing whether a within-subject difference contrast on average differs from zero) generalizes readily to multiple independent variables that are crossed with groups of observations.

Consider a design from cognitive psychology in which the effects of study time and word abstraction on memory are examined. Each participant reads six 20-word lists. Three of the lists consist of abstract nouns (e.g., hope) and three consist of concrete nouns (e.g., chair). This independent variable, list type (Abstract versus Concrete), is crossed with the amount of time that participants are given to study each list: 1 minute, 2 minutes, or 3 minutes. Participants are shown one list for the appropriate amount of time and, after a short filler task, asked to recall as many words on the list as they can. They then repeat this procedure for the other five lists, counterbalancing order across participants. The dependent variable is the number of words they recall from each list. The resulting hypothetical data for five participants are given in Figure 12.17. The last row also gives the means of the observations in each of the six conditions.

If the 30 observations (five participants × six memory scores from each) were independent of each other, then the dependent variable would be regressed on five contrast code predictors, two to code study time (1 minute versus 2 versus 3), one to code list type (abstract versus concrete), and two to code their interactions. But because there are stable individual differences in the ability to memorize these lists of words (notice, for instance, that Participant 4 consistently remembers more words than the other participants), there are positive dependencies in these memory scores. Accordingly, we must estimate and test these various contrasts within groups of linked observations (i.e., within subjects) rather than between them. And to do this, we proceed exactly as we did in the last example, asking whether within-subject contrasts or differences differ from zero. The only difference is that in this design, with six conditions, there are five within-subject contrasts that will need to be estimated and tested.

In Figure 12.18, we give the contrast codes (δ_h) that will be used to test the within-subject effects. Because study time clearly can be ordered, the first and second of these codes (W_{1i} and W_{2i}) code the linear and quadratic effects of study time; W_{3i} codes list type; W_{4i} codes the linear study time × list type interaction; and W_{5i} codes the quadratic study time × list type interaction. Using these contrast codes, we compute for each participant the five within-subject contrast scores, using Equation 12.1:

$$W_i = \frac{\sum_h \delta_h Y_{hi}}{\sqrt{\sum_h \delta_h^2}}$$

	Study time					
	1 minute		*2 minutes*		*3 minutes*	
Participant	*Abstract*	*Concrete*	*Abstract*	*Concrete*	*Abstract*	*Concrete*
1	10	13	12	14	16	17
2	8	12	9	12	11	13
3	12	13	14	14	16	16
4	15	17	16	17	19	20
5	12	13	15	16	16	17
Mean	11.4	13.6	13.2	14.6	15.6	16.6

Figure 12.17 Data for a 3 × 2 within-subject design

	Study time					
	1 minute		*2 minutes*		*3 minutes*	
Composite variable	*Abstract*	*Concrete*	*Abstract*	*Concrete*	*Abstract*	*Concrete*
Linear time: W_{1i}	−1	−1	0	0	+1	+1
Quadratic time: W_{2i}	−1	−1	+2	+2	−1	−1
List type: W_{3i}	−1	+1	−1	+1	−1	+1
Lin × List: W_{4i}	+1	−1	0	0	−1	+1
Quad x List: W_5i	+1	−1	−2	+2	+1	−1

Figure 12.18 Within-subject contrast codes (δ_h)

The resulting values of W_{1i} through W_{5i} for each of the five participants, and their means, are given in Figure 12.19. We have also included in this table each subject's score on W_{0i}. As always, in computing W_{0i} we set all values of δ_h equal to +1; accordingly, it captures the between-subject variation in the data.

Each of the within-subject contrast scores, W_{1i} *through* W_{5i}, codes a within-subject difference in memory scores as a function of the experimental conditions. And for each, we want to ask a Model A/ Model C question to determine whether the mean within-subject difference coded by that contrast differs significantly from zero. Our estimated models and their associated sums of squares are given here:

Model A:	$\hat{W}_{1i} = 3.60$	SSE = 4.70
	$\hat{W}_{2i} = -0.46$	SSE = 4.10
	$\hat{W}_{3i} = 1.88$	SSE = 6.20
	$\hat{W}_{4i} = -0.60$	SSE = 0.70
	$\hat{W}_{5i} = -0.12$	SSE = 0.10
MODEL C:	$\hat{W}_{1i} = 0$	SSE = 69.50
	$\hat{W}_{2i} = 0$	SSE = 5.17
	$\hat{W}_{3i} = 0$	SSE = 23.83
	$\hat{W}_{4i} = 0$	SSE = 2.50
	$\hat{W}_{5i} = 0$	SSE = 0.17

Participant	W_{1i}	W_{2i}	W_{3i}	W_{4i}	W_{5i}	W_{0i}
1	5.00	−1.15	2.45	−1.00	0.00	33.48
2	2.00	−0.58	3.67	−1.00	0.00	26.54
3	3.50	−0.29	0.41	−0.50	−0.29	34.70
4	3.50	−1.44	1.63	−0.50	−0.29	42.46
5	4.00	1.15	1.22	0.00	0.00	36.33
Mean	3.60	−0.46	1.88	−0.60	−0.12	34.70

Figure 12.19 Subject scores on within-subject W_i

In each case, Model A estimates the mean within-subject contrast and Model C fixes that mean within-subject contrast at zero. The sums of squared errors for Model A equal the sum of the squared W_i scores around their mean, the sums of squared errors for Model C equal the sum of the squared W_i scores around zero, and the SSRs in each case equal $n(\bar{W}_i)^2$. To illustrate these SSRs, the mean of W_{1i} is 3.60, which, when squared and multiplied by 5 (n) equals 64.80. This is the SSR for the W_{1i} model comparison, taking the difference between SSE(C) (69.50) and SSE(A) (4.70).

Notice that we are doing a series of Model A/Model C comparisons here, one for each within-subject contrast that we are interested in. And in any one of these model comparisons the number of observations is the number of participants (5), and hence the observations in each of these model comparisons are independent of each other.

In addition to these within-subject contrasts (and associated sums of squares), there is variation between subjects in these scores and this variation is captured by W_{0i}, which has a sum of squares around its mean of 131.00.

From this, we summarize the variation in the 30 Y_{hi} scores (and the five tests of the within-subject mean differences) with the source table in Figure 12.20. The values of the regression coefficients (b) in this table come from the means for each of the W_i, putting those estimates back into the metric of Y_{hi} by dividing each W_i mean by $\sqrt{\sum_h \delta_h^2}$. Thus, for instance, the mean value of W_{1i}, 3.60, is the parameter estimate from its Model A. If we divide that by $\sqrt{\sum_h \delta_h^2}$ which for W_{1i} equals $\sqrt{4}$, the resulting value is 1.80. In terms of $\bar{Y}_h$, this value is half the difference between the mean number of words recalled in the two 3-minute conditions and the mean number recalled in the two 1-minute conditions, i.e., 1.80 = ((15.6 + 16.6)/2 – (11.4 + 13.6)/2)/2. Thus, once these parameter estimates are returned to the Y_{hi} metric, they can be interpreted in the same way that we interpreted the parameter estimates associated with contrast-coded predictors in Chapters 9 and 10.

The results of this analysis suggest that memory for the word lists improves with longer study time (the significant linear effect of time), that people remember words from concrete list better than abstract lists, and that the concrete versus abstract difference is smaller at longer (versus shorter) study times. These derive from the signs of the respective parameter estimates and knowledge of the contrast weights that were used to derive the various W_i. What is perhaps surprising here is that the effects are quite large despite the relatively few numbers of participants. In part this is because we made up these data so that effects would be significant. But in part this is also because we have employed a within-subject design with considerable positive dependencies in the data due to subjects. Had we employed a design in which each observation came from a different subject (thus 30 subjects in total), permitting us to treat them as independent, the substantial between-subject variation in these data would have been included in the error term for the test of the between-subject effects of interest. With the present design, however, that substantial between-subject variation in the data is controlled or removed from the error term(s) used to test each effect. It will frequently be the case that within-subject designs are more powerful than between-subject designs for exactly this reason (i.e., substantial positive nonindependence due to subjects).

Up to this point, every source table that we have presented has relied on a single Model A, which then is compared with numerous Model Cs to test the various effects of interest. As a result, in every source table up to this point there has been only a single error term, involving the sum of squared errors and mean squared error of Model A. Now, however, when testing contrasts that vary within groups of linked observations, as in this within-subject design, each individual contrast, each W_i, is tested with its own unique Models A and C, resulting in unique error terms for each single-degree-of-freedom contrast. Thus, in this situation, there is an additional reason to refrain from omnibus multiple-degree-of-freedom tests. Up to this chapter, we expressed our dislike of such tests because if Model C is rejected in favor of Model A, we are unable to specify what it is about Model A that makes it preferable to Model C. Now, if each single-degree-of-freedom contrast is tested with its own unique Models A and C, it makes little sense to conduct pooled tests, since doing so might force us to pool or combine rather different error terms.

To illustrate the problem, in the present design one might be tempted to pool together the two-degree-of-freedom effects for time and the two interaction degrees of freedom to test omnibus null hypotheses about time differences and time—list type interactions. In Figure 12.21, we have done just this, reporting only the pooled or combined two-degree-of-freedom tests for the time effect and the time × list type interaction. These rows were computed by adding up the sums of squares and degrees of freedom from the respective rows of the source table in Figure 12.20.

The difficulty with these pooled multiple-degree-of-freedom tests is that they cannot be expressed as model comparisons. That is, there is no Model A/Model C comparison that we can write out that represents these tests; rather, they emerge from collapsing or averaging across two different Model A/Model C comparisons, which may have very different MSEs that are inappropriately averaged. Consider, for instance, the two error terms for the two single-degree-of-freedom interaction comparisons that are reported in the full source table of Figure 12.20. The MSE for the Linear time—List type interaction was 0.17. The MSE for the Quadratic time—List type interaction was

Source	*b*	*SS*	*df*	*MS*	*F*	*PRE*
Between subjects		131.00	4	32.75		
Within subjects						
Linear time	1.80	64.80	1	64.80	55.15	.93
Linear time error		4.70	4	1.17		
Quadratic time	–.13	1.07	1	1.07	1.04	.21
Quadratic time error		4.10	4	1.02		
List type	.77	17.63	1	17.63	11.38	.74
List type error		6.20	4	1.55		
Linear time x List type	–.30	1.80	1	1.80	10.29	.72
Linear time x List type error		0.70	4	0.17		
Quadratic time x List type	–.03	0.07	1	0.07	2.67	.41
Quadratic time x List type error		0.10	4	0.03		
Total within		101.17	25			
Total		232.17	29			

Figure 12.20 Source table for data in Figure 12.17

Source	*SS*	*df*	*MS*	*F*	*PRE*
Between subjects	131.00	4	32.75		
Within subjects					
Time	65.87	2	32.93	29.94	.88
Pooled time error	8.80	8	1.10		
List type	17.63	1	17.63	11.38	.74
List type error	6.20	4	1.55		
Time × List type	1.87	2	0.93	9.35	.70
Pooled time × List type error	0.80	8	0.10		
Total within	101.17	25			
Total	232.17				

Figure 12.21 Pooled source table for data in Figure 12.17

0.03. Clearly the magnitudes of these errors are very different, one being more than five times the size of the other, emerging from two very different Model As. Yet in the pooled source table of Figure 12.21, we have ignored this distinction and simply averaged them together.

In ANOVA textbooks that deal with repeated-measures designs (and experimental factors that vary within subjects), an assumption often called *sphericity* or *compound symmetry* is deemed to be very important. This assumption is equivalent to the assumption that when pooling error terms to conduct an omnibus test the various error terms (from different Model As) are homogeneous. By relying on single-degree-of-freedom within-subject contrasts, each tested with its own unique Model A and Model C and hence its own unique error term, this assumption is not necessary. Our regression-based approach forces the researcher to examine single-degree-of-freedom within-subject contrasts. This is not the case if repeated-measures designs are analyzed using widely available programs for repeated-measures ANOVA, which routinely provide omnibus tests that pool potentially rather different error terms.

Nonindependence in Mixed Designs

We have now dealt with multiple independent variables that vary either between or within groups of nonindependent observations. But so far, we have not discussed cases in which the groups of linked observations are crossed with one or more independent variables while simultaneously being nested under levels of others. Such designs are widely used in some behavioral disciplines and are commonly known as mixed designs, with some independent variables varying within and some between groups of linked observations.

To return to an example in which observations are linked because they come from the same groups, rather than the same participant consider a research design in which heterosexual married couples are asked to rate their marital satisfaction on a 1–9 point scale, with higher numbers indicating greater satisfaction. Some of these couples have been married for 15 years and some of them for 30 years. And among each group some had two or more children and others had none. Among the couples with children, those married for 15 years have young children at home and those married for 30 years have grown children who live elsewhere. There are thus three independent variables of potential interest in this design: the gender of the spouse, the length of the marriage, and whether there are children. The gender of the spouse varies within couple. Length of marriage and whether couples have children vary between couples. Observations within couples are likely to show positive dependencies: If one spouse reports high satisfaction, for instance, the other is likely to as well. The hypothetical data from 16 couples are given in Figure 12.22. As this figure makes clear, there are four couples in each of the four cells defined by length of marriage and the presence or absence of children.

It would be incorrect to analyze these data as if they were 32 independent observations because of the dependencies induced by two observations collected from each couple. One of the independent variables of interest varies within couples while the other two factors vary between couples. In the analyses that examine mean differences as a function of these variables and their interactions, each couple should appear in each Model A/Model C comparison only once, to assure independence of observations.

As in the case when independent variables varied between groups of linked observations, we will analyze the effects of length of marriage, the presence or absence of children, and their interaction using only the between-couple variation in the data. Thus, for these three effects we will construct between-couple contrast-coded predictors: X_{1i} codes length (+1 if 30 years; –1 if 15 years); X_{2i} codes whether or not the couple has children (+1 if yes; –1 if no); and X_{3i} is their product, which captures the interaction. And just as we did in the first section of the chapter, when we examined the effects of independent variables that varied between groups of linked observations, we will examine the effects of these between-couple contrast-coded predictors on the between-couple variation in the data, creating a W_{0i} with both values of δ_h equal to +1, thus in essence averaging across the satisfaction scores of the male and female spouses in each couple.

Since gender of spouse varies within couples rather than between, we must follow the strategy developed in the earlier sections of the chapter devoted to the analysis of independent variables that are crossed with groups of linked observations. Namely, we will calculate a W_{1i} score for each couple that captures the gender of spouse difference, using a δ_h value of +1 for the female satisfaction rating and –1 for the male satisfaction rating. To determine whether the male and female mean satisfaction ratings are significantly different from zero, we will compare a Model A that estimates the mean of W_{1i} to a Model C that fixes it at zero, just as we did in the case of crossed independent variables earlier.

The approach just outlined permits us to examine the within-couple effects (captured by W_{1i}, the within-couple contrast) and the between-couple effects (both the length of marriage and children main effects and their interaction, estimated by examining whether the three between-couple contrast-coded predictors, X_{1i} through X_{3i}, are useful predictors of W_{0i}).

Missing from this outlined analysis, however, are tests of the interactions between independent variables that vary within and those that vary between couples. For instance, it may be the case that there is a bigger difference in satisfaction ratings as a function of length of marriage for male

No children						*Children*					
15 Years			*30 Years*			*15 Years*			*30 Years*		
Couple	*M*	*F*	*Couple*	*M*	*F*	*Couple*	*M*	*F*	*Couple*	*M*	*F*
1	5	6	5	6	5	9	5	5	13	7	6
2	7	6	6	6	5	10	4	5	14	8	7
3	7	6	7	6	4	11	5	4	15	7	7
4	4	5	8	8	6	12	6	5	16	8	7
$\bar{Y}$	5.75	5.75		6.50	5.00		5.00	4.75		7.50	6.75

Figure 12.22 Data for hypothetical mixed example

than for female spouses. This is an entirely appropriate question that we might ask of our data, since gender of spouse and length of marriage are crossed with each other (permitting a test of their interaction) even though one factor (gender) varies within couples and one factor (marriage length) varies between them.

Since every interaction can be equivalently expressed in more than one way, the previous question about whether the length of marriage difference in satisfaction is different for male and female spouses is equivalent to asking whether the gender of spouse difference is larger or smaller for couples married for 30 years than for those married 15 years. Rephrasing this interaction question in this way gives a hint about how we might examine it, while overcoming the nonindependence problem inherent in these data. That is, the within-couple contrast code, W_{1i}, captures the gender difference in satisfaction. To ask whether this within-couple gender difference depends on the length of the marriage, we regress it on the between-couple contrast-coded predictor that codes marriage length, X_{1i}, and test whether the slope of this predictor differs from zero. If it does, then that would mean that the magnitude of the gender difference depends on the length of marriage, meaning that the gender difference is different for couples who have been married longer than for couples who have been married less long. This is the classic definition of an interaction.

In general, then, to test the interactions between the within-couple gender difference and the between-couple independent variables, we will regress W_{1i} on the three between-couple contrast-coded predictors, X_{1i}, X_{2i}, and X_{3i}. In such a model, the slope associated with each of these contrast-coded predictors estimates the degree to which the within-couple gender difference, coded by W_{1i}, differs as a function of each of the between-couple independent variables and their interaction. Additionally, because X_{1i}, X_{2i}, and X_{3i} are all contrast-coded predictors with means of zero across the four between-couple conditions, to test whether the average within-couple gender difference across the four between-couple conditions equals zero, one can simply test whether the intercept in this W_{1i} regression model differs from zero.

Let us then specify two between-couple regression models, separately regressing W_{0i} and W_{1i} on the three contrast-coded predictors that code the two between-couple independent variables and their interaction:

$$W_{0i} = \beta_{00} + \beta_{01}X_{1i} + \beta_{02}X_{2i} + \beta_{03}X_{3i} + \varepsilon_{0i}$$
$$W_{1i} = \beta_{10} + \beta_{11}X_{1i} + \beta_{12}X_{2i} + \beta_{13}X_{3i} + \varepsilon_{1i}$$

In the W_{0i} model, we are examining only variation in the data between couples, analyzing a composite score that averages the scores from the two spouses in a couple. In this model, the slopes associated with each of the contrast-coded predictors estimate the degree to which the between-couple independent variables (and their interaction) affect the average responses of each couple. In the W_{1i} model, we are examining gender differences within couples. The intercept in this model, β_{10}, estimates the magnitude of the average gender difference, within couples, on average across the four cells of the between-couple design. And each of the slopes for the between-couple contrast-coded predictors in this model tests whether the gender difference (within couples) in satisfaction depends on the between-couple independent variables (and their interaction). From these data these two models are estimated as

$$\hat{W}_{0i} = 8.31 + 0.80X_{1i} + 0.18X_{2i} + 0.80X_{3i}$$
$$\hat{W}_{1i} = -0.44 - 035X_{1i} + 0.09X_{2i} + 0.18X_{3i}$$

Each of the coefficients in these models can be put back into the Y metric by dividing by the denominator of the respective W, which in both cases is $\sqrt{2}$, yielding the parameter estimates that are presented in the second column of the source table given in Figure 12.23.

Source	b	SS	df	MS	F	PRE
Between couples (W_{01})						
Length (X_{1i})	.56	10.12	1	10.12	9.54	.44
Children (X_{2i})	.12	0.50	1	0.50	0.48	.04
Length × Children (X_{3i})	.56	10.12	1	10.12	9.54	.44
Error between		12.75	12	1.06		
Total between		33.50	15			
Within couples (W_{1i})						
Gender (intercept)	–.31	3.12	1	3.12	8.82	.42
Gender x Length (X_{1i})	–.25	2.00	1	2.00	5.66	.32
Gender x Children (X_{2i})	.06	0.12	1	0.12	0.35	.03
Gender x Length x Children (X_{3i})	.12	0.50	1	0.50	1.42	.11
Error within		4.25	12	0.35		
Total within		10.00	16			
Total		43.50	31			

Figure 12.23 Mixed-model ANOVA source table

The rows of this table provide tests of the various effects, either between couples (from the W_{0i} equation) or within couples (from the W_{1i} equation). And for each of these tests the appropriate error term (either between or within) comes from the sum of squared errors from one of the two models. Thus, each test represents a model comparison, setting a particular parameter equal to zero in the Model C, with which the aforementioned Model A is compared. The between-couple tests examine the significance of each of the slopes associated with the contrast-coded predictors. The within-couple tests examine the significance of the intercept (i.e., whether the gender difference on average differs from zero) as well as the significance of each of the slopes. Let us go through the tests summarized in this table.

First, between couples, there is a significant length of marriage difference in satisfaction ratings. This F and PRE results from a comparison of the W_{0i} Model A with a Model C that fixes the slope of X_{1i} at zero. The estimated slope of X_{1i} in Model A equals 0.80. In the source table, we have converted this back into the original Y metric, dividing 0.80 by $\sqrt{2}$. The result, .56, equals half the difference between the mean satisfaction ratings for couples married for 30 years and those married for 15 years. There also is a significant length of marriage × children interaction in the mean satisfaction ratings given by each couple, tested by comparing the W_{0i} Model A with a Model C that fixes the slope of X_{3i} at zero. Given our codes and the resulting parameter estimates, it appears that only among couples with children does satisfaction increase with the length of the marriage.

Turning to the within-couple effects, first there is a significant gender difference, tested by comparing the W_{1i} Model A to a Model C that fixes the intercept at zero. The coefficient reported for this test in the source table, i.e., –.31, equals half the difference between the mean satisfaction ratings of the female and male spouses. The other significant effect within couples is the gender × length of marriage interaction, tested by comparing the W_{1i} Model A to a Model C that fixes the slope of X_{1i} at zero. It appears that the tendency for males to report higher satisfaction than females is greater among couples who have been married 30 (versus 15) years.

Once again, to overcome the problem of nonindependence due to couples, each couple occurs in each Model A, the Model A for W_{0i} and the Model A for W_{1i}, only once. Thus, the residuals in these models come from couples rather than from observations within couples. Additionally, the variances of these residuals (i.e., the mean square errors within couples and between them) are rather unequal (1.06 between versus .35 within), indicative of the expected positive dependence of observations due to couples. Between-couple effects are appropriately tested by comparing them to the magnitude of the between-couple error variance; within-couple effects (and that includes the interactions of a within-couple difference with between-couple factors) are appropriately tested by comparing them with the magnitude of the within-couple error variance.

Summary

The strategy that we have just employed in analyzing the data from a mixed design, with one independent variable varying within couples (gender) and two varying between them (length of marriage and the presence or absence of children), can readily be generalized to other, more complicated designs in which there are grouped observations, giving rise to potential nonindependence, with some factors varying within groups and others varying between them. Let us review the general strategy.

When there are multiple observations that are likely to be nonindependent because they are grouped in some way (e.g., they come from the same subject, the same couple, the same small group, etc.), one needs to combine those observations, analyzing composite scores to meet the assumption that the residuals are independent of each other. With s linked observations from each group, one forms a total of s composite W_i scores, and these composite W_i scores are used as dependent variables in a series of between-group regression models. The first of these W_i scores, W_{0i}, is defined by setting all values of δ_h equal to 1, adding up the s scores from each group and then dividing by the square root of s. This W_{0i} composite score captures all of the between-group variation in the observations. One regresses this W_{0i} composite score on the between-group contrast-coded predictors (or continuous predictors as well; see Chapter 13) that code any between-group factors of interest. In this way, one tests the main effects and interactions of these between-group factors, just as they were tested in Chapters 9 and 10.

The remaining $s-1$ W_i composite scores are computed by using within-group contrast codes as values of δ_h to code the within-group factors of interest. Thus, each remaining W_i (from the set W_{1i}, W_{2i}, . . ., W_{s-1i}) codes a specific within-group comparison or difference between observations, capturing the main effects and interactions of the within-group factors of interest. Each of these W_i composite scores represents a single-degree-of-freedom within-group comparison. One then regresses these W_i scores on the contrast-coded predictors that code the between-group factors of interest. In each of these regression models, a test of whether the intercept differs from zero is a test of the within-group comparison or contrast coded by that particular W_i. Tests of the regression coefficients for the between-group contrast-coded predictors in each of these models are tests of interactions between the within-group contrast coded by that particular W_i and the between-group contrasts.

Importantly, there is one error term for the test of all between-group independent variables, while each within-group contrast has its own unique error term (from its own Model A) that is used to test that particular within-group contrast as well as the interactions of that contrast with the between-group factors. What may seem strange initially about these tests of interactions of a within-group contrast and one or more between-group factors is that they are not examined using product predictors. Rather, they are examined simply as slopes of the between-group contrast-coded predictors, asking whether a particular within-group difference (i.e., W_i) depends on those predictors.

To this point, we have insisted on computing W_i and then reporting coefficients that have been put back into the Y_{hi} metric by dividing each by $\sqrt{\Sigma_h \delta_h^2}$. Our insistence here has been motivated by the desire to compare the source table (and sums of squares) from the correct analysis that appropriately handles the nonindependence in the observations with an incorrect one that ignores the grouping of observations. Thus, we wanted to show that the incorrect analysis simply combined the within-group and between-group errors into a single error term, ignoring the fact that observations within groups are more or less similar than observations between them. At this point, however, it is important to say that the use of W_i scores is not necessary. Equally appropriately, one can simply analyze the average of each group's s scores, instead of W_{0i}, to capture the between-group variation in the data, and simple within-group differences to capture the within-group contrasts of interest. This analysis of within-group averages and differences will yield the same F and PRE statistics as does the analysis of W_i scores, although the resulting sums of squares will not add up to give the total sums of squares in the Y observations. The additional advantage of using simple averages and differences, over and above computational ease, is that the resulting parameter estimates will be in the correct Y metric rather than having to be converted from the W_i metric.

We give one last example to illustrate all that we have just summarized, this time without data. We give simply the design and outline the resulting source table. Imagine that a study in cognitive psychology were conducted, in which participants read word lists of varying length for a fixed amount of time. One set of lists consists of 10 words, one of 20 words, and one of 30 words. In addition, half of these word lists consist of nouns and half consist of adjectives. These two independent variables vary within subjects, so that each participant reads six different lists, created by crossing the three levels of list length with the two levels of word type (noun versus adjective). In addition, there is an independent variable that varies between participants: study time. Some participants are given 1 minute to study each list, some 2 minutes, and some 3 minutes. Thus, the design is a three (list length) by two (word type) by three (study time) factorial experimental design, with the first two factors varying within subjects and the last varying between them.

Each subject provides six observations, one memory score for each of the six lists of words that they read. These six observations are likely to manifest positive nonindependence due to being grouped within subjects: Some participants exhibit better memory across all lists than do other subjects. Accordingly, with six observations we need six composite scores from each subject. Each of these six composite scores is defined using the values of δ_h in Figure 12.24. The first of these codes, designated Y_{ave}, is the average of each subject's six memory scores across the six lists. It captures all the between-subject variation in the data. The remaining five composite scores code within-subject contrasts or differences and are labeled $Y_{dif1} - Y_{dif5}$. The values of δ_h for these within-subject contrasts indicate that the first codes the linear effect of list length, the second codes the quadratic effect of list

	10 Words		*20 Words*		*30 Words*	
	Noun	*Adjective*	*Noun*	*Adjective*	*Noun*	*Adjective*
Composite variable	δ_h	δ_h	δ_h	δ_h	δ_h	δ_h
Y_{ave}	+1	+1	+1	+1	+1	+1
Y_{dif1}	−1	−1	0	0	+1	+1
Y_{dif2}	−1	−1	+2	+2	−1	−1
Y_{dif3}	−1	+1	−1	+1	−1	+1
Y_{dif4}	+1	−1	0	0	−1	+1
Y_{dif5}	+1	−1	−2	+2	+1	−1

Figure 12.24 Values of δ_h for composite variable in word list memory example

length, the third codes the word type difference, the fourth codes the linear list length by word type interaction, and the fifth codes the quadratic list length by word type interaction.

To code the between-subject independent variable, study time, we construct two between-subject contrast-coded predictors: X_1 with values of λ_i of −1, 0, and +1 to code the linear study time difference; and X_2 with values of λ_i of −1, 2, and −1 to code the quadratic study time difference. Then we estimate six regression models, regressing each Y composite score (Y_{ave}, and $Y_{dif1} - Y_{dif5}$) on the two between-subject contrast-coded predictors, X_1 and X_2. In the model where Y_{ave} is the dependent variable, we are looking only at between-subject variation. Here we are interested in the slopes of the two between-subject contrast-coded predictors, X_1 and X_2, which estimate the effects of study time on average across all six types of memory lists. Then in the five models with dependent variables that code within-subject differences, we are interested in both the intercepts and the slopes associated with the two between-subject contrast-coded predictors. Each intercept estimates the mean within-subject difference coded by the particular Y_{dif} dependent variable, and each of the slopes *for* X_1 and X_2 estimate the degree to which that within-subject difference depends on study time (linearly and quadratically).

Assuming that there are 20 participants in each of the three between-subject conditions (study time), this analysis results in the source table summarized in Figure 12.25, indicating only the sources of variance and corresponding degrees of freedom.

Source	*df*
Between subjects	
Linear time	1
Quadratic time	1
Error between	57
Total between	59
Within subjects	
Linear list length	1
Linear list length x Linear time	1
Linear list length x Quadratic time	1
Linear list length error	57
Quadratic list length	1
Quadratic list length x Linear time	1
Quadratic list length x Quadratic time	1
Quadratic list length error	57
Word type	1
Word type x Linear time	1
Word Type x Quadratic time	1
Word type error	57
Linear list length × Word type	1
Linear list length × Word type × Linear time	1
Linear list length × Word Type × Quadratic time	1
Linear list length × Word type error	57
Quadratic list length × Word type	1
Quadratic list Length × Word type × Linear time	1
Quadratic list length × Word type × Quadratic time	1
Quadratic list length × Word type error	57
Total within	300
Total	359

Figure 12.25 Rows of the source table for the word list memory example

Conclusion

In this chapter, we have considered models for the analysis of categorical independent variables when observations are nonindependent because they are grouped in some way, and the independent variables vary either between groups of linked observations or within them. In the vocabulary of typical experimental design books in psychology, such analyses are called repeated-measures and mixed-model ANOVA. Although our approach to such ANOVA models yields results that are identical to those that would be obtained using more traditional ANOVA programs, the advantage to our approach, once again, is that we only need to estimate Models A and C, using any least-squares regression procedure, once the data transformations to define both the dependent and predictor variables have been done.

In addition to relying on our by-now thoroughly familiar Model A and Model C comparisons, our approach to these repeated-measures situations has the additional advantage that we are readily able to extend these traditional topics by including independent variables that are measured continuously rather than categorically. It is to these extensions that we turn in the next chapter.

Notes

1 The consistency of this sum of squares for the condition, as we have already said, is due to having put $\sqrt{\sum_h \delta_h^2}$ in the denominator of W_{0i} rather than more simply $\sum_h \delta_h^2$. Had we used the simple $\sum_h \delta_h^2$ average for each group, rather than W_{0i}, the sums of squares would not have worked out so nicely, although the resulting F and PRE would have been unaffected.

2 The values of MS_b and MS_w that are used in computing the intraclass correlation are the mean squares between and within groups after having removed any effects due to the independent variables of interest. In the present case, the independent variable varies between groups so its effects are removed from the variation between groups.

3 A test for whether the intraclass correlations differs from zero is provided by the F ratio of the mean square between to the mean square within in the case of positive dependence. In the case of negative dependence, the F ratio is computed as the mean square within divided by the mean square between. While the degree of nonindependence can be tested in this way, we recommend being very conservative. Whenever data are linked, it is generally wise to assume nonindependence.

4 We presume that the two puzzle sets and the order of conditions are both counterbalanced across the two levels of the independent variable. Thus, each puzzle set is done equally often with the experimenter present and absent. In addition, half of the subjects complete the puzzles with the experimenter present first and half complete the puzzles with the experimenter absent first.

5 This equivalence is ensured by the fact that the denominator of W equals $\sqrt{\sum_h \delta_h^2}$ Had we computed a simple difference score for each subject (i.e., the numerator of W_{1i}), the sum of squares would not have been equivalent even though the resulting PRE and F values would have been correct.

13 Incorporating Continuous Predictors With Nonindependent Data

Toward Mixed Models

One of the distinct advantages of our model comparison approach is that we integrate analytic frameworks which are usually treated separately: the analysis of designs with categorical independent variables (ANOVA) and the analysis of correlational data, with more or less continuously measured independent variables (multiple regression/correlation). In terms of model estimation and comparison, the model comparison approach is appropriate regardless of the measurement metric of predictor variables. One simply learns new ways in which parameter estimates can be interpreted (i.e., as mean differences or adjusted mean differences in the case of categorical variables as well as the traditional slope interpretations). This same advantage for our approach extends to the treatment of designs in which observations are nonindependent because they are grouped in some way. As we will show in this chapter, one can readily incorporate more or less continuously varying independent variables into the models underlying the analysis of nonindependent observations.

We start with the simple case where one or more continuously measured independent variables varies between the groups of linked observations; in other words, the groups of linked observations are nested under levels of the continuously varying predictors. We will approach this using the techniques introduced in the previous chapter. We then turn to the case in which continuous predictors vary *within* the groups of linked observations. And, finally, we examine cases in which the independent variables of interest vary both within and between grouped observations. The exposition will permit us to show how our approach finds a ready extension to recent developments in models known variously as multilevel, hierarchical, and mixed effects models. We provide an introduction to such models, involving multiple random effects that can either be hierarchically nested or crossed. We also provide a brief description of alternative ways to handle nonindependent observations when mixed effects models are inappropriate.

Continuous Predictors Varying Between Groups of Linked Observations

Suppose a grocery store chain is interested in alternative marketing strategies for a particular product, which we will refer to as Product X. Each of ten stores in the grocery chain markets Product X in four different ways over the course of four weeks. Two of the ways involve displaying a discount price and the other two involve displaying the product along with a coupon that must be redeemed. Crossed with that, on two weeks Product X is displayed at the end of an aisle in the grocery store and on the other two weeks it is displayed in the middle of an aisle. Thus, there are two crossed independent variables that vary within stores across time: whether the discount price is displayed or a coupon must be redeemed; and whether Product X is displayed at the end of the aisle or in the middle. The dependent variable is the number of units of Product X sold each week by each store, measured four different times. We assume that the specific week for each type of marketing strategy is randomly determined for each store, so that time and the two independent variables are unconfounded. Hypothetical data from the ten stores are given in Figure 13.1.

DOI: 10.4324/9781003438335-21

	Discounted Price		*Coupon*	
Store	*Aisle End*	*Aisle Middle*	*Aisle End*	*Aisle Middle*
1	45	43	42	37
2	56	55	43	43
3	71	68	65	59
4	23	24	22	21
5	49	50	50	48
6	67	64	67	60
7	32	30	29	29
8	51	50	48	43
9	45	44	45	47
10	58	56	57	53
Mean	49.70	48.40	46.80	44.00

Figure 13.1 Hypothetical product × sales data from 10 stores

	Discounted Price		*Coupon*					
Store	*Aisle End*	*Aisle Middle*	*Aisle End*	*Aisle Middle*	Y_{ave}	Y_{dc}	Y_{em}	Y_{int}
1	45	43	42	37	41.75	9	7	−3
2	56	55	43	43	49.25	25	1	1
3	71	68	65	59	65.75	15	9	−3
4	23	24	22	21	22.50	4	0	−2
5	49	50	50	48	49.25	1	1	−3
6	67	64	67	60	64.50	4	10	−4
7	32	30	29	29	30.00	4	2	2
8	51	50	48	43	48.00	10	6	−4
9	45	44	45	47	45.25	−3	−1	3
10	58	56	57	53	56.00	4	6	−2
Mean	49.70	48.40	46.80	44.00	47.23	7.30	4.10	−1.50

Figure 13.2 Product X sales data plus composite scores

With four observations from each store, we presume nonindependence in these data due to store. And that assumption seems reasonable given the large store-to-store differences in the total number of Product X purchases. Compare the difference, for instance, between store 3 and store 4, with the former having many more sales of Product X than the latter.

Our analysis of these data accordingly treats each of the independent variables as within-store factors. Therefore we need to compute four composite scores for each store to enable the appropriate analyses that overcome the dependence issues in these data. These four composite scores, along with the original data, are given in Figure 13.2. The first composite score, Y_{ave}, is the simple average of each store's sales of Product X across the four conditions (and weeks). Its variation from store to store captures the between-store variation in these sales data. The second composite score, Y_{dc} (with $\delta_h = +1$ if Discount and −1 if Coupon), captures the within-store sales difference as a function of whether the discount price was shown or whether a coupon was required. The third composite score, Y_{em} (with $\delta_h = +1$ if Aisle End and −1 if Aisle Middle), captures the within-store sales difference as a function of whether the display was at the end of the aisle or in the middle. And the final composite *score*, Y_{int}, captures the within-store interaction difference, examining whether the discount—coupon difference is different for end of aisle versus middle of aisle displays.

To analyze the effects of the independent variables on the sales data, we examine three different Model A/Model C comparisons, one for each of the three composite difference scores, asking whether each within-store mean difference differs significantly from zero. The model comparisons are as follows:

Model(s) A	*Model(s) C*
$Y_{dc} = \beta_0 + \varepsilon$	$Y_{dc} = 0 + \varepsilon$
$Y_{em} = \beta_0 + \varepsilon$	$Y_{em} = 0 + \varepsilon$
$Y_{int} = \beta_0 + \varepsilon$	$Y_{int} = 0 + \varepsilon$

The model estimates and SSE(A)s and SSE(C)s are as follows:

Model(s) A	*SSE(A)*	*Model(s) C*	*SSE(C)*
$\hat{Y}_{dc} = 7.30$	572.1	$\hat{Y}_{dc} = 0$	1105
$\hat{Y}_{em} = 4.10$	140.9	$\hat{Y}_{em} = 0$	309
$\hat{Y}_{int} = -1.50$	58.5	$\hat{Y}_{int} = 0$	81

The first of these model comparisons suggests that the mean difference due to the discount versus coupon is significant ($F_{1,9} = 8.38$, PRE = .48), such that more sales ensue from advertising the discounted price rather than using coupons. The second suggests that the mean difference due to end versus middle of aisle product placement is also significant ($F_{1,9} = 10.74$, PRE = .54), with greater sales if Product X is placed at the end of the aisle than in the middle. The model comparison for the test of the interaction is not significant, offering no evidence that the discount versus coupon difference in sales depends on the aisle placement of the product.

Now we introduce an additional variable, *GS*, which varies between stores: the gross sales volume (in hundreds of thousands of dollars) per week of the store in question. This variable, which is continuous rather than categorical, is given in the final column of Figure 13.3 where we reproduce the data and the composite scores given in Figure 13.2. What we would now like to examine is how the gross sales of these ten stores (*GS*) affects the results we have so far examined.

We can first ask about the relationship between gross sales (*GS*) and sales of Product X on average between stores. It seems likely that stores with larger total sales volume are likely to sell more of Product X than stores with smaller total sales volume. And indeed, when Y_{ave} is regressed on *GS*, we find a significant between-stores relationship:

Model A: $\hat{Y}_{ave} = -16.91 + 33.40GS$ SSE(A) = 669.5
Model C: $\hat{Y}_{ave} = 47.23$ SSE(C) = 1669.3

This comparison yields an F of 11.95, with 1 and 8 degrees of freedom, and a PRE of .60. Not surprisingly stores with higher gross sales volume sell more of Product X.

We can also ask about interactions between the gross sales volume of the stores and the independent variables that vary within stores. Perhaps, for instance, the end of aisle placement,

	Discounted Price		*Coupon*						
Store	*Aisle End*	*Aisle Middle*	*Aisle End*	*Aisle Middle*	Y_{ave}	Y_{dc}	Y_{em}	Y_{int}	*GS*
1	45	43	42	37	41.75	9	7	−3	2.3
2	56	55	43	43	49.25	25	1	1	1.7
3	71	68	65	59	65.75	15	9	−3	2.2
4	23	24	22	21	22.50	4	0	−2	1.4
5	49	50	50	48	49.25	1	1	−3	1.9
6	67	64	67	60	64.50	4	10	−4	2.4
7	32	30	29	29	30.00	4	2	2	1.6
8	51	50	48	43	48.00	10	6	−4	1.8
9	45	44	45	47	45.25	−3	−1	3	1.9
10	58	56	57	53	56.00	4	6	−2	2.0
Mean	49.70	48.40	46.80	44.00	47.23	7.30	4.10	−1.50	1.92

Figure 13.3 Product X sales data plus composite scores and gross sales (*GS*)

compared to the middle aisle placement, makes an even bigger difference in sales of Product X in stores with greater total sales volume. Or perhaps the difference in the sales of Product X as a function of displaying the discounted price versus giving coupons is different in stores with greater total sales than in those whose total sales is smaller. And finally, even though overall there was no interaction between the aisle placement of Product X and how its discount was managed (display discount versus redeeming a coupon), it may be the case that this interaction depends on the overall gross sales of the store.

To examine these interactions of the within-store categorical independent variables and the between-store continuous gross sales (*GS*) independent variable, we want to regress each within-store difference score on *GS* and test whether each slope associated with *GS* differs significantly from zero. As we saw in Chapter 11, when a within-subject (in this case within-store) difference is regressed on a between-subject (between-store) predictor, the resulting slope informs us about the interaction of the "within" independent variable and the "between" one. This remains true even when the "between" independent variable is continuous, as it is in this example.

The only slight complication in estimating these models is that if we regress the within differences on *GS*, the intercepts will no longer equal the mean of those difference scores. Rather, they will equal the predicted difference for hypothetical stores whose gross sales volume equal zero. To make these intercepts more meaningful, we want to mean deviate or center the predictor variable. The intercepts in the resulting models will then continue to tell us about the overall or mean effects associated with the within-store manipulations.

The resulting three Model As (with *GSc* being the centered *GS* variable, i.e., $GSc = GS - 1.92$) are as follows:

$\hat{Y}_{dc} = 7.30 - 0.18\text{GSc}$	$\text{SSE(A)} = 572.1$
$\hat{Y}_{em} = 4.10 + 9.91\text{GSc}$	$\text{SSE(A)} = 52.9$
$\hat{Y}_{int} = -1.50 - 3.79\text{GSc}$	$\text{SSE(A)} = 45.6$

As must be the case when predictors are centered, the intercepts in these models equal the means of the three composite difference scores that are the dependent variables. These models can be compared with compact ones that fix the intercepts at zero to test the average effect of the within-store manipulations (and their interaction). Tested in the context of these models, these comparisons yield somewhat different results than those already reported for the repeated-measures ANOVA in which *GS* was ignored. Rather than having an error term with 9 degrees of freedom, as in the earlier model comparisons, each of these Model As estimates two parameters and hence their error degrees of freedom equal 8. Additionally, to the extent that the *GSc* predictor variable explains variation in the difference scores, the SSEs for these Model As will be reduced compared to what they were when no predictor variable was included. (See Chapter 6 regarding tests of β_0 in simple regression.)

To test the interactions of *GS* with the within-store manipulations, we want to compare these Model As with Model Cs that omit the predictor variable and only estimate the intercepts. These Model Cs were given earlier, although they were called Model As in the context where we were asking whether the intercept was different from zero (with no predictor in the model)—i.e., testing the effects of the within-store manipulations without the between-store *GSc* predictor.

Only in the case of aisle placement difference score (Y_{em}) is the slope of *GSc* significantly different from zero ($F_{1,8} = 13.31$; PRE = .62). What this means is that the aisle placement effect reported earlier (more sales if Product X is displayed at the end of the aisle than in the middle) depends on the gross sales volume of the store. Since the estimated slope of *GSc* is positive, it appears that in stores with larger gross sales volume, the aisle placement of Product X makes a bigger difference than in stores with smaller gross sales volume. For stores having sales volume at the mean, the difference in Product X sales as a function of aisle placement is estimated as 4.10 units per week. But as total sales volume of the stores goes up the aisle placement makes a bigger difference at higher volume stores. As we saw when interpreting interactions with continuous predictors in Chapter 7, it is typically informative to derive "simple" effects at representative values of a component predictor. In the present case this is equivalent to deriving predicted values of the within-store difference at representative values of *GSc*, for instance one standard deviation above and below its mean. The standard deviation of the *GS* variable is 0.32. Thus, the predicted values of the aisle end versus aisle middle sales difference of Product X at one standard deviation above and below the mean of *GS* equal

$$\hat{Y}_{em,+1sd} = 4.10 + 9.91(0.32) = 7.27$$
$$\hat{Y}_{em,-1sd} = 4.10 + 9.91(-0.32) = 0.93$$

By deviating *GS* from values other than its mean, the significance of the simple aisle end versus aisle middle sales difference could be tested at particular values of *GS* other than its mean. For instance, rather than deviate *GS* from its mean of 1.92, we might deviate it from the value of *GS* that is one standard deviation above its mean (i.e., 1.92 + 0.32). If Y_{em} were then regressed on this new deviated predictor, the intercept would equal the predicted difference for stores one standard deviation above the *GS mean* (i.e., the intercept would equal 7.27, the value computed above), and it could be tested for significance by comparing it, as Model A, to a Model C that fixed the intercept at zero.

As this example illustrates, examining the effects of continuous predictor variables that vary between groups of nonindependent observations (in this case between stores) is entirely consistent with the lessons of Chapter 12. One uses the between-group variables as predictors of the various composite dependent variables. In general, it makes sense to center the continuous predictors so that the intercepts estimate the mean of the within-group contrast scores. For a model in which the composite average is the dependent variable, the slope of a continuous between-group predictor tells us about the "main effect" of that predictor, on average, across levels of the within-group factors. For a model in which a within-group contrast or difference is the dependent variable, the slope for the continuous predictor tells us about the interaction of the within-group difference with the between-group predictor. Multiple between-group continuous predictors (as well as categorical ones and their interactions) may be used simultaneously, both in predicting the composite average scores and in predicting within-group contrast difference scores. Of course, when there are multiple predictors, there is also likely to be redundancy among them. The slope of a predictor in such models would then tell us about the partial effect of a predictor, controlling for the other predictors in the model.

"Continuous" Predictors That Vary Only Within Groups

Suppose we were examining the effects of two different diets on weight gain in eight newly born laboratory animals. Half of the animals were randomly assigned to one diet and half to the other, and their weights were recorded every day for ten days. The hypothetical data are given in Figure 13.4. The observations are linked because they come from the same animal. Day thus varies within animals, whereas Diet varies between animals.

Day really is a categorical variable, with ten levels. One could accordingly define a composite score that is the average of all ten observations for each animal and then nine within-animal contrasts or differences. Given the intrinsic ordering of Day, a set of nine orthogonal polynomial codes would probably be of most interest, coding the linear Day effect, the quadratic Day effect, the cubic Day effect, and so forth. Of course, one would probably not be interested in much beyond the linear effect: asking simply whether the animals' weights increased over the course of the ten days and whether the increase depended on the diet they were fed.[1] Values of the within-animal contrast code (δ_h) that codes the linear component of Day, having ten levels, are −4.5, −3.5, −2.5, −1.5, −0.5, +0.5, +1.5, +2.5, +3.5, and +4.5.

In Figure 13.5 we have reproduced the data along with two composite scores: the average of each animal's weight across the ten days, Y_{ave}, and the linear difference using the values of δ_h given in the previous paragraph, Y_{difLin}. Using the procedures specified in the last chapter, the between-animal main effect of Diet is estimated by regressing Y_{ave} on the contrast-coded Diet variable:

$$\hat{Y}_{ave} = 13.725 - 1.450\text{Diet} \qquad \text{SSE} = 50.755$$

A comparison of this model with a Model C that fixes the slope of Diet at zero yields $F_{1,6} = 1.99$ and PRE = .25. With only eight animals in the dataset, the effect of Diet on the mean weight (i.e., across the ten days) is not significant.

		Day									
Animal	*Diet*	*1*	*2*	*3*	*4*	*5*	*6*	*7*	*8*	*9*	*10*
1	−1	10	12	12	15	13	16	17	19	17	20
2	−1	8	7	8	9	10	10	12	14	13	15
3	−1	12	13	16	18	17	19	22	23	23	26
4	−1	13	12	14	14	15	17	17	18	20	21
5	1	12	12	13	13	14	14	15	14	15	15
6	1	8	7	7	8	9	9	10	10	12	11
7	1	10	10	11	11	12	13	13	14	14	15
8	1	13	12	12	13	15	14	15	15	16	15

Figure 13.4 Day x Diet data

Animal	*Diet*	*Day* *1*	*2*	*3*	*4*	*5*	*6*	*7*	*8*	*9*	*10*	Y_{ave}	Y_{difLin}
1	−1	10	12	12	15	13	16	17	19	17	20	15.1	84.5
2	−1	8	7	8	9	10	10	12	14	13	15	10.6	72.0
3	−1	12	13	16	18	17	19	22	23	23	26	18.9	122.5
4	−1	13	12	14	14	15	17	17	18	20	21	16.1	79.5
5	1	12	12	13	13	14	14	15	14	15	15	13.7	29.5
6	1	8	7	7	8	9	9	10	10	12	11	9.1	41.5
7	1	10	10	11	11	12	13	13	14	14	15	12.3	47.5
8	1	13	12	12	13	15	14	15	15	16	15	14.0	33.0

Figure 13.5 Day × Diet data with computed composite scores

To examine the within-animal linear effect of Day and its interaction with Diet, we regress the linear difference, Y_{difLin}, on Diet (contrast-coded):

$$\hat{Y}_{difLin} = 63.750 - 25.875\text{Diet} \qquad \text{SSE} = 1719.875$$

We are interested in two questions here. First, the intercept estimates the average linear difference across all eight animals. Testing whether its value is significantly different from zero yields $F_{1,6} = 113.42$ and PRE = .95. Thus, unsurprisingly these animals gained weight across the ten days. Second, the slope for the between-animal predictor variable, Diet, estimates the Linear Day × Diet interaction. A test of whether it differs from zero yields $F_{1,6} = 18.69$ and PRE = .76. Thus, it appears that the first four animals, fed with Diet −1, gained weight at a significantly faster rate than did the animals fed with Diet +1.

So far, we have applied the procedures of Chapter 12 for mixed models to these data, treating Day as a categorical within-animal predictor variable. Day has ten levels and could thus be coded into nine orthogonal within-animal contrasts. We have only examined one of these, the linear effect of Day. But one could similarly examine each of the other eight within-animal contrasts and their interactions with Diet if one had reasons to do so. What, then, makes this an analysis of a "continuous" independent variable that varies within groups of linked nonindependent observations?

Let us think about these same data in a different way. In Figure 13.6 we have taken the data matrix of Figure 13.4 and transposed it, so that animals are now the columns rather than the rows. The first column designates the value of the Day variable (for each of its h levels), the second column gives the Y values for the first animal, Y_{h1}, the third column gives the Y values for the second animal, Y_{h2}, and so forth. If we wanted to treat Day_h as a continuous within-animal predictor variable, we could estimate its effects by regressing weight on the values of Day_h (i.e., 1–10) separately for each animal. In other words, for animal 1, we could regress Y_{h1} on Day_h across the ten days. The resulting model for the first animal is

$$\hat{Y}_{h1} = 9.467 + 1.024\ \text{Day}_h$$

In this model, the intercept, 9.467, is the predicted weight for this animal on Day 0, and 1.024 is the expected linear increase in weight for each additional day. To make the intercept more interpretable, we can center Day_h (DayC_h) around its mean, 5.5, with the resulting parameter estimates

$$\hat{Y}_{h1} = 15.100 + 1.024\ \text{DayC}_h$$

Day_h	Y_{h1}	Y_{h2}	Y_{h3}	Y_{h4}	Y_{h5}	Y_{h6}	Y_{h7}	Y_{h8}
1	10	8	12	13	12	8	10	13
2	12	7	13	12	12	7	10	12
3	12	8	16	14	13	7	11	12
4	15	9	18	14	13	8	11	13
5	13	10	17	15	14	9	12	15
6	16	10	19	17	14	9	13	14
7	17	12	22	17	15	10	13	15
8	19	14	23	18	14	10	14	15
9	17	13	23	20	15	12	14	16
10	20	15	26	21	15	11	15	15

Figure 13.6 Transposed Diet x Day data matrix

The slope is, of course, unchanged, and the intercept now equals the mean weight across the ten days for the first animal.

The interpretation of this within-animal slope (i.e., as the expected linear increase in weight for this animal for each additional day) is similar to the meaning of the Y_{difLin} within-animal difference score. The difference score captured the linear difference in weight for each animal across the ten days. In fact, one can show that the within-animal slope, when Y_{hi} is regressed on Day_h, has a numerator that is equal to the within-animal difference score, Y_{difLin}, that we computed earlier. The formula for the least-squares slope in simple regression was given in Chapter 6 as

$$b = \frac{\Sigma\left(X - \bar{X}\right)\left(Y - \bar{Y}\right)}{\Sigma\left(X - \bar{X}\right)^2}$$

In the present case, the values of the predictor variable are 1–10, with a mean of 5.5. Therefore, the values of $(X - \bar{X})$ are –4.5, –3.5, –2.5, –1.5, –0.5, 0.5, 1.5, 2.5, 3.5, and 4.5, which are the same values that we used in constructing the within-animal difference score, Y_{difLin}. Hence, defining these as the values of δ, the within-animal contrast codes, the formula for the slope becomes

$$b = \frac{\Sigma\delta\left(Y - \bar{Y}\right)}{\Sigma\delta^2} = \frac{\Sigma\delta Y - \bar{Y}\Sigma\delta}{\Sigma\delta^2}$$

Since $\Sigma\delta$ equals zero, this expression for the slope reduces to

$$b = \frac{\Sigma\delta Y}{\Sigma\delta^2}$$

The numerator is our within-animal linear difference score, Y_{difLin}, and the denominator is simply the sum of the squared values of δ, which in this case equals 82.5. Hence if we take the value of the slope for this first animal, 1.024, and multiply it by 82.5 we get 84.5, which is that animal's value on the composite linear difference score, Y_{difLin}, given in Figure 13.5. If we were to conduct eight within-animal regressions (i.e., one regression for each animal), regressing each animal's ten Y_{hi} scores on the centered Day_h variable, the intercepts in the resulting models would equal each animal's mean weight (what we earlier called the Y_{ave} composite score), and the slopes would equal each animal's composite linear difference score (Y_{difLin}) divided by a constant: $\Sigma\delta^2$.

In Figures 13.7 and 13.8 we present for each animal the values of the intercepts (b_{0i}) and slopes (b_{1i}) that result from the eight within-animal regressions, taking each animal's ten weight scores, Y_{hi}, and regressing them on the centered Day_h variable. Also included here is the Diet between-animal contrast-coded independent variable.

Now we want to repeat our Model A/Model C comparisons to ask about the main effect of the between-animal variable Diet, the Linear effect of Day (within-animal), and the Diet × Linear Day interaction. But this time, we are going to use each animal's intercept and each animal's slope from the within-animal regressions as our dependent variables, regressing each one on Diet.

Animal	*Diet*	b_{0i} (Y_{ave})	b_{1i} ($Y_{difLin}/\Sigma\delta^2$)
1	−1	15.1	1.024
2	−1	10.6	0.873
3	−1	18.9	1.485
4	−1	16.1	0.964
5	1	13.7	0.358
6	1	9.1	0.503
7	1	12.3	0.576
8	1	14.0	0.400

Figure 13.7 Intercepts and slopes from within-animal regressions

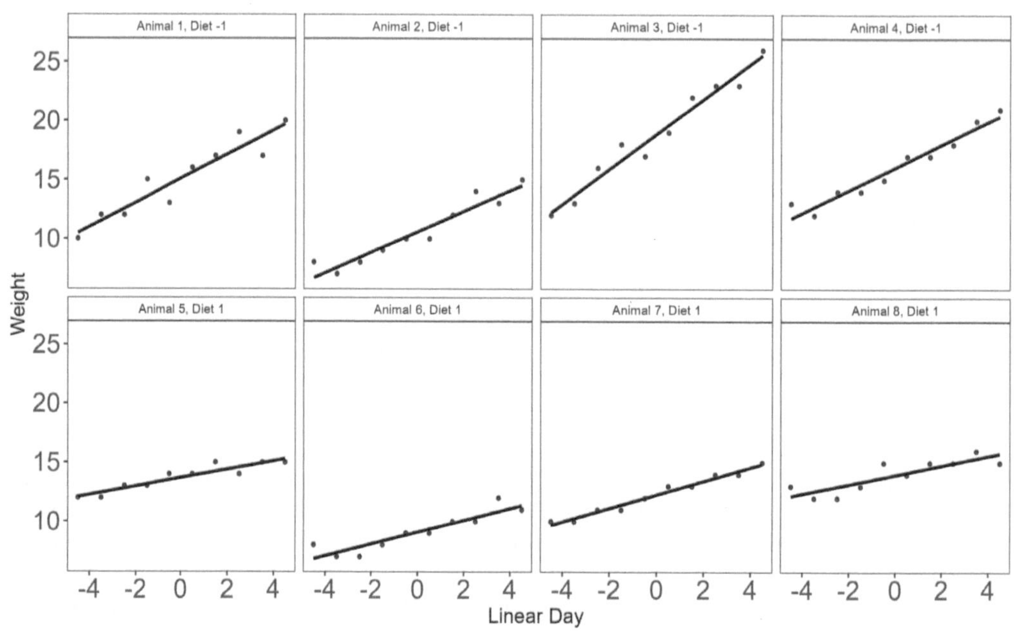

Figure 13.8 Plot of each animal's weight on the y-axis as a function of Day on the x-axis along with the corresponding regression line estimated as a linear function of Day

Since the intercept (b_{0i}) exactly equals each animal's composite average score (Y_{ave}), the first of these regressions is identical to what we obtained previously using Y_{ave} as the dependent variable:

$$\hat{b}_{0i} = 13.725 - 1.450\ \text{Diet}_i \qquad \text{SSE} = 50.755$$

A comparison of this model with a Model C that fixes the slope of Diet at zero once again yields $F_{1,6} = 1.99$ and PRE = .25. As before, this model comparison examines whether the animals given one diet differ in average weight (i.e., averaging across the ten days) from the animals given the other diet.

The second Model A is one in which the within-animal slopes (b_{1i}) are regressed on Diet:

$$\hat{b}_{1i} = 0.773 - 0.314\text{Diet}_i \qquad \text{SSE} = 0.253$$

In a deep sense this model is identical to the one estimated earlier where Y_{difLin} was the dependent variable. Since the value of each animal's estimated slope is simply that animal's value of Y_{difLin} divided by $\sum_h \delta_h^2$, the coefficients in this model equal the estimated coefficients from the earlier Y_{difLin} model divided by this same factor. Accordingly, the intercept in this model estimates the average linear effect of day ($F_{1,6} = 113.42$ and PRE = .95), and the parameter associated with the Diet predictor variable estimates the Linear Day × Diet interaction ($F_{1,6} = 18.69$ and PRE = .76). In other words, these models, using the intercepts and slopes from the within-animal regressions as the dependent variables, estimate exactly the same effects as those estimated using the procedures developed in Chapter 11 for mixed-factor designs. The two procedures for dealing with designs in which some independent variables (e.g., Day) are crossed with groups of linked observations (e.g., from the same animal) and others vary between groups of linked observations (e.g., Diet) are thus equivalent. This will always be the case whenever the intervals between levels of the within-subject factor are equal in magnitude and there are no missing data.

Why have we gone to these lengths to show that the analysis of slopes and intercepts that derive from the within-animal regression models gives us the exact same information as the procedures developed for mixed designs in Chapter 12? The answer is that the analysis of such slopes and intercepts, derived from within groups of linked observations, provides us with tremendous flexibility in examining the effects of independent variables of interest that vary both within and between groups of linked observations. Let us review the procedure once again before examining more complicated scenarios.

With observations that are nonindependent because they are grouped in some way (e.g., they come from the same subject, the same family, the same group, etc.), one can estimate the effects of independent variables that vary within the grouped observations. One can do so by estimating within-group regression models, regressing the dependent variable for each set of linked observations on the independent variables for that group. From such within-group models, assuming that the independent variables have been centered, the intercepts will tell us about the mean of the linked observations for each group and the slopes will estimate the differences in those observations associated with the within-group independent variables of interest. (These may involve partial associations if multiple partially redundant within-group independent variables are used as predictors.) Such independent variables can vary either continuously or categorically and interactions among them may also be included in within-group models, if such interactions are theoretically expected.

The intercepts and slopes from these within group regression models become the data values for further analysis, where the groups are the units of analysis rather than the individual observations.

In essence one regresses the within-group intercepts and slopes, in separate models, on any between-group independent variables that are of interest. In the model where the intercepts are the data variable, assuming centering of within-group predictors, the slopes associated with the between-group predictors estimate the effects of those predictors "on average," averaging across the levels of the within-group predictors. In the slope model(s), where the within-group slopes constitute the data variables, the slopes associated with the between-group predictors estimate interactions between the within-group predictors and the between-group predictors.

This description of the relevant models is both abstract and dense. The important point is that this general approach to the analysis of data that exhibit dependencies because of grouping is a very general and flexible one that has found widespread use in recent years under the name of multilevel modeling, albeit with different estimation procedures. In the following we provide a general introduction to these models.

Multilevel Modeling

To introduce these models initially in a rather simplified context, imagine that we have data from students who are grouped in classrooms in a large school. From each student we have a standardized achievement test score, and we want to analyze these scores as a function of different independent variables, some of which vary within the classrooms and some of which vary between them. We assume that classrooms induce dependence in the test scores, caused by factors such as within-classroom interaction, common experiences, etc. We will use Y_{hi} to represent the test score of the *h*th student in the *i*th classroom, thus keeping the same subscripts in this example that we have used all along: The *i* subscript refers to the group of linked observations, and the *h* subscript refers to the individual observations within the group.

Let's further imagine that the students and teachers at this school each come from one of two households, both alike in dignity: the Montagues or the Capulets. We have one independent variable, the student's family, that varies from student to student within classrooms, and another independent variable, the teacher's family, that varies between classrooms (but does not vary within classroom—assuming each class is taught by one teacher). The first of these we will refer to as SF_{hi} (contrast-coded with values of +1 if the student is a Montague and –1 if the student is a Capulet). We will refer to the teacher family variable as TF_i. It is contrast-coded in a parallel way. Notice that the student family variable has both *h* and *i* subscripts because it varies from student to student within classrooms, while the teacher variable only has an *i* subscript since it varies only between classrooms. For context, there are several questions we might ask. Do Montague students perform better (or worse) than Capulet students? Are Montagues better (or worse) teachers (or more precisely, do the students of Montague teachers perform better [or worse] than the students of Capulet teachers)? And do students perform better on the test if they are taught by someone from their own family (perhaps because they are more receptive to instruction or because of bias during testing)?

We start by estimating the effect of the student-family variable on achievement test scores by estimating slopes and intercepts from the following within-classroom model:

$$Y_{hi} = \beta_{0i} + \beta_{1i}\text{SF}_{hi} + \varepsilon_{hi}$$

Notice that the parameters in this model have *i* subscripts, meaning that for each classroom, we estimate an intercept and a slope. The intercept is the mean achievement test score in a given classroom, averaging across the Montague and Capulet student group means in that classroom. The slope is the difference (actually half the difference given our coding) between the mean test score of the Montague students and the mean test score of the Capulet students in the classroom. In the language of multilevel models, the within-classroom model is referred to as the level-one model, meaning that it captures the mean and the effect of student family at the lowest level of the nested data structure, across the students who are nested within a classroom.

The parameters from this level-one model are in turn modeled at level-two, between classrooms, with a level-two model for the level-one intercepts and a level-two model for the slopes:

$$\beta_{0i} = \alpha_{00} + \alpha_{01}\text{TF}_i + \mu_{0i} \qquad \text{(model of the level-1 intercepts)}$$
$$\beta_{1i} = \alpha_{10} + \alpha_{11}\text{TF}_i + \mu_{1i} \qquad \text{(model of the level-1 slopes)}$$

Both of these level-one parameters are modeled as a function of the level-two independent variable, teacher family. In other words, the intercepts and slopes for each classroom are each regressed separately on teacher family. To differentiate the parameters in these level-two models from those in the level-one model, we use α instead of β, and each parameter has two subscripts. The first subscript indicates whether the parameter *comes from* a model of the level-one intercepts (0) or from a model of the level-one slopes (1). The second subscript indicates whether the parameter *represents* an intercept or a slope in the level-two model. So, for instance, the parameter α_{01} *comes from* the model of level-one intercepts (so we use a leading 0 in the subscript, 01) and it *represents* the slope of the *TF* predictor at level 2 (we use a trailing 1 in the subscript, 01). Thus, α_{01} refers to the slope associated with teacher family in the model that regresses the level-one intercepts on teacher family. Finally, to differentiate the errors in these level-two models from those in the level-one model we have used μ_{0i} and μ_{1i} for the residuals in the level-one intercepts and slopes, respectively.

In Figure 13.9 we provide interpretations of the parameters and error terms for these models.

The level-two models can be used to test specific null hypotheses by comparison of these models with appropriate compact models. For instance, a test of whether α_{01} differs from zero examines whether Montague and Capulet teachers differ in terms of their classrooms' average test score (or more precisely the average of the Montague student average and the Capulet student average in each

β_{0i}	Mean test score for the *i*th classroom, on average across the Montague and Capulet students.
β_{1i}	Half the difference between the average of the Montague students and the average of the Capulet students in the *i*th classroom.
ε_{hi}	Individual test score residual around the predicted student family mean for the *i*th classroom.
α_{00}	Mean of the classroom average test scores, on average across the Montague and Capulet teachers.
α_{01}	Half the difference in classroom average test scores between those classes with Montague teachers and those with Capulet teachers.
μ_{0i}	Residual in classroom mean test scores over and above that predicted by the model.
α_{10}	Mean of the within-classroom student-family effect, averaging across the Montague and Capulet teachers.
α_{11}	Half the difference between the Montague and Capulet teachers in the within-classroom difference in test scores due to student family.
μ_{1i}	Residual in classroom student-family difference in test scores over and above that predicted by the model.

Figure 13.9 Interpretations of parameters and error terms for models predicting achievement test scores

class). A test of whether α_{10} differs from zero tests the mean score difference between Montague and Capulet students, collapsing across Montague and Capulet teachers. And a test of whether α_{11} differs from zero tests the student family by teacher family interaction: Does the student family difference within classrooms depend on the teacher's family? For example, are student scores higher when the student and teacher come from the same family rather than from different families?

These models could be estimated sequentially, first at level-one, estimating each classroom's intercept and slope, and then at level-two, estimating the level-two intercepts and slopes. But doing this is not optimal, particularly if classrooms differ in size and in variability. One would prefer a simultaneous estimation procedure that somehow produces best estimates, given the overall variability in the data attributable to the various sources. The full multilevel model that we would like to estimate is found by substituting the level-two models for the parameters into the level-one model. We reproduce the models here, and then we do the substitution in the level-one model.

Level-one:	$Y_{hi} = \beta_{0i} + \beta_{1i}SF_{hi} + \varepsilon_{hi}$
Level-two:	$\beta_{0i} = \alpha_{00} + \alpha_{01}TF_i + \mu_{0i}$
	$\beta_{1i} = \alpha_{10} + \alpha_{11}TF_i + \mu_{1i}$

Substituting level-two models for the parameters in the level-one model:

$$Y_{hi} = (\alpha_{00} + \alpha_{01}TF_i + \mu_{0i}) + (\alpha_{10} + \alpha_{11}TF_i + \mu_{1i})\, SF_{hi} + \varepsilon_{hi}$$

and then rearranging

$$Y_{hi} = \alpha_{00} + \alpha_{01}TF_i + \alpha_{10}SF_{hi} + \alpha_{11}TF_i{\times}SF_{hi} + \mu_{0i} + \mu_{1i}SF_{hi} + \varepsilon_{hi}$$

What we now have is a single model that integrates the level-one and level-two models that we would like to estimate. In some ways, this integrated multilevel model should look very familiar; in other ways, it looks rather different. First, the part of the model with the α parameters seems to make sense. We are estimating an intercept, which is like the grand mean (given contrast coding), then the effect of teacher family, then that of student family, and finally their interaction. These α effects are called "fixed" effects. They reflect the MODEL or, equivalently, the part of the DATA that we understand through our analysis. Fixed effects allow us to make predictions (what score would we predict for a new Montague student in a class with a Capulet teacher?), and they are routinely tested for statistical significance, yielding something like a t statistic and a p-value. But then the model gets rather strange since we have three different residual or error terms, which are defined in Figure 13.9. First there is residual variance among the classroom means, μ_{0i}. Second, from classroom to classroom, there is residual variance in the slope associated with student family, $\mu_{1i}SF_{hi}$. Both of these residuals are manifested at the level of the classroom. Finally, there is residual error at the level of the individual student test scores over and above the predicted values from the model, ε_{hi}. These different error terms are called "random" effects. They are ERROR—the part of the DATA that we do not understand. Though the variance attributable to random effects is estimated by the analysis, those estimates do not allow us to make predictions for a new student, and they are not routinely tested for significance. These three errors likely have different variances, meaning that they cannot be simply lumped together into one general error term.

Because we do not have just one error term to minimize, the estimation of this model cannot be done using those least squares procedures that have been used throughout the book to this point. Rather we have multiple error terms, and we need to estimate them (and the model parameters) in some optimal sense. The optimal approach finds values of the parameters and residual variance that maximize the joint likelihood of the data, assuming a normal distribution. Back in the conclusion of Section A, we provided a brief introduction to maximum likelihood estimation. We discussed the idea that, in the context of most of the book's models, least-squares estimates and maximum likelihood estimates were one and the same. But in the present context, with multiple error variances, no least-squares estimate is obtainable. Instead, through an iterative approach, we can find the parameter and residual variance values that maximize the joint likelihood of the data.[2] Mixed models procedures available in many statistical packages like R, SAS, and SPSS will do the estimation.

An equivalent way to write the mixed model is informative in specifying the random components of the model:

$$Y_{hi} = (\alpha_{00} + \mu_{0i}) + (\alpha_{10} + \mu_{1i})\, SF_{hi} + \alpha_{01} TF_i + \alpha_{11} TF_i{\times}SF_{hi} + \varepsilon_{hi}$$

According to this specification, the model's intercept (in the first set of parentheses) has two components—the overall mean, α_{00}, and a "random" component that allows the mean score to vary from classroom to classroom, μ_{0i}. Likewise, the slope for the *SF* predictor (in the second set of parentheses) has two components, its average slope across classrooms, α_{10}, and another "random" component that allows the slope of *SF* to vary from classroom to classroom, μ_{1i}. The slopes for *TF* and the *SF* by *TF* interaction do not have random components, since *TF* varies only between classrooms and not within them. That is, since each classroom has either a Montague teacher or a Capulet teacher but not both, we cannot estimate the effect of TF within a particular classroom. To allow random variation in a slope across classrooms, one must be able to estimate the slope for each classroom. If the predictor, like TF, does not vary within classroom, its slope is not estimable.

To do the mixed-model estimation, programs generally require that the data be in "long" form, meaning that each row of data is an individual observation (in this case from an individual student within a classroom). The data file in this case would consist of as many rows of data as there are students, specifying for each an individual Y_{hi} score, a student identifier, *h*, a classroom identifier, *i*, the student's family, SF_{hi}, and the family of the teacher in that classroom, TF_i. Then the mixed-model specification indicates that Y_{hi} scores are to be modeled as a function of student family, teacher family, and their interaction, with random intercepts and random student–family slopes for classrooms. The output generally gives the parameter estimates and the random error variances (in this case the classroom intercept variance, the classroom slope variance for student family, and the level-one residual variance). It is also possible to estimate the covariance between the random intercepts and random slopes for each classroom. These variances (and the covariance) can frequently be of substantive interest. For instance, it might be interesting to know whether classrooms vary in the magnitude of the student-family difference, or whether classrooms with higher average scores show smaller student-family differences, a relationship that is reflected in the intercept–slope covariance.

Because there are multiple error variances in the model, calculations of degrees of freedom and standard errors for parameter estimates are complex and beyond the scope of what we cover. Most mixed-model programs provide estimated degrees of freedom and approximate *t*-tests for individual parameters under what is called the Satterthwaite approximation. Additionally, model comparisons can be made by comparing deviance statistics of nested models, with differences in deviance values distributed approximately as chi-square statistics. Details on these issues can be found in texts devoted exclusively to multilevel mixed models (Brauer & Curtin, 2018; Hox, 2010; Snijders & Bosker, 2012).

There are distinct advantages of using a general multilevel modeling approach to the estimation of models for nonindependent data (compared with the least-squares approaches outlined earlier in this chapter and the preceding one). First, one need not make restrictive assumptions about the numbers of observations in the groups that give rise to the nonindependence. For instance, in the present context, the number of students in each classroom could vary considerably, whereas our previous approach assumed a constant number of observations in each group (and only one observation for each level of the within-group predictors). Second, and relatedly, estimation is possible in the presence of missing data, whereas with repeated measures ANOVA approaches, if a group of data is missing one or more observations typically the whole group must be omitted from the analysis. Third, as we next illustrate, these models accommodate continuous predictors that vary both within and between groups, at both level one and level two. And finally, a simultaneous estimation of the level-one and level-two models is both efficient and statistically optimal.

A Second Multilevel Model With Continuous Predictors

Suppose a researcher were interested in the relationship between how much one exercised and one's mood. Ideally, if she were interested in seeing whether there was a causal effect of exercise on mood, she would experimentally manipulate exercise level, randomly assigning participants to different levels and then later measuring their mood. More practically and to get an initial assessment of how exercise and mood are related, she might do a longitudinal observational study, measuring both exercise level and mood repeatedly over time in some sample of participants. Figure 13.10 presents some of the data that might have been collected.

In these hypothetical data, occasions of measurement are sampled within participants with the number of occasions varying from participant to participant. On each occasion, participants indicate their daily level of exercise (1–10 self-rating scale) and the positivity of their mood after exercising (again 1–10 self-rating scale). In practice, many different participants and many different occasions within participants would be sampled.

We would like to build a model in which the exercise variable is used to predict mood. Observations are grouped in the sense that numerous occasions come from the same participant and dependency due to that grouping is likely. Further, the independent variable, exercise, varies both within and between participants. To illustrate what this means, consider participants 1 and 2 in Figure 13.10. Clearly within these two participants, exercise varies from occasion to occasion. Additionally, on average, participant 1 exercises less than participant 2 (participant 1 average: 2.67;

Participant	*Occasion*	*Exercise*	*Mood*
1	1	3	5
1	2	1	4
1	3	4	4
2	1	5	6
2	2	7	5
2	3	3	8
2	4	5	5
3	1	1	3
3	2	3	7

Figure 13.10 Hypothetical exercise and mood data

participant 2 average: 5.00). Exercise thus varies between participants as well as within participants. To fully understand the relationship between exercise and mood in these data, we need to examine that relationship both within and between participants.

We start the specification of the multilevel model by writing out the model for relating exercise to mood within each participant, i.e., the level-one within-participant model:

$$M_{hi} = \beta_{0i} + \beta_{1i}E_{hi} + \varepsilon_{hi}$$

With occasions grouped or nested within participants subscript h now refers to the occasion and subscript i refers to the participant. According to this model, each participant has an intercept and a slope. Because the exercise variable has not been centered, each participant's estimated intercept will be the predicted mood on days with no exercise. An individual participant's slope will estimate the difference in mood as a function of having exercised one additional unit on a given occasion. Given that we will be predicting these level-one, within-participant parameters estimates at level two, between participants, it makes sense to center the level-one predictor and, importantly, to center it around each participant's own mean level of exercise. That is what we have done in the following revised level-one model, where $\bar{E}_i$ equals the mean exercise score of the ith participant across all observed occasions of that participant.

$$M_{hi} = \beta_{0i} + \mu_{1i}(E_{hi} - \bar{E}_i) + \varepsilon_{hi}$$

In this revised level-one within-participant model, each participant's estimated intercept will equal that participant's mean mood score across occasions.[3]

At level-two, we model the level-one parameters as a function of the participant's mean exercise level:

$$\beta_{0i} = \alpha_{00} + \alpha_{01}\bar{E}_i + \mu_{0i}$$
$$\beta_{1i} = \alpha_{10} + \alpha_{11}\bar{E}_i + \mu_{1i}$$

In order to make the intercepts in the level-two models more interpretable, we will also center the predictor in these models, centering each participant's mean exercise level around the grand mean, $\bar{E}.$:

$$\beta_{0i} = \alpha_{00} + \alpha_{01}(\bar{E}_i - \bar{E}.) + \mu_{0i}$$
$$\beta_{1i} = \alpha_{10} + \alpha_{11}(\bar{E}_i - \bar{E}.) + \mu_{1i}$$

Thus, in the level-one model, the predictor is the difference between a participant's exercise score on a given occasion and his or her mean exercise level, while in the level-two model, the predictor is the difference between each participant's mean exercise level and the grand mean of exercise.

We can now integrate the level-one and level-two models (both with their centered predictors) into a single multilevel model, substituting the level-two models for the parameters in the level-one model:

$$M_{hi} = (\alpha_{00} + \alpha_{01}(\bar{E}_i - \bar{E}.) + \mu_{0i}) + (\alpha_{10} + \alpha_{11}(\bar{E}_i - \bar{E}.) + \mu_{1i})(E_{hi} - \bar{E}_i) + \varepsilon_{hi}$$

Rearranging terms, to group together the parameters (the "fixed effects") and the random error terms, yields

$$M_{hi} = \alpha_{00} + \alpha_{01}(\bar{E}_i - \bar{E}.) + \alpha_{10}(\bar{E}_{hi} - \bar{E}_i) + \alpha_{11}(\bar{E}_i - \bar{E}.)(E_{hi} - \bar{E}_i) + \mu_{01} + \mu_{1i}(E_{hi} - \bar{E}_i) + \varepsilon_{hi}$$

Thus, the model to be estimated includes fixed parameters for the grand mean, α_{00}; the between-participant effect of mean exercise level, α_{01} (do people who habitually exercise more than the average have higher average moods?), the within-participant mood effect, α_{10} (on occasions when exercise levels are higher or lower than is habitual do people report higher moods?); and their interaction, α_{11} (does the relationship between mood and variation in exercise across occasions depend on a person's habitual level of exercise?). In addition to these fixed effects, the model specifies random variation in participant means (i.e., intercepts) and in their slopes for day-to-day fluctuations in exercise levels (and presumably their covariance). Output from most mixed-model procedures would generally include the estimated fixed parameters (and their associated standard errors) as well as the variances of the random components.

The interaction term in the model is included as a result of including the between-participant predictor in the level-two model for the slope. If one respecified the level-two slope model without the predictor

$$\beta_{0i} = \alpha_{00} + \alpha_{01}(\bar{E}_i - \bar{E}.) + \mu_{0i}$$
$$\beta_{1i} = \alpha_{10} + \mu_{1i}$$

one would be omitting the interaction from the full mixed model, which becomes

$$M_{hi} = \alpha_{00} + \alpha_{01}(\bar{E}_i - \bar{E}.) + \alpha_{10}(E_{hi} - \bar{E}_i) + \mu_{01} + \mu_{1i}(E_{hi} - \bar{E}_i) + \varepsilon_{hi}$$

It is entirely possible, given the correlational nature of the observational data that underlies this example, that the two exercise effects (one due to within-participant fluctuations and one due to between-participant variation) have very different associated slopes. For instance, day-to-day fluctuations in exercise might be negatively predictive of day-to-day fluctuations in mood while nevertheless chronic exercisers are chronically in better moods.

Figure 13.11 presents a hypothetical plot of possible results from this mixed model. Each thin line is the level-one model for a given participant, showing the participant mean (the dot)

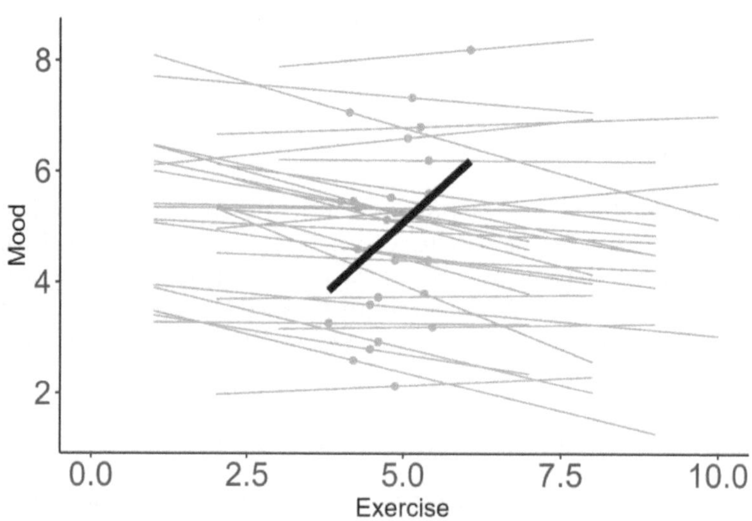

Figure 13.11 Exercise-mood hypothetical relationships from multilevel model

and the within-participant relationship between exercise and mood. Many of these lines have negative slopes. At the same time, on average across the participants, it seems from these hypothetical data that participants who exercise more on average have higher mood scores on average, as indicated by the thick line in the figure, which represents the level-two model for participant means.

The heavy line depicts the between-participant relationship between mean exercise level and mean mood. Lighter lines depict within-participant relationships between exercise and mood on different occasions. Points represent the participant means on both exercise and mood.

Crossed Random Effects Models

We have used two data examples in the previous section to illustrate multilevel models. The first involved students (level one) nested within classrooms (level two). The second involved occasions (level one) nested within participants (level two). Because these models incorporate error variances at both levels, appropriate analyses permit generalization of effects with hypothetical new samples at both levels. In other words, if we found a significant effect of exercise on mood, either within participants across occasions or between participants, and if the assumptions underlying the analysis were met, then we could reasonably generalize conclusions to other samples of participants and other samples of occasions that might have been used, assuming these were sampled in the same way as those included in the original data set. This sort of generalization has been the goal of null hypothesis testing throughout the models used in this book. We want to know whether parameter estimates might reasonably be expected to differ from a priori values (usually zero) if other error tickets (i.e., other samples of errors) were used. But in the present context, there are error tickets accruing from multiple "random" factors at both levels one and two.

Frequently in psychology experiments, random factors are crossed with each other rather than being nested one within the other, as in the previous examples. Suppose, for instance, that one were interested in whether stereotypic judgments (e.g., "how dangerous is this individual?") of Black male faces differed significantly from such judgments of White male faces. One might ask a sample of participants to judge a sample of faces, half of which were Black and half White, with the goal of determining whether, on average, Black faces are judged differently from White faces. Just as the participants should properly be considered a sample of possible participants, so too the faces used should be considered just a sample of possible faces that might have been used to represent the two ethnic categories.

Consider the hypothetical data in Figure 13.12. Each row of data represents the judgments of ten faces by an individual participant, with the first five judgments of Black faces and the second five of White faces. In this example, faces are crossed with participants, meaning that every face is judged by every participant. In this sense, this design, with two sampled factors, is different from those considered earlier where the two sampled factors (for instance students in classrooms) were nested, one under the other.

The standard data analytic approach to this design would be to compute two mean scores for each participant, one for the judgments of the Black faces and one for the judgments of White faces, and then to ask whether there is a significant difference between these two means on average across participants, using a within-participant repeated measures ANOVA. In essence, this standard analysis treats participant as a random factor but ignores variation in the data that accrues from individual faces because it averages across the faces within ethnicity. For instance, in the data of Figure 13.12, within the Black faces, it seems that ratings of Face4 are rather high on average and ratings of Face5 rather low on average. This difference, due to faces, is lost if one simply computes and analyzes average ratings for each participant.

Participant	*Black Faces*					*White Faces*				
	Face1	*Face2*	*Face3*	*Face4*	*Face5*	*Face6*	*Face7*	*Face8*	*Face9*	*Face10*
1	7	8	5	9	4	4	5	3	7	8
2	6	5	4	8	3	5	6	4	6	7
3	4	6	2	7	1	4	3	2	4	4
4	6	7	3	7	4	6	4	3	6	7

Figure 13.12 Hypothetical data from participants judging Black and White faces on a stereotypic dimension of judgment

Clark (1973) and Judd et al. (2012) have shown that this standard analytic approach, ignoring the variation due to faces, results in substantially inflated test statistics, if the goal is to reach conclusions about the ethnicity difference that generalize to studies involving different samples of both participants and faces. In other words, "significant" effects due to ethnicity may fail to replicate in studies that involve other faces sampled in the same way from the two ethnic categories. In essence both participants and faces induce random variation (dependencies) in the data, and these must be taken into account in the analysis if one is to reach appropriate statistical conclusions.

To understand this bias if the variation due to faces is ignored, simply imagine that different samples of faces were used. In one study, all the Black faces, by luck of the draw, were like faces 2 and 4 with relatively high ratings. And in another study, all the Black faces, again simply by chance, were more like faces 3 and 5, with relatively low ratings. Because of these chance differences each of these studies may have reached very different conclusions if they failed to account for the sampling variability inherent in the data due to the mean face differences.

The appropriate analysis involves a mixed model that has "error" terms deriving from both random factors, participants and faces. However, these two random factors are not nested, as in the previous examples, so one cannot specify the appropriate mixed model by talking about random factors and models at levels one and two. Instead, the mixed model specifies the fixed parameters that one wishes to estimate along with all other components of random variation that are estimable in the data. For the current design, the fixed effects are the grand mean and the mean difference due to face ethnicity. The grand mean is the intercept in a linear model with ethnicity contrast-coded and the mean difference due to face ethnicity is the slope for the ethnicity predictor. Participants and faces are both random factors, for which one can estimate a mean: In the matrix of ratings of Figure 13.12, the row means are the participant means and the column means are the face means. Additionally, for participants, but not for faces, one can calculate an ethnicity slope or difference. That is, because participants are crossed with ethnicity, an ethnicity slope can be estimated for each participant. But this is not true for the random factor of face since each face occurs in only one ethnicity condition rather than in both.

Hence, for this design, the full mixed model that should be estimated, including all variance components attributable to both random factors (participants and faces), is

$$Y_{ij} = \beta_0 + \beta_1 Eth_j + \mu_{0i} + \mu_{0j} + \mu_{1i} Eth_j + \varepsilon_{hi}$$
$$Y_{ij} = (\beta_0 + \mu_{0i} + \mu_{0j}) + (\beta_1 + \mu_{1j})\, Eth_j + \varepsilon_{hi}$$

In this model, the i subscript is used to refer to participant and j to face. Y_{ij} is the ith participant's judgment of the jth face, and Eth_j is a contrast-coded predictor representing ethnicity of the face. We have given the model in two versions, first with the fixed effects followed by the four random effects, and second with the intercept components in the first parentheses and with the slope components in the second parentheses. The model specifies an overall intercept (or mean) with variations around that mean due to participant and face means; then an overall difference due to ethnicity, allowing that slope to vary randomly from participant to participant; and finally, random error at the level of the individual observation. One can estimate this mixed model using maximum likelihood mixed modeling procedures, again with the data in "long" form (i.e., one rating of one face by one participant per row of data) rather than in the form given in Figure 13.12. The output from the estimation procedure would include the two parameter estimates (and associated standard errors) and variance estimates attributable to all four random components of variance (along with the covariance between participant intercepts and slopes).

With nested designs (e.g., occasions nested under participants), we saw that independent variables could vary either between or within levels of the grouping factor. This is also true in the case of designs with two random factors that are crossed (as in the participant by face example). Independent variables can vary between and/or within the levels of each random factor. In the design of Figure 13.12 and the mixed-model specification just given, the independent variable, ethnicity, varies within participants but between faces. Many other designs are possible and the appropriate mixed-model specification for each is given by Judd et al. (2012). The basic rule is that random intercepts are generally estimable for both random factors while random slopes are estimable for only those independent variables that vary within the levels of a particular random factor.

Alternatives to Multilevel Models

We have discussed crossed and nested random effects models. In each example dataset, there were relatively few units of analysis. For the purposes of demonstration, in our analysis of whether judgments of Black and White faces differed, there were only four participants, who provided ten observations each (five for Black faces and five for White faces). Similarly, in our analysis of the effects of exercise on mood, we hypothetically sampled only four participants, who provided three observations each. Were we to actually estimate these models, we would likely need many more observations. Multilevel models in which observations are nested within a higher order unit often require 20–50 higher order units to yield unbiased estimates, although there is no definitive cutoff (McNeish & Stapleton, 2016).

Recruiting sufficient samples for multilevel models is not always possible. Consider the example involving Montague and Capulet students. Let us assume that we only collected data from four Montague-teacher classrooms and four Capulet-teacher classrooms. For the moment, we will ignore teacher family and focus only on the within-classroom effect of student family. We likely do not have enough classes to generate unbiased estimates, so a multilevel model is inappropriate. However, neither can we ignore nonindependence due to classroom and estimate a regression model that ignores the grouping variable. So what can we do?

One possibility is to build a model in which we account for between-classroom variation as we did in Chapters 9, 10, and 11, addressing categorical predictors. Recall that, in those chapters, we created a set of dummy or contrast codes to assess the difference between the means of the different groups. A similar approach allows us to differentiate within- and between-classroom variation. Such models are referred to as "fixed-effects models." By statistically controlling for between-classroom variation, we can estimate the effect of student family on achievement, knowing that our estimates reflect only within-classroom, between-student differences in academic performance.

We would first create m-1 categorical predictors, such as contrast codes, where m is the number of classrooms. In the current example, seven contrast codes account for all the between-classroom differences in academic performance. We would then estimate Model A by regressing academic performance on these seven categorical predictors (which capture differences between classrooms) and one contrast-coded predictor reflecting differences between Montague and Capulet students. A comparison of Model A to a Model C that includes only the seven categorical predictors for classroom tests whether the partial slope associated with student family differs from zero, holding constant between-classroom differences in academic performance. This analysis is nothing new. We presented similar models in Chapter 11, which involved controlling for category (in that case, groups of students defined by college and first-generation status) while testing a focal predictor (aptitude). These fixed-effects models are conceptually equivalent, but here we emphasize their role in addressing nonindependence. An important drawback of fixed-effects models is that the results do not generalize to levels of the grouping variable other than those we have sampled. Whereas a multilevel model would treat both student and classroom as random factors, a fixed-effect analysis (as the name suggests) treats classrooms as a fixed factor. Accordingly, any significant effects would not be expected to generalize to new samples of classrooms.

There are other occasions when using mixed effects models is undesirable. For example, there may be too few observations nested under a higher order unit, the errors may violate assumptions of normality, homoscedasticity, or independence (above and beyond that which is accounted for by the model). Statistical programs that estimate clustered standard errors (also referred to as empirical standard errors) can reduce the effect of these violations on the standard errors of the regression estimates, yielding more accurate estimates of F, PRE, and p. We recommend several articles that provide detailed information on clustered standard errors, fixed-effects models, and other alternatives to mixed effects models that address nonindependence (McNeish, 2023; McNeish et al., 2017; McNeish & Kelley, 2019). Each of these alternatives has benefits and drawbacks. As always, researchers should carefully consider the analytic strategy that best tests their hypotheses given the constraints of their data.

Summary and Conclusion

We began this chapter by showing how continuous predictors that vary between groups of linked observations can be incorporated into models that examine differences within groups. The strategy was a simple extension of the analyses described in Chapters 6 and 12. That is, variables that assess within-group differences are regressed on the continuous predictors. Tests of the intercepts in these models provide tests of the within-group differences so long as the continuous predictors are centered. Tests of the slopes tell us whether the magnitudes and directions of the within-group differences depend on the level of the continuous predictor(s).

We then examined how to incorporate continuous predictors that vary only within groups of linked observations. In this chapter, for example, lab animals' weights were recorded daily for ten days, and we examined whether the animals gained weight during the ten days and whether that weight gain depended on diet—a categorical variable that varied between subjects (i.e., animals). We described two ways to analyze these data, noting that they yield the same conclusions whenever the within-subject intervals (e.g., day) are equal and there are no missing data. The first method involved computing a composite variable reflecting the linear change in weight for each rat, using the methods described in Chapter 12. The second involved estimating a separate regression equation for each rat. That is, for each rat, weight was regressed on a continuous day variable having values of 1 to 10. The intercepts and slopes from these equations then became the dependent variables in further analyses. This second method, commonly known as multilevel modeling, is a simple extension of our modeling approach to repeated-measures ANOVA.

We then extended our discussion of multilevel modeling, noting that they are special cases of what are called linear mixed models. Such models allow simultaneous estimation of effects at multiple levels, with appropriate error variances accruing from random factors at the different levels. Estimation is based on maximum likelihood procedures rather than least-squares, allowing efficient and optimal estimation.

Multilevel models offer distinct advantages over traditional repeated measures analyses. First, they can incorporate continuous as well as categorical independent variables and these independent variables may be partially redundant with each other.

Second, as we have indicated, multilevel models easily accommodate missing data, unequal within-subject intervals, and unequal numbers of observations in the groups of nonindependent observations. In the weight-gain example, for instance, some animals may have data for only eight or nine days rather than all ten days.

Third, the predictor variables (whether continuous or categorical) can vary both within groups of linked observations and between them. Consider again the example in which exercise was used to predict mood. The amount of exercise varied within participants from day to day. It also varied between participants; some participants exercised more overall than did others. Multilevel modeling provides a way of simultaneously estimating both within participant and between participant effects.

Fourth, we introduced crossed random effects models in which multiple random factors are crossed with each other rather than being nested, as in multilevel models. In a crossed random effects model, for example, we described how both participants and stimuli (i.e., faces) could, and indeed, should, be treated as random factors. The goal of research is to reach conclusions that generalize to a broader group of participants and to a broader group of stimuli; crossed random effects models permit that sort of generalization.

Finally, we note that there are times when mixed effects models may not be appropriate, for example, when there are too few higher order units to produce unbiased estimates. We provide a brief explanation of fixed-effects models (which are equivalent to models discussed in Chapter 11) and summarize other ways to analyze nonindependent data.

We have sought to illustrate the flexibility available through multilevel analysis and to encourage researchers to think creatively about the sorts of questions they can ask about grouped or linked observations. Although we have highlighted the considerable advantages of multilevel programs, we believe that the intelligent data analyst can accomplish a great deal with a simple general-purpose multiple regression program. Thinking carefully about the model comparisons that make the most theoretical sense prior to embarking on the further complexities of multilevel modeling programs will provide a more thorough understanding of the data and the models that are theoretically reasonable.

Notes

1 As discussed in the previous chapter, because each within-animal contrast has its own unique error term, the test of any one of them is completely independent of the test of any and all others. Thus, we can test only the linear within-animal contrasts in which we are interested, ignoring the other within-animal contrasts.

2 In fact, for most multilevel models the estimation involves a variation of maximum likelihood that is called 'restricted' maximum likelihood. What this means is that the parameters and the residual variances are estimated sequentially rather than simultaneously.

3 Centering around each participant's own mean rather than the grand mean is important so that the level-one intercepts are the participant's own mean mood rather than the participant's predicted mood at the mean exercise level of all participants and occasions.

14 Outliers and Ill-Mannered Error

Until Chapter 12, we generally assumed that error in our basic equation

DATA = MODEL + ERROR

was well-behaved. In the two previous chapters, we examined the consequences of nonindependent errors and developed an analytic strategy to address the problem. In this chapter, we examine other ways in which error may be ill mannered. In particular, we consider violations of our original assumption that ε_i is normally distributed with homogeneous variance. These problems are generally not as serious as those associated with nonindependence. However, in many instances the problems caused by non-normality and heterogeneous variance are substantial and, in any case, an examination of the assumptions almost always leads to a better understanding of one's data and the construction of better models.

There are two broad classifications of violations of assumptions about error: systematic and idiosyncratic. As we shall see, the systematic violations are often best identified visually by examining a variety of specialized graphs. Violations that are sufficient to make us doubt either the estimated model or the appropriateness of the probability values for the test statistics usually stand out in these specialized graphs. Our remedy, just as it was for nonindependence, is to transform the data so that we may use our full range of techniques for building models. Idiosyncratic violations arise from wild observations, commonly referred to as *outliers*, that for one reason or another are dramatically inconsistent with the other observations. The presence of outliers frequently causes violations of normality and homogeneous variance assumptions. Hence, we begin by considering methods for detecting and resolving the problems of idiosyncratic outliers and then turn to the visual methods for detecting systematic violations. As a practical matter, it is usually best to address the idiosyncratic outlier issues first, and then consider possible systematic violations of the normality and homogeneous variance assumptions.

Outliers

Outliers are extreme observations that for one reason or another do not belong with the other observations in the data. There are many ways in which outliers can be introduced into the data. The first cause we consider is scientifically uninteresting but quite troublesome and common. Data recording and data entry errors can put wild values into the data and the predictor variables used in models. The use of computers for data analysis creates many opportunities to introduce such outliers in our data and at the same time reduces our chances of finding them, unless we look carefully. Consider, for example, a study involving participants' heights, measured in inches, and weights, measured in pounds. While completing an online survey, one participant accidentally reverses the

DOI: 10.4324/9781003438335-22

numbers for height and weight. As researchers, if we analyzed the data by hand, we would likely notice such a reversal. However, when analyzing data on a computer we are further removed from our data and therefore less likely to notice such a mistake.

A second cause of outliers is that the observations are not a single homogenous set to which a single model will apply, but rather a heterogeneous set of two or more types of observations, one of which is much more frequent. The infrequent cases of the other types will appear as outliers. Discovering that there are really two kinds of things in our data when only one was expected is almost always interesting scientifically. An example is identifying children who thrive and excel despite being raised in adverse environments. Examining other characteristics of extreme observations, especially if they are outliers, may provide clues for building more complex models. In this case, we again need statistical techniques for identifying extreme or outlier observations, not necessarily so that they can be removed or corrected but so that they can be examined with greater care to extract their full informational value.

A third cause of outliers is what statisticians refer to as error distributions with "thick tails" in which extreme errors occur with greater frequency than expected for a normal distribution. We examine the normality assumption in greater detail later. For now, it suffices to note that just a few observations from the thick tails of a non-normal error distribution might appear as outliers.

Throughout the book, we have used the minimization of SSE as the method for estimating model parameters and comparing models. As we noted in Chapter 2, SSE is especially sensitive to outliers. A single outlier can dramatically alter estimates of group means or regression parameters. Hence, an outlier can bias the parameter estimates and/or inflate SSE. A cursory examination of the usual output from statistical regression programs would not suggest the presence of such an error. Allowing an outlier to "grab" the model parameters is obviously undesirable. Perhaps more insidious, even if the model is not distorted, inflation of SSE can increase the difficulty of testing complex models by reducing power. If outliers amplify the total error, they may overshadow any reductions in error achieved by increasing model complexity. We therefore need statistical techniques for identifying erroneous observations so that they can be removed or corrected in the data.

As an example of the deleterious effects of an outlier, in this case, a data recording error, let us again consider the following two sets of numbers that we examined in Chapter 2:

Set 1: 1, 3, 5, 9, 14	$\bar{Y} = 6.4$	$\text{MSE} = s^2 = 26.8$
Set 2: 1, 3, 5, 9, 140	$\bar{Y} = 31.6$	$\text{MSE} = s^2 = 3681$

Set 2 is identical to Set 1 except that a data entry error has added an extra digit to 14 to make it 140. As we noted before, this error has a dramatic impact on the mean, which is our estimate of β_0 for the simple model. In this case, the estimate changes from 6.4, which is in the middle of the data values, to 31.6, which is not very representative of any of the data values. The effects on SSE and, consequently, MSE are at least as bad. The inflated MSE makes any hypothesis testing extremely difficult. Examine the 95% confidence interval for β_0 for both sets of numbers:

Set 1: [0, 12.8]
Set 2: [–43.7, 106.9]

The confidence interval for Set 2 is much wider than that for Set 1; in fact, the confidence interval for Set 1 is entirely included within the confidence interval for Set 2. Hence, many hypotheses about β_0 that could be easily rejected for Set 1 will not be rejected for Set 2. The data entry outlier thus not only produces a misleading parameter estimate but also reduces the statistical power of any inferences. The outlier has so inflated the SSE that it is much more difficult to detect any proportional reduction in error produced by using β_0 instead of B_0 in a model for these data. The same would be true for more complex models.

Detecting Outliers

There are three separate outlier questions. Given that our analyses nearly always have predictor variable(s) X_{i1} ($X_{i2}, \ldots, X_{ip-1}$) and always have a data or outcome variable Y_i, two outlier questions are obvious. Is the value for the predictor variable (or the set of predictor values) unusual? Is the value for the outcome variable unusual? The third question asks about the influence of the observation on all the parameter estimates, which involves both unusual predictor and outcome values. That is, does the observation distort or have undue influence on the overall model? Each of these three outlier issues is considered in turn after we consider an example, which we use throughout as an illustration.

An Outlier Example

We now consider a detailed example to demonstrate the serious consequences of even a single outlier and to motivate the detection techniques to be presented later. Figure 14.1 displays the SAT verbal scores and high school ranks for 13 students. The data values on the left side of the figure are the correct values. The values on the right side are identical except that we have simulated an outlier (such as could be caused by a data entry error) by reversing the SAT and high school rank values for the sixth student. Below each dataset are the results of simple regression analyses to determine whether high school rank is a useful predictor of SAT verbal scores. Note that PRE and, consequently, F are slightly larger for the dataset that contains the data reversal for the sixth

	Original data		*With reversal*	
Student	*SAT*	*HSRank*	*SAT*	*HSRank*
1	42	90	42	90
2	48	87	48	87
3	58	85	58	85
4	45	79	45	79
5	45	90	45	90
6	48	86	86	48
7	51	83	51	83
8	56	99	56	99
9	51	81	51	81
10	58	94	58	94
11	42	86	42	86
12	55	99	55	99
13	61	99	61	99
	$b_0 = 5.95$ $b_1 = 0.50$ PRE = 0.29 $F_{1,11} = 4.53$		$b_0 = 96.55$ $b_1 = -0.50$ PRE = 0.33 $F_{1,11} = 5.35$	

Figure 14.1 SAT verbal and high school rank scores with and without data error (reversal of scores for Student 6)

student and the parameter estimates are dramatically different. The intercept changes from about 6 to 97, and the slope changes sign from +0.5 to –0.5. The PRE and F are statistically significant (at $\alpha = .05$) for the dataset with the data entry error, so we would conclude that as high school rank increases, SAT verbal scores decrease! This nonsensical result might cause us to check the data further, but nothing in the basic regression results or in a quick scan of the data in tabular form indicates that a serious problem exists.

How might we detect the outlier in the second dataset? It is almost always easier to see outliers in graphs than in tables. Hence, the first step is to examine the scatterplot for SAT and HSRank. The scatterplots along with the best-fitting lines for the datasets with and without the reversal of the scores for the sixth observation are displayed in Figure 14.2. The vertical axes in both scatterplots are drawn to the same scale so that the size of an error (i.e., deviation from the least-squares regression line) is comparable.

The scatterplot for the original data in the first half of Figure 14.2 has the pattern we would expect when the model is appropriate. The data values are clustered around a straight line that has a slope of 0.5, as revealed by the regression analysis in Figure 14.1. Note that none of the observations is unusually far from the line or from the other observations. Thus, no outlier is apparent in the first scatterplot. In contrast, an outlier stands out clearly in the second scatterplot for the dataset with the reversed observation. Again, none of the observations is particularly far from the best-fitting line (hence the significant value of PRE = .33). However, the outlier observation in the upper left-hand corner of the second scatterplot is unusual in three ways. First, the value of this outlying observation on the predictor variable HSRank is unusually small. If we examine the values on the predictor variable for all observations, there is only one large gap in those values—between the predictor value for this unusual observation (48) and the next lowest predictor value (79). Second, the value on the data variable (SAT) for the outlier is unusually large. Again, there is only one large gap between the data value for the unusual observation (86) and the next highest data value (61). Third, if the outlier case were omitted from the analysis, then the best-fitting line would have a positive slope (as in the first scatterplot) rather than a negative slope. This makes the outlier case unusual because it does not appear that omitting any other observation would have nearly as large an impact on the slope for the regression line. These three ways in which the outlier is unusual suggest that to detect outliers we will want to ask the following three questions for each

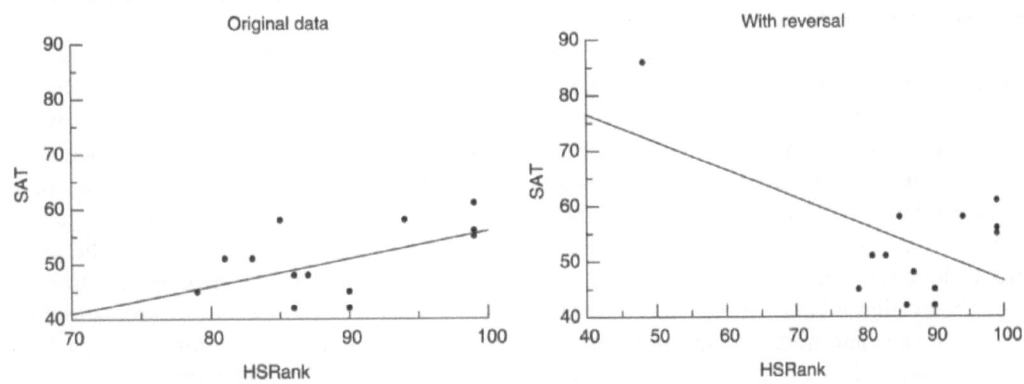

Figure 14.2 Scatterplots and best-fitting lines for data without (left) and with (right) reversal of scores for sixth observation

observation (note: we use the subscript i to represent a general observation and subscript k to represent an observation being considered as an outlier):

1 Is X_{k1} or, more generally, is the set of predictor values $X_{k1}, X_{k2}, \ldots, X_{k(p-1)}$ unusual?
2 Is Y_k unusual?
3 Would omission of the observation produce a dramatic change in the parameter estimates $b_0, b_1, b_2, \ldots, b_{p-1}$, or equivalently in the predictions $\hat{Y}_i$?

If the answer to any of these questions is yes, then we have an outlier requiring attention. For the HSRank and SAT data it appears as if the answer to all three questions is yes for the artificial outlier we produced by reversing the scores.

We now consider formal techniques for answering the three outlier questions. The statistical literature abounds with various mathematical indices and graphical procedures for each question. Except in the most unusual circumstances, most of these procedures give essentially the same answers, so we consider only a small subset of the possible techniques. We have selected those techniques that are most consistent with the general approach to data analysis via model comparisons.

Is the Predictor Value X_k Unusual?

All observations contribute equally to estimating the mean: That is, each observation has a weight of $1/n$. In contrast, each observation does not contribute equally to the estimate of the slope in regression. To gain insight into this issue and its importance, it is useful to reconsider an expression from Chapter 6 for estimating the slope of the best-fitting least-squares line. That is,

$$b_1 = \Sigma w_i \left[\frac{Y_i - \bar{Y}}{X_i - \bar{X}} \right] \text{ where } w_i = \frac{\left(X_i - \bar{X}\right)^2}{\Sigma\left(X_i - \bar{X}\right)^2}$$

The term in brackets is the slope "suggested" by each observation or data point—the change in Y between the point and the mean for the variable Y divided by the change in X between the point and the mean for the variable X—that is, the slope of a line between the data point and the mean point. The estimated slope for the regression is simply the weighted average of all the individual slopes suggested by each point. Figure 14.3 illustrates the individual slopes for each data point in the original data of Figure 14.1.

The weight (w_i) given to each data point when calculating the overall slope estimate is based on the unusualness of the data point in terms of the predictor variable X. The further the predictor variable is from the mean, the greater its weight. This makes sense because the slopes of lines with a long horizontal component in Figure 14.3 are unlikely to be affected much by small changes, perhaps caused by error, in the outcome variable Y. Hence, our confidence in the suggested slope is greater for extreme values of X. In contrast, for X values close to the mean, small changes in Y could dramatically alter the data point's suggested slope. Hence, our confidence in the estimated slope is much less for values of X that do not differ much from the mean. One can think of each observation as having a vote concerning the overall estimate of the slope, but the votes of some observations count more than the votes of others. If all the observations are telling more or less the same story, then all the observations ought to be voting for essentially the same slope. In the case of a perfect relationship, each data point's suggested slope would be identical and would equal the overall slope. In that case, it would not matter that some votes count more than others because all the votes would agree perfectly. However, when the votes disagree, the differential weights matter. In general, it is undesirable for a single observation to have a very large percentage of the

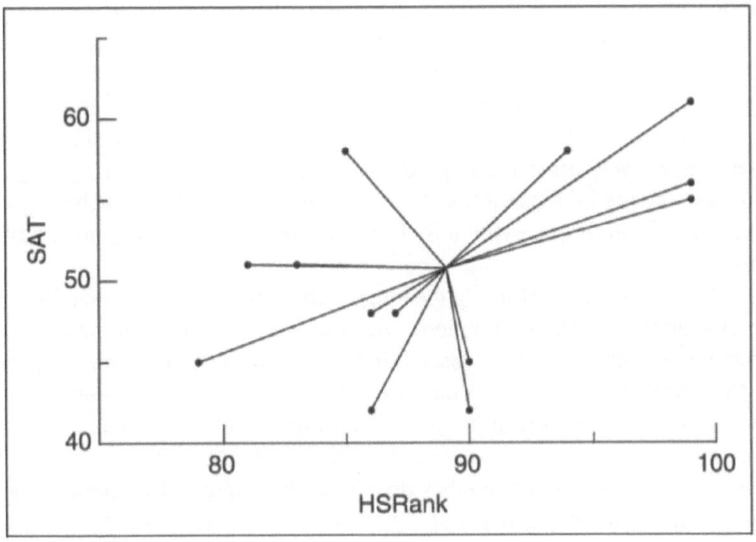

Figure 14.3 Individual slopes for each data point for original data in Figure 14.1

total weight because the slope votes of the other data points are ignored in calculating the "overall" slope. In that case, the "overall" slope is really a description of only one data point. This occurs when one observation has a very unusual predictor value relative to the other observations.

Most modern regression programs report the lever for each observation in the data (sometimes unhelpfully referred to as the diagonal of the hat matrix) h_i, which represents the weight or leverage that the observation has in determining the overall model. The lever is thus a measure of the importance of that observation's votes for the parameters in the model. For simple regression, the lever is defined as

$$h_i = \frac{1}{n} + \frac{\left(X_i - \bar{X}\right)^2}{\sum_{j=1}^{n}\left(X_j - \bar{X}\right)^2}$$

For simple regression, the model consists of two parts—the mean (or intercept) and the slope. Each observation has equal weight in the voting for the mean (i.e., $1/n$) and a weight proportional to the squared deviation of the predictor variable from its mean in the voting for the slope (i.e., w_i). The lever, as previously defined, is simply the sum of those two weights.

In some datasets, a few observations have unusually high levers—much greater than the other observations. This situation raises concerns because it implies that the estimated model primarily reflects the observations with high levers. It is inappropriate and misleading to report a regression equation as if it applies to all n observations when it only applies to the few observations that have particularly high levers.

In judging the magnitude of levers, it is useful to note that the sum of the levers necessarily equals the number of parameters, which is two for the case of simple regression. Hence, the average lever equals $2/n$ for simple regression. If the lever of an observation is near 1.0, it implies that one of the two parameters of the simple regression model is allocated to predict that single observation; this is clearly undesirable. In Figure 14.4 there are columns for the lever values for the original data of Figure 14.1 and for the lever values when there is the data entry error. For a simple regression based on 13 observations, the average lever equals $2/n = .15$. For the original data, which conform to our notions of a typical regression, the levers range around this average from .08 to .26, as expected. However, for the data where we introduced an outlier by reversing the predictor and outcome variables for one observation, the levers range from .08 to .15 except for one extreme value of .76. This implies that of the two parameters in the simple regression, 0.76 of a parameter is effectively allocated to predict a single observation, leaving only 1.24 effective parameters to make the predictions for the other 12 observations. Having such a high proportion of the model focused on a single observation is clearly inconsistent with the goal of having the model describe all the data.

Most researchers know that restricting the range (more properly, restricting the variance) of the predictor variable attenuates the correlation. The converse applies when there is a single outlier with an extreme predictor value: that observation artificially increases the range (variance) and thus inflates the correlation. Allowing one observation to inflate the correlation, or even to create one when otherwise there would be no relationship, increases the chances of making Type I errors—rejecting the null hypothesis when the null hypothesis is in fact correct. Observations with unusual predictor values, assuming they do not also have unusual outcome values, often make Model As appear better than they actually are. In those cases, the story told by the regression model really only pertains to that one observation, and it is very misleading to pretend that the regression story applies to all the data.

The generalization of levers from models with one predictor to models with multiple predictors is the same as the generalization of parameter estimates in Chapter 7. If there is no redundancy among the predictor variables, then additional terms for each predictor variable are added to the lever equation. If there is redundancy, then the computations are best left to the computer algorithms. Redundancy also introduces a new wrinkle conceptually for the lever. An observation's predictor values may not be unusual with respect to individual predictor variables. However, the pattern across all the predictors might be unusual. For example, if we were using actual height and weight as predictors in a sample of adolescents, a height of 5′ 9″ would not be unusual, and a weight of 95 pounds would not be particularly unusual, but the combination would be. The generalization of the lever detects this kind of unusualness quite well.

	Original data			*Data with reversed observation*		
Observation	*Lever*	*RStudent*	*Cook's D*	*Lever*	*RStudent*	*Cook's D*
1	.08	−1.91	0.125	.08	−1.03	0.04
2	.08	−0.31	0.005	.08	−0.53	0.01
3	.11	1.96	0.182	.08	0.38	0.01
4	.26	−0.14	0.004	.10	−1.35	0.10
5	.08	−1.18	0.056	.08	−0.70	0.02
6*	.09	−0.22	0.003	.76	4.63	11.86
7	.14	0.61	0.033	.08	−0.43	0.01
8	.25	0.05	0.001	.15	0.95	0.08
9	.19	0.84	0.086	.09	−0.54	0.02
10	.12	0.89	0.055	.11	0.86	0.05
11	.09	−1.41	0.094	.08	−1.27	0.06
12	.25	−0.15	0.004	.15	0.83	0.06
13	.25	1.09	0.195	.15	1.60	0.21

Figure 14.4 Outlier indices for the data of Figure 14.1

Is the Data Value Y_k Unusual?

If we want to identify those Y_i that are unusual, we must ask "unusual with respect to what?" The answer in the model comparison approach is that we want to find those Y_i that are unusual with respect to the model we are considering. The errors:

$$e_i = Y_i - \hat{Y}_i$$

tell us how unusual the *i*th observation is with respect to the model. If the absolute value of e_i is small, then the model makes a good prediction for that Y_i, so it is not unusual. On the other hand, if e_i is large, then the model makes a bad prediction for that Y_i, so it is unusual. The individual error e_i is often referred to as the *residual* because it is the part of the original observation Y_i that is "left over" after the prediction $\hat{Y}_i$ has been subtracted. Hence, we often refer to an examination of the individual error terms as an "analysis of the residuals."

It is generally difficult to identify outliers just by examining the absolute magnitude of e_i for two reasons. First, the importance of an error of a given magnitude is relative to the other errors and the magnitude of the prediction. We therefore need a way to transform the residuals to a common scale on which we can judge small and large values. We might, for instance, standardize them by dividing each one by the root-mean-square error.

A second problem is that there is a paradox in using the standardized residual for identifying unusual values of Y_i. Extreme data values tend to "grab" the model so that the estimated parameters minimize e_i for those extreme values. We are asking whether a particular data value, Y_k, is unusual with respect to the model, but the extreme Y_k has itself been used to determine the model. If it has seriously altered or biased the parameter estimates in the model, then Y_k might not be unusual in terms of the magnitude of its residual. Instead, we ought to determine whether Y_k is unusual with respect to a model determined by all the other observations *except* Y_k.

Mathematical statisticians have developed a multitude of transformations of the residuals e_i to solve these two problems in interpreting the magnitude of particular residuals. Most regression programs in computer statistical packages can produce a large variety of these transformed residuals. Some of these transformed residuals only solve the scaling problem, others eliminate the paradox by removing the effect of the *k*th observation when considering the *k*th residual, and some do both. We consider only one of these many residual indices, the *studentized deleted residual*, for three reasons. First, the studentized deleted residual both solves the scaling problem and eliminates the paradox, so there are good theoretical reasons for choosing it as an index of whether Y_k is unusual. Second, the studentized deleted residual has a natural interpretation in the context of the model comparison approach used in this book. We will demonstrate that the square of the studentized deleted residual is simply the F for comparing appropriately chosen Models C and A. Third, it is very unlikely that the studentized deleted residual would fail to detect an unusual Y_k that could be detected by any of the other transformed residuals.

We develop the studentized deleted residual by considering a specific model for an outlier. If an observation is so extreme that it is unlike the other observations, then we should be able to reduce the error appreciably by adding a specific parameter to the model just for that one observation. Designating a parameter for a single observation will ensure that there will be no error in our prediction for that observation. The other parameters are thereby freed to describe the remainder of the data. Because all error due to the outlier is eliminated and the other parameters generally provide a better fit to the remaining data, our overall measure of error will inevitably be less. In effect, we will be considering whether the outlier is fundamentally different from the other data observations. We will refer to an original multiple regression model augmented by the addition of a specific parameter for the suspected outlier as an *outlier model*. To screen data for outliers, we will test the outlier model for each observation. That is, we will test one observation at a time to see whether it is so extreme that we should allocate a specific parameter to it.

As with all other models, the question inherent in the outlier model is whether the additional complexity is worth it. Adding a parameter for the suspected outlier to the model will inevitably reduce the error, but will the reduction in error justify the increased complexity of the model? We can use PRE and F to answer this question. To be more formal, the model comparison for the outlier model is

$$\text{MODEL A: } Y_i = \beta_0 + \beta_1 X_{i1} + \ldots + \beta_{p-1} X_{i,p-1} + \beta_p X_{ip} + \varepsilon_i$$
$$\text{MODEL C: } Y_i = \beta_0 + \beta_1 X_{i1} + \ldots + \beta_{p-1} X_{i,p-1} + \varepsilon_i$$
$$H_0: \beta_p = 0$$

$$\text{where } X_{ip} = \begin{Bmatrix} 1 & \text{for } i = k \\ 0 & \text{for } i \neq k \end{Bmatrix}$$

Model C is the original regression model and Model A is the outlier model having an additional parameter that estimates the effect of a single observation. If the original model really describes all the observations, then it should not be useful to isolate a single observation with its own parameter.

As an example, consider the sixth observation in the data of Figure 14.1 with the data reversal error. Let $X_i = 1$ if $i = 6$ and 0 otherwise. Then the estimation of the aforementioned models reveals

$$\text{MODEL A: } \hat{SAT}_i = 6.71 + 0.50\text{HSRank}_i + 55.49X_i$$
$$\text{MODEL C: } \hat{SAT}_i = 96.55–0.50\text{HSRank}_i$$
$$\text{PRE} = .68 \; F_{1,10} = 21.4 \; p < .01$$

Thus, devoting one parameter to the sixth observation or, equivalently, omitting the sixth observation from the data used to fit the model reduces the error by 68%. With 11 degrees of freedom for error left after estimating b_0 and b_1, we would expect that the omission of any one observation would, on average, reduce the error by only about 1/11 = 9%. The very much larger proportional reduction in error obtained by omitting the sixth observation suggests that it is a very unusual observation with respect to the model for the data. This is confirmed by the very large value of $F_{1,10} = 21.4$. Note also that the sign of the coefficient for HSRank changed between Models C and A. Indeed, omitting the sixth observation caused a dramatic difference in the estimation of both the intercept and the coefficient for HSRank. We return to this fact later.

It is interesting to consider what would happen if we were to test the outlier model for an observation that is not an outlier. Suppose, for example, that we tested the outlier model for the first observation in the data with the reversal error in Figure 14.1. Estimating the coefficients in the compact and augmented models yields (where now $X_i = 1$ if $i = 1$ and 0 otherwise)

$$\text{MODEL A: } \hat{SAT}_i = 95.71–0.48\text{HSRank}_i - 10.67X_i$$
$$\text{MODEL C: } \hat{SAT}_i = 96.55–0.50\text{HSRank}_i$$
$$\text{PRE} = .096 \; F_{1,10} = 1.06 \; p = .33$$

Devoting a special parameter to the first observation is clearly not worth the added complexity. The proportional reduction in error of 9.6% is little more than we would expect by chance (9%), which is confirmed by the low, nonsignificant value of F. Note that in contrast to omitting the sixth observation, omitting the first observation produces minimal changes in b_0 and b_1.

These results suggest that we can use the outlier model as a means of screening our data for possible outliers. We simply test the outlier model n times, each time letting a different observation be the "odd one out." Although the outlier model is conceptually clear, it would be tedious to specify and test n different models for each regression. Fortunately, this is not necessary because Belsley et al. (1980) have shown that F for testing the outlier model for the kth observation is given by

$$F = \frac{e_k^2(n - PA - 1)}{\text{SSE}(1 - h_k) - e_k^2}$$

All of the components for computing F for the outlier model are therefore available either from the original regression or from the levers h_k. Most computer programs for multiple regression will provide t (the square root of this F) as the *studentized deleted residual* for each observation. In the model comparison approach, it makes sense to think of the studentized deleted residual as a test of the outlier model. The studentized deleted residuals for the observations in our example are listed in Figure 14.4 under its common abbreviation "RStudent." For the original data without an outlier, none of the values would be statistically significant using $\alpha = .05$. However, in the dataset with the reversed observation, the sixth observation has a very large value of RStudent = 4.63 (which squared equals 21.1, identical to the F for testing the outlier model for this observation). If the model is describing all the data, then there should not be such a large value for RStudent. The sixth observation is definitely an outlier.

We have now answered the question of whether an observation's Y value is unusual with respect to a model of the other observations. The studentized deleted residual is a t statistic testing the outlier model. If the resulting t is large, then we conclude that allocating an additional parameter for that particular observation is worthwhile. This conclusion is equivalent to saying that that particular observation's Y value is quite unusual with respect to the other observations. In other words, one model does not apply to all the observations.

We must be careful about using the outlier model to screen the data for outliers. If we examined the studentized deleted residuals for all observations, we would be performing the equivalent of n statistical tests on the same set of observations. If we use $\alpha = .05$ to determine the critical values of PRE and F, or equivalently RStudent, for rejecting the original model in favor of the outlier model, then we have a 5% chance of making a Type I error for *each* of the n model tests we perform. The probability that we do *not* make a Type I error in n trials equals $.95^n$. For our example with 13 observations, the probability that we do not make a Type I error is $.95^{13} = .51$. So the probability of making at least one Type I error when screening the data with the outlier model equals $1 - .51 = .49$. This is unacceptably large. One solution is to use the *Bonferroni inequality*. According to the Bonferroni inequality, the probability of making at least one Type I error will be less than α if the critical value for each test is chosen so that $\alpha' = \alpha/n$. For our example with 13 observations, $\alpha' = .05/13 = .004$, which would imply a critical value for F of about 13 or 14 (equivalent to a t of 3.6 or 3.7). The F (or RStudent) for the sixth observation easily exceeds this more conservative standard and so would be declared a statistically significant outlier.

In general, we do not recommend that the squared studentized deleted residual or the F that results from evaluating the outlier model be used routinely as a formal statistical test unless one has external information questioning the validity or reliability of a particular observation. Instead we suggest that large values of RStudent be used to indicate observations requiring closer scrutiny. An easy rule of thumb applies to RStudent. For reasonably large n, only about 1% of the studentized deleted residuals should be less than −3 or greater than +3. So if the absolute value of RStudent is > 3, careful attention to that observation is required. Absolute values of studentized deleted residuals of > 4 ought to be extremely rare and even after the Bonferroni adjustment will be statistically significant. Hence, all alarm bells ought to sound. An RStudent that large is a clear indication of an unusual value of Y_k. Using a model for which a studentized deleted residual is > 4 could be very misleading.

Would Omission of Y_k Dramatically Change the Predictions?

While considering the graphs in Figure 14.2 and examining the results of the outlier model for the sixth observation, we noted that omitting that observation causes a dramatic change in the estimate of b_1 from –0.5 when the sixth observation is included to +0.5 when it is omitted. It did not appear that omitting any other observation would have an appreciable effect on the estimate b_1 and certainly would not change the sign of the estimate. This suggests that we search for outliers by considering the change in the parameter estimates resulting from the deletion of each observation from the analysis. The deletion of a "typical" observation should have little or no effect on the parameter estimates, but the deletion of an unusual observation that may have "grabbed" the estimates might result in a large change in the parameter estimates. As we noted when defining SSR conceptually, differences in parameter estimates are reflected directly in different predicted values. Hence, to assess whether the set of parameters changes dramatically as a function of one observation, we can compare the predictions for the model based on the entire set of observations to the predictions for the model estimated when one observation is omitted. Let $\hat{Y}_{i,[k]}$ represent the predicted value for the ith observation from a model estimated with the kth observation omitted. Cook's D (1977, 1979) compares the two sets of predictions as

$$D_k = \frac{\sum_i \left(\hat{Y}_i - \hat{Y}_{i,[k]}\right)^2}{PA(MSE)}$$

A large value of Cook's D indicates that the model predictions are very different depending on whether the kth observation is included or not. For any particular observation that was of concern, it would be relatively simple to compute the two regressions with and without the kth observation and then compute the sum of squared differences for the two sets of predictions. However, this again would be tedious as a screening procedure for outliers if we had to perform n separate regressions, each time omitting a different observation. Fortunately, as was the case for the outlier model and studentized deleted residuals, this is not necessary because the following equivalent formula for Cook's D shows that it can be calculated from basic components available from the original regression. That is

$$D_k = \frac{e_k^2}{PA(MSE)}\left[\frac{h_k}{(1-h_k)^2}\right]$$

For example, for the sixth observation the only quantity we have not determined is MSE, but this is easily obtained as SSE/(n – PA) = 1085.3/11 = 98.66, so

$$D_6 = \frac{13.3^2}{2(98.66)}\left[\frac{.76}{(1-.76)^2}\right] = 11.83$$

This value (within rounding error) as well as the values of Cook's D for all the observations are listed in Figure 14.4. The only very large value of Cook's D is the one for the sixth observation in the dataset in which its predictor and outcome values are reversed. As we noted before, that observation grabs the model and dramatically changes the predictions for all the other observations, which is clearly undesirable in a model that purports to be a description of all the data.

There are only informal guidelines for thresholds above which values of Cook's D should require attention. One suggested rule is to consider any value greater than 1 or 2 as indicating that an observation requires a careful look. Another suggestion is to look for gaps between the largest values of Cook's D. Usually, as in Figure 14.4, a value truly requiring attention is obvious.

Many regression programs will report Cook's D, so the aforementioned computational formula will seldom be needed. However, the formula is conceptually important because it demonstrates that Cook's D is a multiplicative function of the squared error for that observation and the lever for that observation. Thus, the most influential observations in terms of their effects on the parameter estimates will be those having large squared errors (an unusual Y_k value) *and* high leverage (an unusual set of predictor values). Conversely, if either Y_k is not unusual with respect to the model for the other observations (i.e., low value of RStudent) *or* if the set of predictors for the kth observation is not unusual with respect to the sets of predictors for the other observations (i.e., a low value of the lever h_k), then Cook's D will not be large. But also note that if Y_k is not quite unusual enough to attract our attention on its own, and if h_k is also not quite large enough to be considered a high lever on its own, then Cook's D, because it is a multiplicative function of both values, can often be large enough to require attention.

Dealing With Outliers

Once we have identified outliers in a batch of data (by examining the levers, studentized deleted residuals, and Cook's D statistics), we have the problem of what remedial action, if any, should be taken. Although the tools to detect outliers are not particularly controversial, decisions about how to proceed in the face of clearly outlying observations have been a more contentious subject. If an outlying observation clearly originated because of data recording or coding errors, as in the example we have been using, then it clearly makes sense to correct the observation, if possible, and re-do the analysis with the corrected observation included. In many cases, however, it may not be clear why a particular observation is an outlier and, as a result, no clear corrective strategy is available. In this case, controversies may arise about how one should proceed.

In the early days, before the development of sound statistical procedures for identifying outliers, it was reasonable to question analyses that left out problematic observations based on seemingly ad hoc rules. One naturally feared, in the absence of clear rules for identifying outliers, that observations might simply have been omitted because they were problematic for the researcher in arguing for a favored hypothesis. Because of such fears, a tradition emerged in psychological research in which it was generally considered unethical to report analyses based on partial datasets that omitted problematic observations.

We believe that the development of sound statistical procedures for identifying outliers has rendered this point of view partially obsolete. We would argue that it is unethical to include clearly outlying observations that "grab" a reported analysis, so that the resulting conclusions misrepresent the majority of the observations in a dataset. The task of data analysis is to build a story of what the data have to tell. If that story really derives from only a few overly influential observations, largely ignoring most of the other observations, then that story is a misrepresentation.

At the very least, in the presence of clear outliers, we encourage researchers to analyze their data twice, once including the outliers and once not. If the resulting story is generally the same regardless of the inclusion of the outliers, then it is simpler to include the unusual cases. On the other hand, if the story is dramatically different, and the outlier statistics indicate one or more problematic cases, then we would recommend reporting the story without the outliers, while clearly indicating the steps that were undertaken for outlier detection. For journal articles, outlier detection procedures (and other analyses) may often be described in detail in supplementary materials.

Ensuring that procedures are transparent to reviewers and readers fosters more critical evaluation of research findings and facilitates others' replication efforts. However, an interesting question, worthy of further pursuit, is why the unusual cases occurred in the first place and what they may indicate about the theoretical issues that are being explored. Although the majority of observations may have a single story to tell, those few that make for a very different story can often be intriguing cases to follow further, to identify the limits of the story that one has told and how it needs to be expanded for other populations.

Systematic Violations of Error Assumptions

Outliers are idiosyncratic observations, arising for unknown reasons. In addition to problems that arise idiosyncratically, there may be systematic tendencies of errors to violate the assumptions underlying statistical inference. In Chapters 12 and 13 we dealt with the most consequential of these systematic violations, data in which errors or residuals are not independent of each other. The other two assumptions that may be seriously violated are the assumptions that errors are normally distributed and come from a single distribution, thus having a constant variance.

It is essential to realize that the assumptions of normality and constant variance pertain only to the errors in a given model and not to any of the variables themselves, neither the data variable, Y, nor the predictors, X, in the model. We clearly have extensively used predictors whose values are not normally distributed and there is nothing in all we have done that makes any distributional assumptions about the predictors in a model. As far as the data variable, Y, is concerned, here too there are no assumptions about the distribution of this variable, as long as its errors in the models predicting it are normally distributed with constant variance. To illustrate the distinction, consider the case of a categorical predictor variable, with two levels, used to predict some Y. If the mean difference between the two groups was substantial enough, the overall distribution of the Y variable, ignoring the two groups, would be far from normal (i.e., bimodal). Yet within each group the distribution of Y might very well be nicely behaved (i.e., normal in each group, with the same variance). These within-group distributions are the distributions of the errors in the model that predicts the two group means. Sometimes, of course, the Y variable might be distributed in such a way that there is no possible way that its errors are normally distributed. We consider some of these models, that is, models in which the Y variable is dichotomous, in Chapter 15 The important point is that it is the distribution of the errors that is crucial, not the distribution of Y.

The assumptions that errors are normally distributed with a constant variance are often seen as "robust" assumptions, in the sense that the accepted advice is that violations of the assumptions do not result in major statistical errors. While these assumptions are clearly more robust than the independence assumption considered in the last two chapters, violations of normality and constant variance can result in both Type I and Type II errors. Therefore it makes sense for researchers to attend to these issues, particularly in cases where there is good reason to suspect violations—cases that we detail shortly.

As outlined earlier, the detection of outliers involves motivated statistics and model comparisons that give clear indexes of the degree to which particular cases might pose analytic problems. The detection of non-normality and heterogeneous variances of errors is a much less precise science, and we will largely emphasize data plots and inspections of them to detect violation problems. Although there are statistical tests of departures from normality and of heterogeneous variance, we do not emphasize them because they often involve other assumptions that may be problematic. It is our experience that plots of data reveal when these assumptions are violated substantially enough that corrective action is warranted. In this sense, we regard these assumptions as robust: If their violations are likely to cause problems, they should be apparent based on relatively inexact methods of data inspection.

Diagnosing Non-Normal Error Distributions

As already noted, the assumption that errors are normally distributed is a relatively robust assumption. But the validity of this conclusion depends on how the errors deviate from normality. Departures from the normal distribution occur primarily in two ways. Recall that the normal distribution is symmetric, with tails of equal thickness both above and below its central tendency. First, error distributions may depart from normality because they are skewed in one direction or the other, either with a thicker positive than negative tail or the other way around. Second, error distributions may depart from normal because they are too peaked, with tails that are too thin, or insufficiently peaked with tails that are too thick. Although it is important to correct error distributions that are skewed, the most serious problems are encountered when distributions of errors have substantially thicker tails than the normal distribution. Figure 14.5 portrays two distributions. On the left is what is called a Cauchy distribution, resulting from the ratio of two normally distributed variables. On the right is a uniform distribution. Although the Cauchy distribution looks roughly normal, its tails are actually considerably thicker than those of a normal distribution. In contrast, the uniform distribution has tails that are considerably thinner than the normal distribution. Because least-squares estimation particularly attends to extreme errors (by minimizing the sum of *squared* errors), thick-tailed distributions, like the Cauchy distribution, are potentially much more problematic than thin-tailed non-normal distributions, such as the uniform one.

In our judgment, the most useful tool for detecting whether the distribution of errors departs from a normal one is the normal quantile-quantile plot, provided by all major software programs in procedures that plot univariate distributions of data.[1] The normal quantile-quantile plot displays the percentile distribution of the data (i.e., the errors) against the percentile distribution that would be expected if the distribution were normally distributed. Thus, the data values are rank-ordered and the percentile of each score in the distribution, based on the number of observations, is computed. These rank-ordered percentile scores for each observation constitute the values on the vertical axis of the normal quantile-quantile plot. Along the other axis are the z scores for each percentile coming from the normal distribution. Thus, each observation is located by a point in the two-dimensional plot, with its

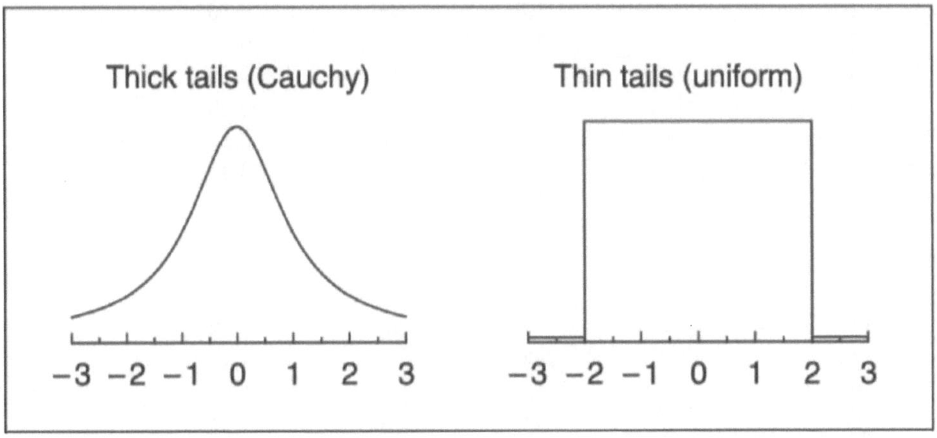

Figure 14.5 Cauchy distribution with thick tails (left) and uniform distribution with thin tails (right)

value on the vertical axis being its percentile in the actual distribution of errors and its value on the horizontal axis being what its z score would be if the percentiles came from a normal distribution.

To illustrate, the values in Figure 14.6 were generated by sampling from a normal distribution with a mean of 0 and a standard deviation of 10. We may think of them as values for the error for some model. In this sample, the mean of the 25 scores is actually 0.2, and the sample standard deviation is 11.5. The rank-ordered data values or raw scores are given in the first column of Figure 14.6. These are the scores that are plotted along the vertical axis of the normal quantile-quantile plot for each observation. In the second column, the value i is given for each score, which is simply a variable that indicates the rank order of the score (hence it goes from 1 to 25). In the third column, the percentile in the sample for each score is given, using the formula $f = (i - .5)/n$. The .5 is included in this formula just to avoid percentiles of 0 and 1.0. Then in the final column of Figure 14.6, the z score associated with the given percentile, coming from the normal distribution, is given. Thus, for instance, the third lowest error of –14 in the distribution is the 10th percentile score in the distribution of the data (i.e., $f = (3 - .5)/25 = .10$). And in a normal distribution of data, the z score corresponding to the 10th percentile score is –1.28. The values in this final column are plotted along the horizontal axis of the normal quantile-quantile plot and thus each observation is located by a point in the plot corresponding to its percentile in

Empirical error values	*i*	*fi*	*Normal quantile z(fi)*
–21	1	.02	–2.05
–16	2	.06	–1.55
–14	3	.10	–1.28
–12	4	.14	–1.08
–12	5	.18	–0.92
–11	6	.22	–0.77
–10	7	.26	–0.64
–8	8	.30	–0.52
–8	9	.34	–0.41
–6	10	.38	–0.31
1	11	.42	–0.20
1	12	.46	–0.10
2	13	.50	0
4	14	.54	0.10
4	15	.58	0.20
5	16	.62	0.31
5	17	.66	0.41
5	18	.70	0.52
7	19	.74	0.64
7	20	.78	0.77
11	21	.82	0.92
13	22	.86	1.08
15	23	.90	1.28
20	24	.94	1.55
22	25	.98	2.05

Figure 14.6 Empirical errors randomly sampled from a normal distribution with their normal quantiles

the sample (vertical axis) and what its z score would be if the distribution were normal (horizontal axis). The resulting normal quantile-quantile plot for these data is given in Figure 14.7.

If the data were perfectly normally distributed they would lie along a straight line in the plot. Since these data were in fact sampled from a normal distribution, the fit of the points to a straight line is exceedingly good. In Figure 14.8 we give four normal quantile-quantile plots for errors sampled from four known distributions. The first row shows another example for the normal distribution, with the original distribution on the left and the normal quantile-quantile plot on the right. The second row depicts a sample from a uniform distribution, with each error value between –2 and +2 being equally likely. In its normal quantile-quantile plot, the plotted points seem to fall along a straight line in the middle of the distribution but then the slope becomes very flat at the ends. The flat slope at the ends indicates that the most extreme scores in the sample (having the highest and lowest percentiles) are not as extreme as they would be if the data came from a normal distribution. This distribution has thinner tails than data from a normal distribution. The third row contains the Cauchy distribution. Note that in the plot on the left side its tails are thicker than those for the normal distribution in the first row. Here, once again, the slope of the points in the middle of the distribution is as it should be, but now at the ends the slope gets very steep, indicating that the data values at the high and low end of the distribution are more extreme than they would be if they came from a normal distribution of data. Normal quantile-quantile plots that look like this are a cause of serious concern. Finally, the last row in Figure 14.8 depicts errors sampled from a transformed beta distribution, which in this case has a positive skew. As the normal quantile-quantile plot makes clear, this distribution has a thicker positive tail than a normal distribution (the slope of the plot becomes very steep at the upper end) and a thinner negative tail than a normal distribution (the slope becomes very flat at the lower end). Again, it is the thick tail on the positive end that would be cause for concern.

Diagnosing Errors With Heterogeneous Variances

As was the case when examining whether errors are normally distributed, we have found that graphical inspection methods are generally most useful for examining whether they manifest a constant variance. Most frequently, violations of this assumption occur when the variance of the errors depends

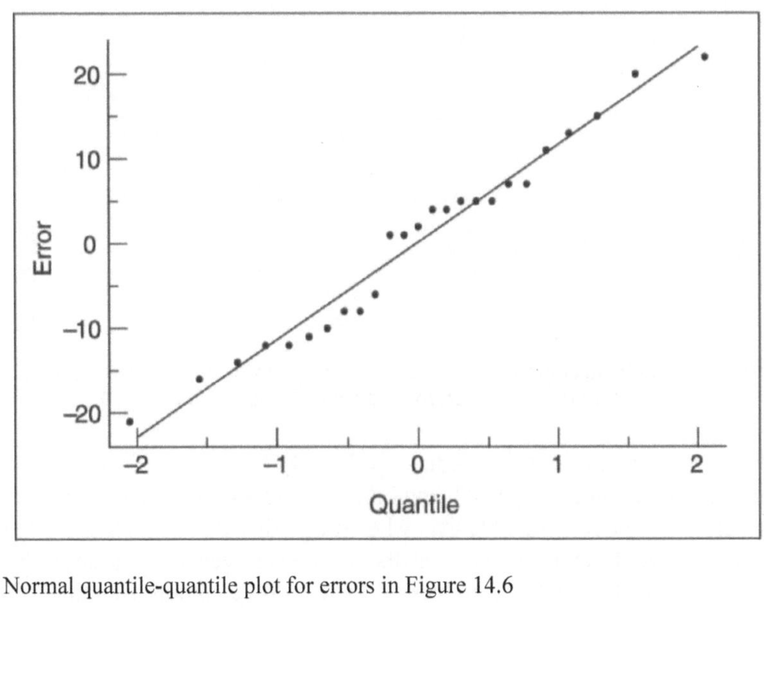

Figure 14.7 Normal quantile-quantile plot for errors in Figure 14.6

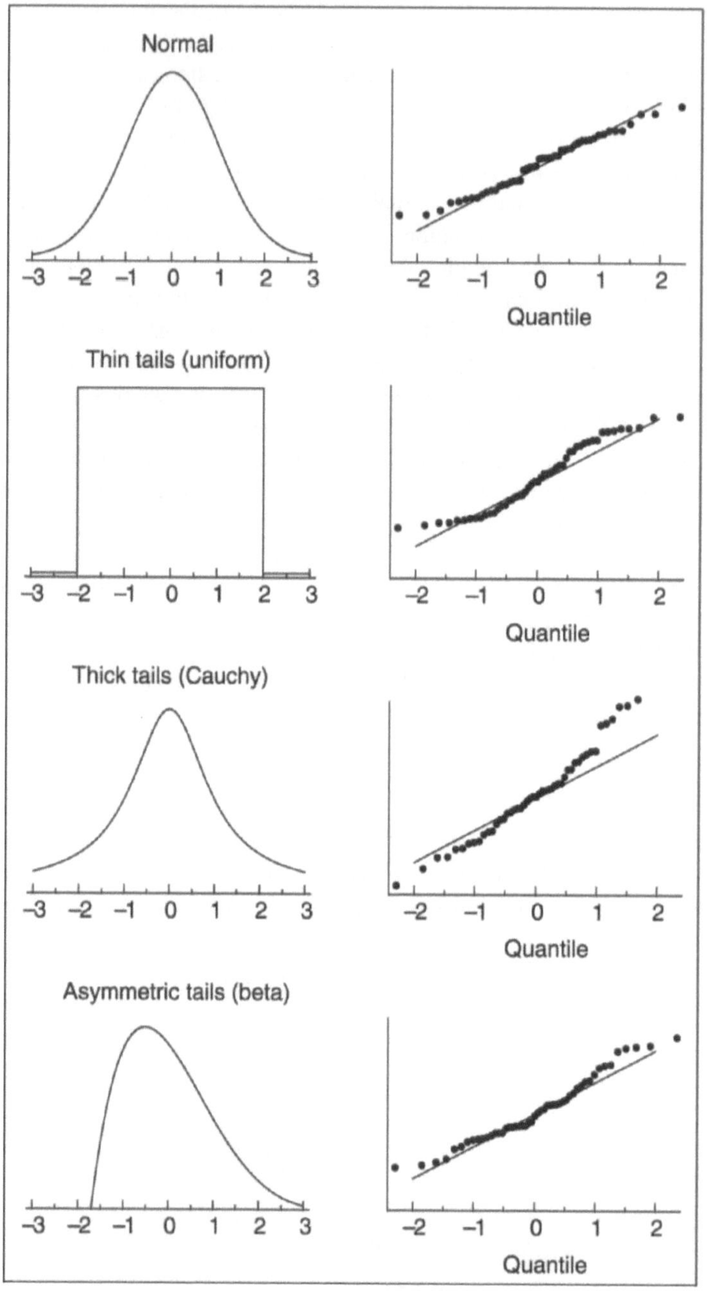

Figure 14.8 Normal quantile-quantile plots for errors sampled from normal (first row), uniform (second row), Cauchy (third row), and transformed beta with skew (fourth row) distributions

systematically on the magnitude of the predicted values, $\hat{Y}_i$. For instance, if we were predicting the sales price of makes of automobiles, then being off by \$1000 may be a much more substantial error when the predicted sales price is \$20,000 than when the predicted sales price is \$100,000.

A useful plot for determining whether errors show heterogeneous variances is to graph the errors as a function of the predicted values, $\hat{Y}_i$. The most common violation of the constant variance assumption, consistent with the example given in the previous paragraph, manifests as a funnel-shaped plot in which the errors become larger in absolute value (both positive and negative errors) as the predicted values become larger. That is, the errors are more spread out at higher predicted values. This is perhaps even easier to see if, on the vertical axis, one plots not the errors themselves, but their absolute value (or more frequently, the square root of their absolute value—such a plot is called a "spread-location plot"). In this plot, if errors are more variable or spread out at higher predicted values, then the absolute values of the errors should become larger at higher predicted values, and the resulting plot will have a generally positive slope.

Figure 14.9 illustrates these kinds of plots with data where the errors have a constant variance (on the left) and with data where the errors were generated to be proportional to the size of the

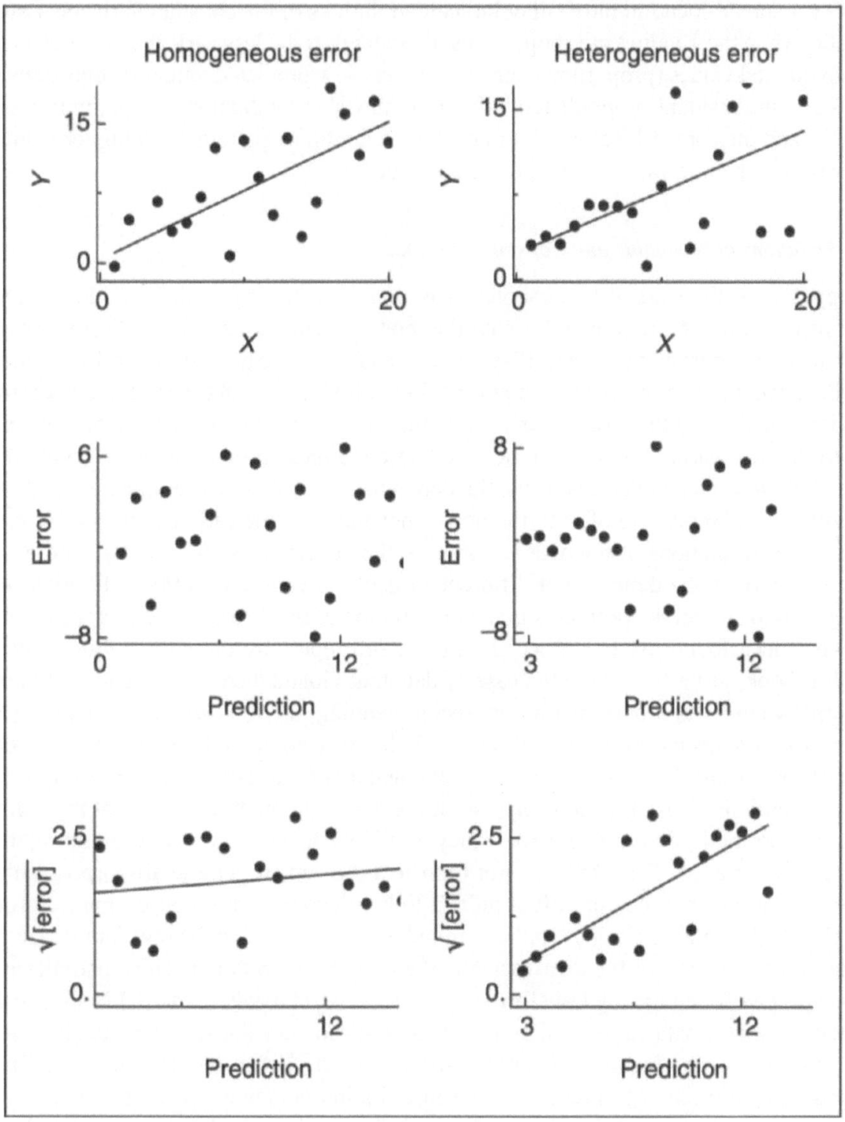

Figure 14.9 Bivariate scatterplots (first row), plots of residuals by predicted values (second row), and spread-location plots (third row) for homogeneous (left) and heterogeneous (right) errors

predicted values (on the right). The top row of scatterplots is simply the bivariate scatterplot, using a single predictor variable (X) to predict Y. From these plots, it is very hard indeed to see any problems. But things become clearer in the residual-by-predicted plots in the second row of the figure. On the left, the plotted points show a nice cloud-like structure with no noticeable funnel shape. However, a funnel shape is apparent in the right-side plot where it seems that the errors are more spread out at higher values of the predicted value. The third row presents the spread-location plots. Here, it is evident that as the predicted value increases, the average absolute error (actually its square root) increases for the heterogeneous errors in the right column. In contrast, the absolute error for homogeneous errors shows very little increase as the predicted values increase.

Although this sort of funnel-shaped pattern is typical, other patterns of residual-by-predicted value plots (or spread-location plots) may indicate violations of the constant variance assumption. For instance, when the Y values are proportions, there tends to be less variance in the errors at both very high predicted values (proportions near 1) and very low predicted values (proportions near 0). In such a case, the residual-by-predicted value plot may show a diamond shape, with less spread at both ends, and the spread-location plot may show a nonlinear pattern, with higher values of the absolute errors in the mid-range of the predicted values.

Resolving Violations of Normality and Constant Variance

As we suggested at the start of this chapter, it is important to deal with individual problematic observations (i.e., outliers) before addressing the more systematic violations of error assumptions (i.e., normality and constant variance). If systematic problems are diagnosed, various remedial actions can be taken. In general, we take an approach that is similar conceptually to our approach for dealing with violations of the independence assumption. There, we transformed our data variable, referring to the new variable(s) as W. Once we had done these transformations, combining non-independent observations in various ways, the dependence problem was eliminated, and we could proceed with our full repertoire of analytic models developed in the earlier chapters. Here, we also recommend transformations of the data variable, of the general form $W_i = f(Y_i)$, where $f()$ represents some function, to be defined. For different sorts of systematic violations, different functions will be required to correct the problems in the distribution of the errors. Once an appropriate transformation has been found, we can proceed with the full repertoire of analytic models developed earlier in this book, just as we did in the case of data that violate the independence assumption.

An alternative to the approach that we are recommending, which is widely used, involves using what are called *nonparametric* statistical tests. We do not recommend nonparametric procedures for two primary reasons. First, it is commonly assumed that such tests make no assumptions about the error distributions. This, unfortunately, is not the case. Although it is true that they effectively deal with non-normally distributed errors, they routinely make very strong assumptions about homogeneity of variance. Thus, they do not represent the sort of generic all-purpose solution to assumption violations that they are often judged to be. Second, many nonparametric procedures are equivalent to the parametric ones we have relied upon, done not on the raw data variable but on a particular transformation of the data variable, the rank transformation. Thus, parametric procedures typically involve analyzing transformed scores, $W_i = f(Y_i)$, where the $f(\,)$ is simply the rank transform, ordering the data values and giving the lowest score a value of 1, the next lowest a value of 2, and so forth. In our judgment, other transformations are better than the rank transformation, both because they are likely to preserve more information contained in the raw data and, more importantly, because they can bring the data more in line with normal distributions with constant variance. The rank transformation results in data that are uniformly, rather than normally, distributed. Because most nonparametric procedures are simply parametric procedures applied to data that have been rank-transformed, it becomes obvious that transformations that are more likely to address problems of non-normality and heterogeneous variances should be used instead.

Nonlinear but monotonic transformations of the data variable can often lead to more normally distributed errors with constant variances. Additionally, data that have been so transformed to alleviate assumption violations may often yield the benefit of simpler models. That is, many times higher order interactions or nonlinear effects in data are eliminated once appropriate transformations have been applied to data that otherwise violate the assumptions we have been considering. As always, we prefer models that are parsimonious.

For certain kinds of data, having certain characteristic distributions, there are widely accepted transformations that generally have the desired effect of correcting assumption violations. For instance, in the case of data variables that are counts (e.g., the number of times someone does something or the number of children in a family), the appropriate transformation is generally a square root transformation, $W_i = \sqrt{Y_i}$. For data that are reaction times, the appropriate transformation is generally the log transformation, $W_i = \log(Y_i)$. For data that are proportions, the logit transformation is often the most appropriate, $W_i = \log[Y_i/(1 - Y_i)]$. And for data that are correlations, Fisher's Z transformation, $W_i = 1/2 \log[(1 + Y_i)/(1 - Y_i)]$, is generally most appropriate. Of course, although these transformations are routinely applied in these cases, we recommend checking whether the chosen transformation has indeed corrected the original distributional problems, by re-examining normal quantile-quantile plots and residual-by-predicted plots, but this time on the residuals from the transformed data variable.

To correct problems of skew in other sorts of distributions, we recommend exploring transformations that derive from what has come to be known as the ladder of powers, $W = Y_i^{\lambda}$. For this approach, one may need to add a constant to Y_i such that all values of Y_i are > 1. If $\lambda = 1$, then of course $W_i = Y_i$ and no transformation is effected. If $\lambda > 1$, then larger values of Y_i become even larger, so that the upper end of the distribution becomes more spread out, relative to the lower end, thus reducing negative skew. The higher the value of λ, assuming $\lambda > 1$, the more dramatic the transformation. For values of $\lambda < 1$, the opposite happens: The lower end of the distribution becomes more spread out, relative to the upper end, thus reducing positive skew. $\lambda = 0$ yields a value of 1 for all cases and so is not used; $\log(Y_i)$ is the transformation that fits in the family of power transformations at that point. Thus, to reduce positive skew, one might start with the square root transformation, $W_i = \sqrt{Y_i} = Y_i^{+.5}$. If positive skew remains, then one might start again, and use the log transformation of Y_i. Even more extreme would be the inverse of the square root $W_i = Y_i^{-.5}$, and then the inverse $W_i = Y_i^{-1}$, and so forth. Finding the correct value of λ to reduce skew and render the distribution more like a normal distribution may be a process of trial and error. One works either up or down the ladder of powers, depending on whether the data variable is positively or negatively skewed, examining the normal quantile-quantile plot for each transformation until the one that yields the most approximately normal distribution is identified.

Importantly, in addition to violating the assumption of normality of errors, skewed error distributions are also routinely characterized by heterogeneous variances. In the case of positive skew, for instance, when dealing with reaction times, it is generally the case that errors are more variable at longer times. Power transformations that reduce skew are thus likely to correct both the non-normality and heterogeneous variance problems.

Example

A detailed example will help to elucidate the diagnostic and remedial tools described in this chapter. These data pertain to the relationship between the number of grants in the physical and social sciences awarded by the National Science Foundation (NSF) to various universities. These data were assembled by a colleague whose university's administration was concerned about the low number of NSF grants they received. They had decided to remedy the situation by providing extra resources to departments in the physical sciences. This colleague hoped to convince his administration that it was equally important to support social science departments because there was a strong relationship between the number of grants received in the physical and social sciences at peer institutions. The data analyzed here are the number of NSF grants of each type for the given state's two primary public universities and their self-identified peer institutions. Each university is categorized as the state's flagship or land grant university. Figure 14.10 shows the data and best-fitting regression lines for the flagship and land grant universities. Despite the apparent divergence of the slopes, the difference in slopes (i.e., a test of the interaction between university category and number of social science grants) is not significant: $F(1,15) = 1.9$, PRE = .11, $p = .19$. However, the plot reveals a potential outlier—a flagship university that has by far the greatest number of physical science grants—that may be distorting the regression line for the flagship universities. A visual check of the regression assumptions as well as an outlier analysis is needed.

Figure 14.11 displays the normal quantile plot (on the left) and the spread-location plot (on the right). One point, which corresponds to the apparent outlier in Figure 14.10, is far away from the normality line in the normal quantile plot; its steep slope relative to the other points identifies this as a "thick-tail" normality violation that could have substantial impact on the analysis. The spread-location plot clearly shows that the square root of the absolute value of the residuals is increasing with the size of the predicted values from the regression; thus the assumption of homogeneity of variance is violated. The studentized deleted residual for the unusual observation is 6.78, its lever is .32, and its Cook's D is 1.33, which is about twice as large as the next value of Cook's D. Its lever does not identify it as a particularly unusual observation in terms of its predictor values. However, the large studentized deleted residual ($p < .0001$) suggests that this university is telling a very different story about the relationship between the number of physical and social science NSF grants; the large value of Cook's D indicates that it is having a disproportionate effect on the overall model. If the outlier is omitted, the interaction is significant: $F(1,14) = 15.38$, PRE = .53,

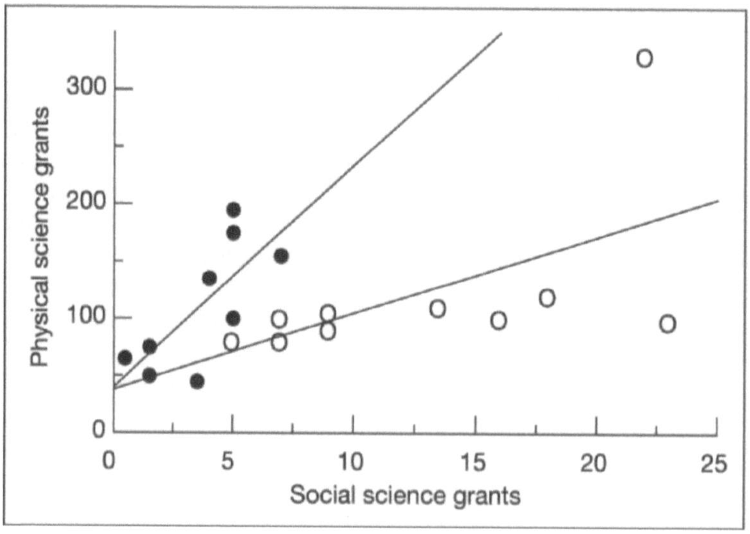

Figure 14.10 Relationship between NSF physical and social science grants at flagship (○) and land grant (●) peer institutions

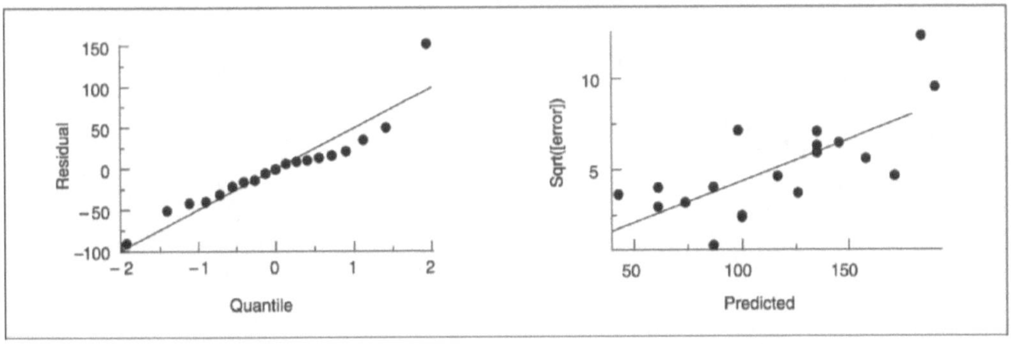

Figure 14.11 Normal quantile plot (left) and spread-location plot (right) for grants data

$p = .0015$. In other words, the story we tell about whether the relationship is different for flagship and land grant universities depends entirely on whether we include this one university. We should not allow one observation to dominate the story we tell about *all* the data! In this particular case, the unusual university has by far the highest total number of grants. And its number of physical science and social science grants fits neither the pattern of the land grant universities nor the pattern of the other flagship universities. In retrospect, this might not be a legitimate peer institution; it may have been included, wishfully, as a peer only because it was in the same intercollegiate athletic conference. For all these reasons, it is appropriate to conduct an analysis with that outlier removed. If the analysis changes appreciably without that university, that of course would not prove any of our earlier post-hoc suppositions. Instead, those suppositions might provide hypotheses to be explored in a larger study including all the state universities.

Removing the one clear outlier does not repair the violation of homogeneity of variance apparent in the right panel of Figure 14.11 (the new graph is not presented here). Hence, a transformation may be appropriate. Also, there are a priori reasons for anticipating the need for a transformation of these data. The scale for the number of grants is not likely to be linear with an underlying scale of institution quality. For example, is the functional difference between, say, 5 and 10 grants equivalent to the difference between, say, 105 and 110 grants? We are likely to judge the second difference to be negligible while considering the first difference to be quite large. Analyzing the raw data implicitly treats these two differences as if they were equal. Also, as suggested earlier, counts are likely to require a square root (or similar) transformation for models of counts to have errors with homogeneous variance. Although one need not transform on both sides, it seems appropriate in this case to transform both the criterion (number of physical science grants) and the predictor (number of social science grants). Figure 14.12 shows the normal quantile plot of the residuals and the spread-location plot after the square root transformation has been applied with the outlier omitted (it remains an outlier, although not as extreme, after the transformation). Both plots suggest that the analysis of the transformed data without the outlier reasonably satisfies the normality and homogeneity of variance assumptions. Any weaker power transformation (i.e., a power between .5 and 1) leaves a positive slope in the spread-location plot, whereas any stronger transformation (such as the log or the inverse) induces a negative slope. Hence, the square root transformation (power = .5) is best for correcting the variance problems in these data.

An analysis of the transformed data reveals a second problematical observation—a land grant university that has 56 NSF grants in the physical sciences but none in the social sciences. Some land grant universities have remained closer to their roots as agricultural and mechanical universities and so do not have the full complement of social science departments; this university may be

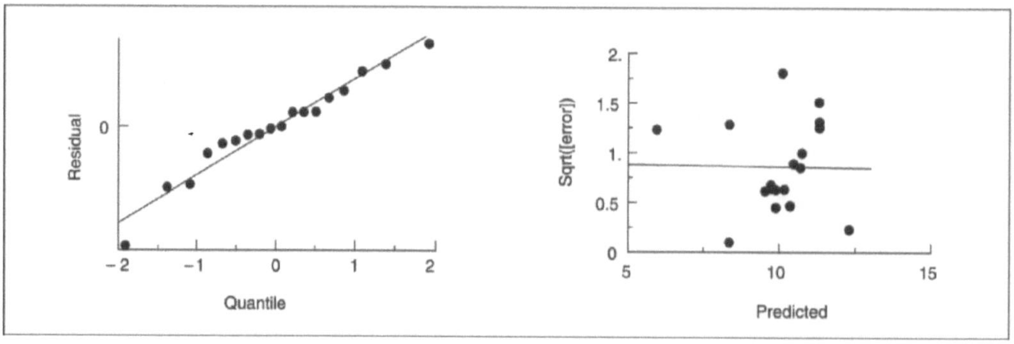

Figure 14.12 Normal quantile and spread-location plots after square root transformation of grants data

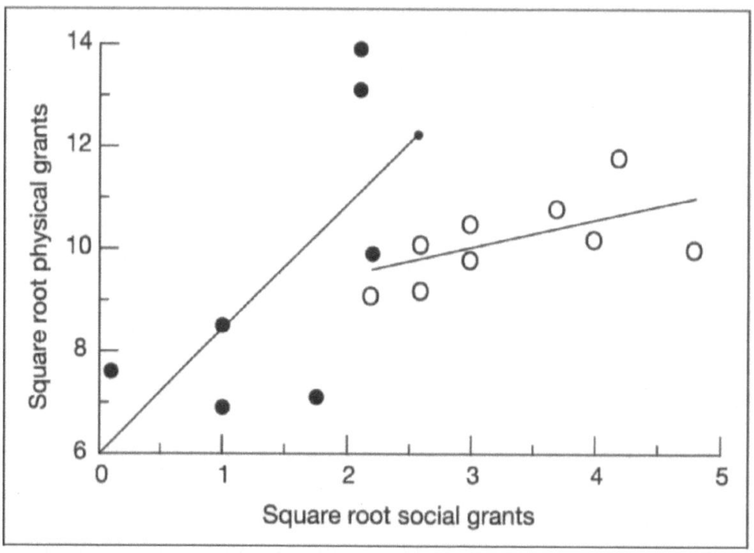

Figure 14.13 Final analysis of square-root-transformed data with one outlier omitted for flagship (○) and land grant (●) peer institutions

such an instance. Fortunately, the story does not change appreciably if this observation is included or omitted. So we will stop with the analysis of the square-root-transformed data with the first outlier omitted; this final analysis is depicted in Figure 14.13. There is no evidence for a relationship between the number of physical and social grants at flagship universities (slope = 0.46, $F(1,14) = 0.6$, PRE = .04, $p = .45$), but there is a relationship for land grant universities (slope = 2.4, $F(1,14) = 15.65$, PRE = .53, $p = .0014$). The difference between the two slopes is statistically significant (slope difference = 1.93, $F(1,14) = 5.15$, PRE = .27, $p = .04$). (Omitting the second problematic observation would strengthen the land grant relationship and enhance the slope difference.) Transforming the original data and omitting the outlier yielded not only an analysis that satisfied the important statistical assumptions but also a clear and consistent story for these data. Including the outlier and not transforming the data produced an analysis that violated the major assumptions underlying the analysis and produced a muddled and inconsistent story for the data.

Summary

In this chapter we have examined two important topics. The first concerns how one identifies and deals with observations that are outliers. We have seen that outliers can be defined in three ways: observations with unusual values on the outcome variable, observations with unusual values on the predictors, and observations that are unusual on both. Generally the first sort of outlier results in inflated error terms and Type II errors; the second tends to result in an overabundance of Type I errors; and the third sort is especially pernicious with outlying observations having the potential to seriously bias parameter estimates. We recommend examining the studentized deleted residual to identify outliers of the first sort, the lever to identify outliers of the second sort, and Cook's D to identify outliers of the third sort. Outliers that result from obvious errors in recording or coding data should be corrected if possible. Otherwise, analyses should be conducted both with and without them. One should not let a few pernicious observations dictate the story and distort it from what it would be without them.

The second half of this chapter discussed graphic displays that can be used to detect violations of the assumptions that residuals are normally distributed with a constant variance. These displays include the normal quantile-quantile plot for detecting non-normality and residual-by-predicted and spread-location plots for detecting heterogenous errors. We concluded by discussing transformations that can be used to deal with the problems of errors that display these assumption violations.

Note

1 But remember that it is the distribution of the errors that one wants to plot, not the distribution of the Y variable.

15 Logistic Regression

Categorical Dependent Variables

In the previous chapter, we introduced a variety of transformations, which enable us to reshape data that are systematically skewed and/or heteroscedastic. For example, if we measure the number of posts each participant makes to social media in one week (a count variable) and attempt to model that number as a function of the participant's extraversion, we face a challenge because the data (or more precisely the errors) are likely to be skewed, and their variance is likely to increase as $\hat{Y}$ increases. A square root or log transformation of Y can reduce both the skew and the heteroscedasticity of the errors. These transformations provide a useful, flexible strategy for dealing with data that would otherwise violate our assumptions about error. Once the data have been transformed appropriately, we can fit a model to the now-well-behaved scores of each participant.

In this chapter, we introduce a related but distinct approach to dealing with non-normal and heteroscedastic distributions of Y. Instead of changing the nature of our outcome variable through some transformation (e.g., square root or logarithmic transformations), we change the way we think about predicted values. The approach we will discuss involves a *generalized* linear model. The general linear model is actually a special case of the *generalized* linear model. The approach assumes that our observations are sampled from some distribution, which need not be normal. As is the case with other distributions, such as F distributions, or chi-square distributions, these distributions have a defined shape, allowing us to estimate the probability that a certain score might occur. For our example involving count data, the number of social media posts among participants with a particular level of extraversion might be sampled from a Poisson distribution. Rather than modeling the observations directly, we instead model the way the underlying *distribution* changes as a function of extraversion. A variety of different generalized linear models have been developed. We will not offer a full treatment in this chapter, but we will discuss one of the most important generalized linear models: the logistic model.

In prior chapters, we considered models with both categorical and continuous predictor variables, but the dependent variable was always continuous. However, sometimes the dependent variable of interest is categorical. For example, we might want to predict whether participants won or lost a contest, have or do not have an ailment, voted or did not vote in the last election. There might even be more than two categories. Our outcomes might distinguish between participants who voted Republican, those who voted Democrat, and those who voted Independent. Continuous dependent variables are generally preferred to categorical dependent variables because more information is available if we model scores (e.g., exam scores) on a continuous scale rather than a categorical scale (e.g., pass/fail). However, sometimes a continuous measure is not available, and we must model a categorical outcome variable instead.

There are special problems when the dependent variable is dichotomous (having two outcome values). By way of example, let us consider a researcher interested in whether the extent of news consumption (say, measured in hours of video and audio exposure per week) is related to whether

DOI: 10.4324/9781003438335-23

a participant votes in a national election. The researcher predicts that participants who consume more news are more likely to vote. Just as we provided numerical codes for categorical predictor variables, we can assign numerical codes to these two outcomes: 0 if someone does not vote and 1 if they do. However, the procedures we have developed for testing model comparisons do not work for categorical dependent variables for several reasons. A naïve model would be:

$$Y_i = b_0 + b_1\text{Hours}_i + \varepsilon_i$$

As we showed in Chapter 6, this equation generates a line. For any number of hours of news consumption, from Hours=0 to Hours=1 to Hours=15, that line yields a particular predicted value, $\hat{Y}$. Let us suppose that, when Hours_i equals its mean, $\hat{Y}_i = 0.74$. If the only possible values for Y are 0 and 1, how could we interpret $\hat{Y}_i = 0.74$? One might argue that 0.74 is interpretable as the probability of voting for the average participant, but clearly it is impossible for any individual participant to obtain a score that corresponds to the predicted value. More problematic, for some value of Hours, this model might yield $\hat{Y}_i = 1.22$. Even if we allow that the predicted value can represent a probability, this number makes little sense. The probability of voting may increase with news consumption, but that probability can never exceed 1 no matter how much news a participant consumes, and it can never drop below 0 no matter how little news a participant consumes. Thus, the predicted values cannot increase (or decrease) linearly and indefinitely. In general, models that predict probabilities as a function of some continuous predictor are characterized by a nonlinear "S" shape. As news consumption decreases, the probability of voting may generally decrease, too, but rather than decreasing indefinitely, the probability of voting should asymptotically approach zero as news consumption drops further and further. Similarly, as news consumption increases, the probability of voting may generally increase, but it should asymptotically approach one. Visual examples of this "S" shape for probabilities appear in many of the figures that follow.

Furthermore, with possible values restricted to 0 and 1, it is not possible for the errors ($e_i = Y_i - \hat{Y}_i$) to have a normal distribution, as assumed for the statistical tests to be valid. If $\hat{Y}_i = 0.74$ for some set of participants, their errors can have only two values: for anyone who votes, error will be 1–0.74 = 0.26; for those who do not vote, error will be 0–0.74 = −0.74. The distribution of these errors is not normal.

An additional problem with modeling Y_is comprised of 0s and 1s involves heteroscedasticity. In the case of a dichotomous outcome, variability in the data depends on the probability of the event. A probability of .5 has errors of greater magnitude than does a probability of .1 or .9. Consider an event for which both the probability of voting (we can call this p) and the probability of not voting (which is $1 - p$) equal 0.50. In this case, there are equal numbers of voters (coded as 1) and non-voters (coded as 0). All participants have outcomes that differ from the overall probability by +/− 0.50. This pattern therefore has relatively high variance, 0.25. By contrast, consider a situation in which p equals 0.90. In this case, there are nine voters for every one non-voter. Although each non-voter (with a score of 0) differs dramatically from the predicted value, with e = –0.90, there are many more voters (with scores of 1) who are very close to the expected value, with e = 0.10. All told, the variance is much smaller in this situation, 0.09.

Part of the solution to the problems of prediction, non-normal errors, and heteroscedasticty is to avoid modeling the data (the 0s and 1s) themselves. Instead we assume that each outcome is sampled from a distribution, and we attempt to model aspects of that distribution. In this case, we might assume that the distribution is a Bernoulli distribution (a kind of binomial distribution). We can conceptualize it as the distribution created if each participant flips a coin to choose whether or not to vote, acknowledging that the coin each participant uses may be biased—some participants' coins are fair with a probability of .5, some participants' coins favor voting ($p > .5$), and other participants' coins favor not voting ($p < .5$). We can then ask, does the bias of each participant's coin depend on news consumption?

The other part of the solution is to avoid modeling the probability, itself, because we know that probability involves a nonlinear "S" shape. Instead, we model something that behaves in a linear fashion. We might consider modeling the *odds* of voting. The odds of an event are defined as the probability that the event occurs, p, divided by the probability that the event does not occur, $1 - p$.

$$odds = \frac{p}{1-p}$$

Odds have a minimum of zero (when p is zero) and approach positive infinity as the quantity, $1 - p$, approaches zero. Odds near 1.0 indicate that the two probabilities are approximately equal, while extreme odds (very small or very large) indicate that one of the two outcomes is much more likely than the other. However, odds are also bounded at zero (not allowing meaningful negative values), and they too behave in a nonlinear fashion. An "S"-shaped probability function yields an accelerating pattern of odds. Our solution is to model the logarithm of the odds, which is called the *logit*. The formula is

$$logit = log\left(\frac{p}{1-p}\right) = log(odds)$$

An "S"-shaped probability function translates to a linear logit function, which enables us to model the logit as a linear function of predictor variables. Elegantly, unlike probability and odds, the logit is not constrained to a particular range. It can assume any real number value, and those logits translate to probabilities ranging from 0 to 1.

Because our approach involves modeling a distribution hypothesized to generate the original outcomes, rather than the original outcomes themselves, the logic of minimizing the sum of squared errors is no longer appropriate. Instead, we use *maximum likelihood* procedures to estimate the models, and our aggregate measure of error will be the model's likelihood. Maximum likelihood was introduced at the conclusion of Section A and discussed in Chapter 13. We will also use the χ^2 ("chi-squared") statistic instead of t and F to evaluate the model comparisons. Fortunately, the meta-logic remains the same. We express a model comparison in terms of Models A and C for the *logits* and then use χ^2 and its probability (assuming Model C to be true) to evaluate whether the extra complexity of Model A is justified in terms of reducing error. Because this technique models *logits*, it is referred to as "logistic regression."

Categorical Predictor

Consider Figure 15.1, which displays the results of a health department survey after an outbreak of an intestinal illness in a particular community. Of 111 people surveyed, 55 reported being sick with the illness (which is coded as 1), and 56 reported being well (which is coded as 0). So the probability of being sick (i.e., having a score of 1 on the dependent variable) is $p = 55/111 = 0.495$, and the probability of being well (having a score of 0 on the dependent variable) is $1 - p = 56/111 = 0.505$.

The odds of being sick is the ratio of these two probabilities,

$$odds = \frac{p}{1-p} = \frac{0.495}{0.505} = 0.980$$

Odds near 1.0 imply that the two probabilities are approximately equal, whereas those more extreme (very small or very large) imply that one of the two outcomes is much more likely than the other.

		Drinking water source		
		Tap	*Bottled*	*Total*
Health status	Sick	49	6	55
	Well	32	24	56
	Total	81	30	111

Figure 15.1 Number of respondents who were sick or well and their drinking water source

The logit is the logarithm of the odds,

$$logit = log(odds) = log(0.980) = -0.020$$

Modeling the logit, we can specify the same kinds of models we used in earlier chapters to ask questions about our data. For example, ignoring the source of water to simply test the overall likelihood of being sick, we might compare the following Model A with Model C1 or C2 to ask whether the chances of being sick differ from 50% or differ from the from the historical average of 2%, respectively. In the first case, we would need to recognize that if the probability is 50%, then the odds = 0.5/0.5 = 1, and the logit = log(1) = 0 (see Model C1). In the second case, we would need to recognize that if the probability is 2%, the odds = .02/.98 = .020, and the logit = log(.020) = −3.892 (see Model C2).

Model A: $logit_{sick} = \beta_0 + \varepsilon_i$
Model C1: $logit_{sick} = 0 + \varepsilon_i$
Model C2: $logit_{sick} = -3.892 + \varepsilon_i$

Now, turning to the effects of water source, we might ask whether being sick is a function of the source of the person's primary water source (bottle or tap). We might code the variable Water as +1 for bottled water and −1 for tap. Comparing the following models asks whether the odds of being sick are different for the two groups:

Model A: $logit_{sick} = \beta_0 + \beta_1 Water_i + \varepsilon_i$
Model C: $logit_{sick} = \beta_0 + \varepsilon_i$

Indeed, all the modeling techniques we have used before are applicable here. For example, to ask whether the relationship between water source and being sick depends on the person's age, we could compare these two models:

Model A: $logit_{sick} = \beta_0 + \beta_1 Water_i + \beta_2 Age_i + \beta_3 Water_i \times Age_i + \varepsilon_i$
Model C: $logit_{sick} = \beta_0 + \beta_1 Water_i + \beta_2 Age_i + \varepsilon_i$

For the simple example of predicting sickness from the source of drinking water (a categorical predictor), we can work out the details of the model comparison. This is not possible for more complicated models for which we will need to rely on the computer to do the estimation. For the simple case, the model comparison is

Model A: $logit_{sick} = \beta_0 + \beta_1 Water_i + \varepsilon_i$

Model C: $logit_{sick} = \beta_0 + \varepsilon_i$
H_0: $\beta_1 = 0$

To compare these two models we need to find the parameter estimates giving the maximum likelihoods for each model and then compare the maximum likelihoods of both models. Conceptually, this is essentially the same as finding the parameter estimates giving the minimum sum of squared errors and then comparing those two sums of squared errors. Only the underlying mathematics and estimation algorithms differ. Just as using a full set of codes yielded the cell means as the predicted values in ANOVA, so too using a full set of codes in logistic regression yields the logits of the cell probabilities as the predicted values.

For Model C, the predicted probability of being sick is $\hat{p} = 55/111 = 0.495$, for which the predicted log odds are $b_0 = \widehat{logit}_{sick} = \log[0.495/(1\text{–}0.495)] = \text{–}0.018$.

For Model A, for those drinking tap water (–1), the predicted probability of being sick is $\hat{p} = 49/81 = 0.605$, corresponding to $\widehat{logit}_{sick} = \log[0.605/(1\text{–}0.605)] = 0.426$. For those drinking bottled water (+1), the predicted probability of being sick is $\hat{p} = 6/30 = 0.2$, corresponding to $\widehat{logit}_{sick} = \log[0.2/(1\text{–}0.2)] = -1.386$. Then we can compute the parameter estimates for Model A: b_0 is the average of the two predicted cell values, and b_1 is the distance of each cell value from this average, or, equivalently, half the distance between the two cell values (assuming codes of +1 and –1). That is,

$$b_0 = \frac{0.426 - 1.386}{2} = -0.48$$

$$b_1 = -0.426 - 0.48 = -1.386 - (-0.48) = \frac{-0.426 + (-1.386)}{2} = -0.906$$

To summarize,

Model A: $\hat{logit}_{sick} = \text{–}0.48\text{–}0.906 Water_i$
Model C: $\hat{logit}_{sick} = \text{–}0.018$

In this simple case, these are the maximum likelihood estimates for the two models. Now it remains to compute ERROR for each model, which will not involve SSE, but rather the maximum likelihoods. These likelihoods will serve as an aggregate measure of the degree to which each model fails to fit the data, and by comparing these ERRORs, we can compare the two models. For binary outcomes with probability *p* of an event (being sick in our example) and *k* events in *N* cases, the likelihood equals

$$L = p^k(1 - p)^{N\text{–}k}$$

This is usually a very small number, so we normally use what is called the "deviance," that is, the extent to which the estimated model deviates from a model that perfectly predicts the data:

$$\text{–}2\log(L) = \text{–}2\log(p^k(1 - p)^{N\text{–}k})$$

both because it will be a more reasonably sized number, and because it will approach having a chi-squared distribution as the number of observations *N* increases. We can then calculate the deviance for Model C, which is given by

$$\text{–}2\log(L) = \text{–}2\log(p^k(1 - p)^{N\text{–}k}) = -2\log(0.495^{55} \times 0.505^{56}) = 153.87$$

For Model A we multiply the likelihoods of the two groups (tap water and bottled water) to obtain the deviance as

$$-2\log(L) = -2\log[(0.2^6 \times 0.8^{24}) \times (0.605^{49} \times 0.398^{32})] = 138.719$$

Just as with the case for computing the sums of squared errors, we will generally depend on computer programs to compute these deviances or model fits. For more complicated models, we *must* rely on the computer algorithms. Figure 15.2 shows the output for the logistic regression model for the water data. Note that the coefficients for the logistic model are exactly those we computed for Model A earlier. Also, the "Null Deviance" equals 153.87, corresponding to Model C so we will denote it as DEV(C), and the "Residual Deviance" equals 138.72, corresponding to Model A, which we will denote as DEV(A). The degrees of freedom are computed exactly as they were for models compared with sums of squared errors. For Model C, there is one parameter so $df(\text{C}) = N - 1 = 111-1 = 110$. Model A includes one additional parameter so $df(\text{A}) = N - 2 = 111-2 = 109$. Finally, we are ready to compare Model C and Model A based on the following information:

$$\text{Model A: } \widehat{logit}_{sick} = -0.480 - 0.906\,Wateri \quad \text{PA=2} \quad DEV(\text{A})=138.72$$
$$\text{Model C: } \widehat{logit}_{sick} = -0.018 \quad \text{PC=1} \quad DEV(\text{C})=153.87$$

Very similar to how we compute SSR when the dependent variable was continuous, we compute the reduction in the deviance as

$$DEV(\text{C}) = DEV(\text{A}) = 153.87-138.72 = 15.15$$

Mathematical statisticians have shown that, as N increases, this difference asymptotically approaches a chi-square distribution with degrees of freedom equal to $PA - PC$. In earlier chapters, we used the F distribution to determine whether the extra complexity of Model A afforded a meaningful improvement in our understanding of the data. In the case of logistic regression, we use the chi-square distribution in a similar fashion to decide whether Model A's complexity is worth the price. Figure A.5 in the Appendix provides the critical chi-square values for a variety of alpha levels and values of $PA - PC$. In this case, we want to compare the obtained chi-square value of 15.15 to the critical value for $\alpha = 0.05$ and $PA - PC = 1$. That critical value is 3.84 and arises so frequently in logistic regression that it should be memorized. Looking across the row in Figure A.5 we can see that if the null hypothesis were true ($\beta_1 = 0$), obtaining a chi-square value in the data as high as 15.15 would be extremely rare. We therefore reject Model C in favor of Model A: The source of the drinking water is significantly related to the probability of being ill, with those drinking tap water more likely to be ill.

	Estimate	*Standard Error*	*Wald Chi-Sq*	*Pr(>|z|)*
(Intercept)	–0.4801	0.2549	3.546	0.059679
water	–0.9062	0.2549	12.634	0.000379
Null deviance:	153.87 on 110 degrees of freedom			
Residual deviance:	138.72 on 109 degrees of freedom			

Figure 15.2 Basic output provided by most logistic regression programs

Just as we computed PRE by comparing the achieved improvement SSR = SSE(C) –SSE(A), we can compare the deviances. In this case, the proportional reduction in deviance (PRD) equals

$$\text{PRD} = \frac{DEV(\text{C}) - DEV(\text{A})}{DEV(\text{C})} = \frac{153.87 - 138.72}{153.87} = \frac{15.15}{153.87} = 0.098$$

There are other ways of generalizing the *PRE* concept to logistic regression, and they have slight differences. Statisticians often refer to one of these measures as being "pseudo-R^2." However, for our purposes, *PRD* provides a useful measure, like *PRE*, of how much better Model A is than Model C.

Now that we know that Model A, predicting different sickness rates conditional on water source, is better than Model C, we need to interpret Model A. We know the predicted values for Model A, and we used them previously to derive the parameter estimates. Such derivations will not be possible for more complicated models so we will go through the exercise of interpreting the coefficients as an example of what to do with more complicated models. First, based on the parameter estimates for Model A, we can compute the predicted *logits* of being sick for both water groups:

Bottle: $\hat{logit} = -0.48 - 0.906 Water_{bottle} = -0.48 - 0.906(+1) = -1.386$
Tap: $\hat{logit} = -0.48 - 0.906 Water_{tap} = -0.48 - 0.906(-1) = 0.426$

Most of us do not think very well in terms of *logits* so the next step is to convert the predicted *logits* to predicted odds using

Bottle: $\text{odds}_{\text{sick}} = e^{\text{logit}} = e^{-1.386} = 0.25$
Tap: $\text{odds}_{\text{sick}} = e^{\text{logit}} = e^{0.426} = 1.53$

That is, the odds of someone being sick who had been drinking bottled water was only 1 in 4, while the odds for someone who had been drinking tap water was about 1.5 to 1 or 3 to 2. For every two people in the tap water group who were well, there were three people who were sick. It is also often useful to report the odds ratio:

$$\text{OR} = \frac{odds_{tap}}{odds_{bottle}} = \frac{1.53}{0.25} = 6.12$$

In many disciplines, such as medicine and public health, results are discussed in terms of odds and odds ratios. This ratio indicates that someone drinking tap water was 6.12 times as likely to be sick as someone drinking bottled water. Finally, we can return to probabilities using these computations:

$$\text{Bottle:}\quad \hat{p} = \frac{odds_{sick}}{1 + odds_{sick}} = \frac{0.25}{1 + 0.25} = 0.200$$

$$\text{Tap:}\quad \hat{p} = \frac{odds_{sick}}{1 + odds_{sick}} = \frac{1.53}{1 + 1.53} = 0.605$$

Knowing that Model A, which allowed separate probabilities of being sick conditional on water source, is better than Model C, which had a single probability of being sick, we also know that the probabilities, odds, and *logits* for the two conditions are each significantly different. This analysis clearly points to tap water as being the source of the illness; however, with nonexperimental data we cannot infer causality. Still, in this case, the survey results motivated the community to check the water treatment plant where they found a malfunction.

Continuous Predictor

Logistic regression permits all the modeling techniques we discussed for least-squares regression. Models may contain continuous variables, coded-categorical variables, and products of those variables. We will first consider a single continuous predictor and then illustrate more complex models. In all these models we will depend on the computer to perform the complex maximum likelihood estimations.

Colleges and universities are concerned about their graduation rates. A success is usually defined as graduation within six years of entry into the college or university. To make admission decisions, many schools compute a predicted grade point average (PGPA) for the student's first year, using test scores, quality of high school, high school GPA, etc., as predictors in a multiple regression model of the kind we have seen before. Although PGPA is designed only to predict first-year success, it seems plausible that it might also predict graduation success. We can test this prediction by comparing these two models:

Model A: $Grad_i = \beta_0 + \beta_1 PGPA_i + \varepsilon_i$
Model C: $Grad_i = \beta_0 + \varepsilon_i$

The question is whether knowing a student's PGPA would improve the predicted probability that the student would graduate. Figure 15.3 displays common statistics from a logistic regression of a large dataset (15,790 observations) from a major research university, summarized in our terms as:

Model A: $\widehat{logit}_{grad_i}$=-3.170+1.330$PGPA_i$ $\quad PA = 2 \quad DEV(\text{A}) = 19398$

Model C: $\widehat{logit}_{grad_i}$=0.695 $\quad PC = 1 \quad DEV(\text{C}) = 20080$

The overall graduation probability in these data is 0.667. The question is whether knowing a student's PGPA improves the predicted probability that the student would graduate. The answer is yes—Model A reduces error relative to Model C: $\chi^2 = 20080 - 19398 = 682, p < .0001$, but not by a large proportion:

$$\text{PRD} = \frac{20080 - 19398}{20080} = \frac{682}{20080} = 0.034$$

The interpretation of the Model A in terms of *logits* is exactly the same as for simple regression. The intercept –3.17 is the predicted *logit* when PGPA = 0 and the slope 1.33 is the expected increase in the *logit* for each unit increase (i.e., each grade level) in PGPA. In more familiar

	Estimate	*Standard Error*	*Wald Chi-Sq*	*Pr(>|z|)*
(Intercept)	–3.16996	0.15229	433.47	<2e–16
PGPA	1.32960	0.05246	642.12	<2e–16
Null deviance:	20080 on 15789 degrees of freedom			
Residual deviance:	19398 on 15788 degrees of freedom			

Figure 15.3 Basic output provided by most logistic regression programs using predicted grade point average (GPA) to predict graduation within six years

terms, $e^{-3.17} = 0.04$ is the odds that a person with PGPA = 0 will graduate and the predicted probability of graduating within six years is $\hat{p} = 0.04/(1+0.04) = 0.04$. It seems unlikely that a student with PGPA = 0 would be admitted so, just as with least-squares models, it is useful to mean deviate the predictor because it renders the intercept more interpretable. The new model, using $PGPAdev_i = PGPA_i - \overline{PGPA}$, is

$$\text{Model A: } \widehat{logit}_{grad_i} = 0.732 + 1.330 PGPAdev_i \qquad PA = 2 \quad DEV(\text{A}) = 19398$$

$$\text{Model C: } \widehat{logit}_{grad_i} = 0.695 \qquad PC = 1 \quad DEV(\text{C}) = 20080$$

The intercept in both models estimates the logit for the average student in the sample. Unfortunately, the estimates of the intercepts in these two models are not identical as they would be with least-squares regression models. The slight difference is because maximum likelihood algorithms do not find an exact solution but instead search until the solutions converge to a result that is "close enough" to the maximum likelihood. In this case, the differences are trivial: From Model A, the odds of graduating for a student with the average PGPA score are $e^{0.732} = 2.08$, with a predicted graduation probability of $\hat{p} = 2.08/(1+2.08) = 0.675$. From Model C, the odds are $e^{0.695} = 2.00$ with a predicted graduation probability of $\hat{p} = 0.667$. Such slight differences are common when using logistic regression and its estimation procedures.

The slope of 1.33 was the same whether using the PGPA model or the mean-deviated model. It says that for every one-unit increase (i.e., a one-level grade increase, C to B, B to A, etc.) in PGPA, the predicted *logit* increases by 1.33 units. Converting to odds, this is the same as saying the odds of graduating increase by a factor of $e^{1.33} = 3.78$ for each grade-level increase in PGPA. When working with odds, addition is replaced by multiplication. Hence, an increase of two grade-levels in PGPA would imply an increase in the odds of graduation by a factor of $3.78^2 = 14.29$. We cannot make easy statements about the increase in the probability of graduation as PGPA increases because it is necessarily nonlinear. It is easiest to see all these relationships graphically. Figure 15.4 shows graphs of Models A and C for *logits,* odds, and probabilities. Clearly, the predicted *logit,* the odds, and the probability of graduating all increase as PGPA increases.

For least-squares models, the statistics for each coefficient (e.g., standard error and associated t statistic) represented a model comparison in which that parameter was set to zero in Model C. For logistic regression, the idea is similar, but the resulting statistics differ. If we compute the ratio of the coefficient to its standard error, we obtain a z statistic. If we square that z, we obtain a chi-squared test for the coefficient. Either the z or the chi-squared statistic can be used to test the coefficient. For PGPA,

$$\chi_1^2 = z^2 = \left[\frac{b_1}{s_{b_1}}\right]^2 = \left[\frac{1.3296}{0.05246}\right]^2 = 642.12$$

The difference in the deviances of the two models ($\chi^2 = 20080 - 19398 = 682$) and the squared ratio of the coefficient to its standard error ($\chi_1^2 = 642.12$) estimate the same chi-squared value asymptotically. However, they seldom converge exactly. Even with almost 16,000 observations, the two estimates for these data differ slightly. Almost always they yield the same conclusion in terms of deciding whether the coefficient is significantly different from zero. To distinguish the two, the squared ratio of the coefficient and its standard error are known as the "Wald chi-square." The ratio itself is similarly known as "Wald's z."

The standard error of the coefficient is also useful in computing confidence intervals in logistic regression. Instead of using the critical value of t as in the least-squares confidence intervals,

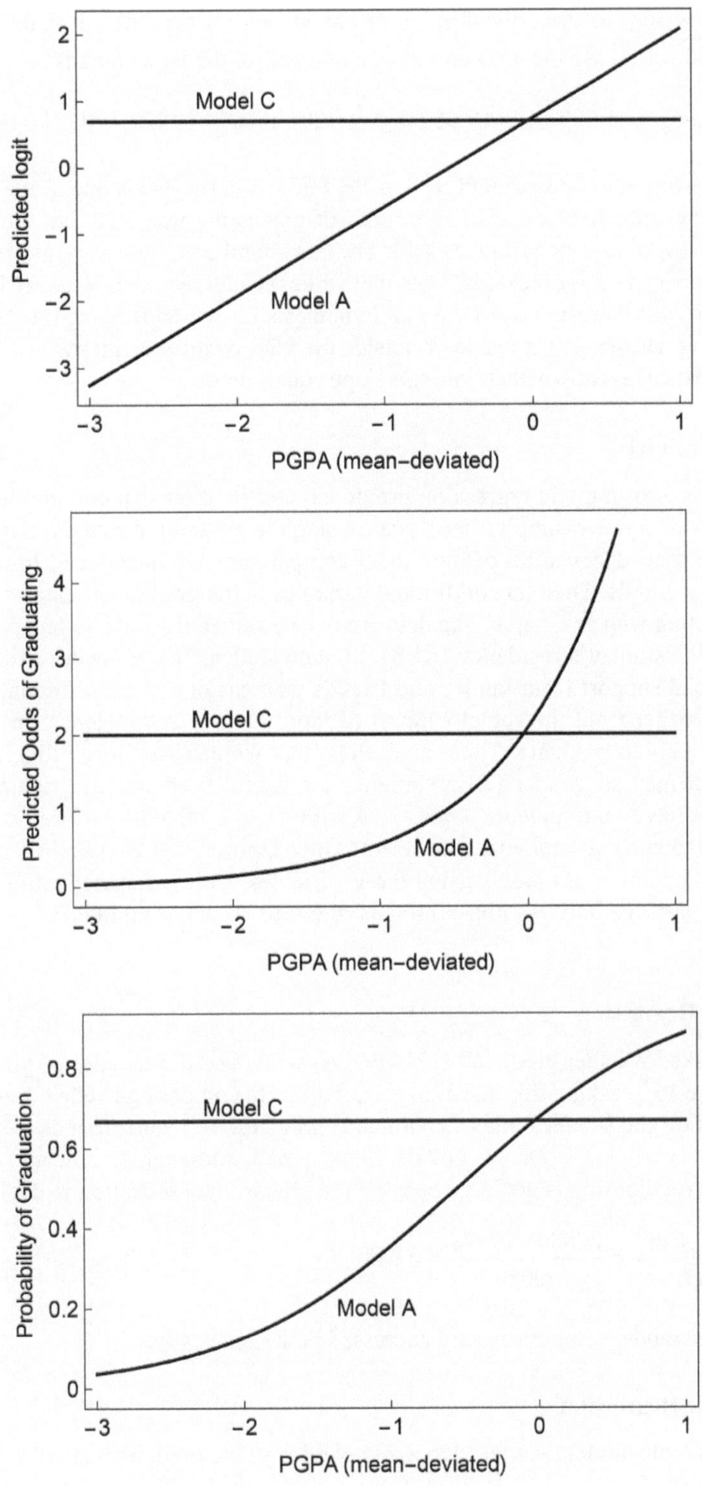

Figure 15.4 Plots of the graduation models in terms of logits, odds, and probabilities

we use the square root of the critical value of chi-squared or the critical value of z, which is $\sqrt{\chi^2_{1,\alpha=.05}} = \sqrt{3.84} = 1.96$. So the 95% confidence interval for the logit slope is

$$\mathrm{CI}_{b_j} = b_j \pm 1.96 s_{b_j} = 1.329 \pm 1.96(0.05246) = 1.329 \pm 0.103 = [1.226,\ 1.432]$$

which can be converted to odds using $[e^{1.226}, e^{1.432}] = [3.41, 4.19]$. For a one grade-level increase in PGPA, the best estimate of the change in odds of graduating was 3.78, but those odds might reasonably be as low as 3.41 or as high as 4.19. The large number of observations provides a rather narrow confidence interval for the odds. Note that while the null hypothesis for the logit is that the slope equals zero, this corresponds to the null hypothesis for the odds being one because $e^0 = 1$. Thus, determining whether 1.0 is inside or outside the 95% confidence interval of the odds is the same as the statistical test of whether the *logit* slope equals zero.

More Complex Models

With the basic tools of logistic regression previously outlined we can compare more complex models. We provide a few examples here, again using the graduation rate data. It is possible to generalize these procedures to all of the model comparisons we introduced in our discussion of least-squares methods. There are additional variables in the graduation dataset that might be useful for predicting who graduates. The data are from a state university so one potentially useful predictor is the student's residency (RES). In-state students have lower tuition and greater proximity to social support from family and friends whereas out-of-state students have higher tuition and are farther from the social support of family and friends. Hence, in-state students might be more likely to graduate. There are beliefs that women are more diligent in their college classes than men so gender (GEN) might be a useful predictor of graduation. College administrators believe that students who arrive with a plan, including a preferred major, are more likely to graduate than students who select "Open Option" (OPEN) as their major at entry. Figure 15.5 lists a number of models using these variables, their parameter estimates, and their deviances. Comparisons between these models allow us to ask a number of questions about graduation rates.

Simple Logistic Regression

Previously we asked whether predicted GPA (PGPA) was a useful predictor of graduation rates. In terms of Figure 15.5, asking this question is equivalent to comparing Models 1 and 2. Model 1 is the simple model, and Model 2 adds a continuous predictor, so it corresponds to simple regression. In this case, $\chi^2 = Dev_1 - Dev_2 = 20080–19398 = 682$. Although significant ($p < .001$), the magnitude of the relationship is not large because the proportional reduction in deviance is small:

$$\mathrm{PRD} = \frac{Dev_1 - Dev_2}{Dev_1} = \frac{20080 - 19398}{20080} = 0.034$$

More complicated model comparisons are addressed in the same way.

Multiple Logistic Regression

Comparing the simple model (Model 1) to a model with all the predictors (PGPA*dev*, Res, Gen, and Open in Model 3) corresponds to multiple regression. In this comparison, $\chi^2 = Dev_1 - Dev_3 = 20080–19353 = 727$. Again, this improvement of Model 3 over Model 1 is significant

Model 1:

$$logit_i = \beta_0 + \varepsilon_i \quad PA = 1$$

$$\widehat{logit_t} = 0.698$$

$$Dev = 20080$$

Model 2:

$$logit_i = \beta_0 + \beta_1 PGPA\ Dev_i + \varepsilon_i \quad PA = 2$$

$$\widehat{logit_t} = 0.732 + 1.33 PGPA\ Dev_i$$

$$Dev = 19398$$

Model 3:

$$logit_i = \beta_0 + \beta_1 PGPA\ Dev + \beta_2 Res_i + \beta_3 Gen_i + \beta_4 Open_i + \varepsilon_i \quad PA = 5$$

$$\widehat{logit_t} = 0.716 + 1.24 PGPA\ Dev + 0.046 Res_i + 0.107 Gen_i + 0.01 Open_i$$

$$Dev = 19353$$

Model 4:

$$logit_i = \beta_0 + \beta_1 PGPA\ Dev + \beta_2 Res_i + \beta_3 Gen_i + \varepsilon_i \quad PA = 4$$

$$\widehat{logit_t} = 0.716 + 1.24 PGPA\ Dev_i + 0.045 Res_i + 0.108 Gen_i$$

$$Dev = 19353$$

Model 5:

$$logit_i = \beta_0 + \beta_1 PGPA\ Dev + \beta_2 Res_i + \beta_3 Gen_i + \beta_4 PGPA\ Dev{*}Res_i + \varepsilon_i \quad PA = 5$$

$$\widehat{logit_t} = 0.688 + 1.22 PGPA\ Dev_i + 0.067 Res_i + 0.108 Gen_i + 0.387 PGPA\ Dev_i{*}Res_i$$

$$Dev = 19302$$

Figure 15.5 Logistic models predicting graduation rates

($p < .001$, and the degrees of freedom = $PA - PC = 4$). However, what would correspond to R^2 is not large:

$$\text{PRD} = \frac{20080 - 19353}{20080} = 0.036$$

Thus, we can conclude that, as a set, all the predictors provide a significant reduction in deviance over the simple model.

Evaluating One Predictor Among Many

As with models of continuous outcome variables, we often want to ask whether a particular predictor variable is needed over and above or when controlling for other predictor variables. For example, we can compare Model 4, which does not include the *Open* predictor, to Model 3, which includes *Open* and additional predictors. This comparison asks whether *Open* is a useful predictor over and above the other predictors. With 1 degree of freedom, $\chi^2 = Dev_4 - Dev_3 = 19353–19353 = 0$. *Open* is not a useful predictor over and above predicted GPA, residency status, and gender.

Interactions

Just as we used model comparisons to evaluate interactions (products of two variables) and polynomials (products of a variable with itself) in least-squares models, we can evaluate them in logistic regression. Predicted GPA is expected to be a more potent predictor for state residents than for out-of-state students. Comparing Model 4, which includes the predictors without the interaction, to Model 5, which adds the product between *PGPADev* and *Res,* assesses whether, controlling for gender, predicted GPA is indeed a more potent predictor for in-state than out-of-state students. Note that the same model comparison rule applies: All the components of the interaction product must be included in both models. In this case, with 1 degree of freedom, $\chi^2 = Dev_4 - Dev_5 = 19353–19302 = 51$, $p < .001$. Thus, the model that includes the interaction is a significantly better model than the one that does not.

We interpret the interaction, in terms of logits, exactly as we did for least-squares models. *Res* was coded +1/2 for in-state students and –1/2 for out-of-state students. Substituting these values into the predicted equation for Model 5 yields

In-State:

$$\widehat{logit}_i = 0.688 + 1.22 PGPADev_i + 0.045\left(\frac{1}{2}\right) + 0.108 Gen_i + 0.387\left(\frac{1}{2}\right) PGPADev_i$$

$$\widehat{logit}_i = 0.711 + 1.41 PGPADev_i + 0.108 Gen_i$$

Out-of-State:

$$\widehat{logit}_i = 0.688 + 1.22 PGPADev_i + 0.045\left(-\frac{1}{2}\right) + 0.108 Gen_i + 0.387\left(-\frac{1}{2}\right) PGPADev_i$$

$$\widehat{logit}_i = 0.665 + 1.03 PGPADev_i + 0.108 Gen_i$$

The slope relating *PGPADev* to the logit (log odds of graduating) is larger for the in-state students (1.42) than for the out-of-state students (1.03). We know this difference is significant because of the comparison of the deviances of Models 4 and 5, which tests the interaction. The comparison of the coefficients shows how much more potent predicted grade point average is for in-state than out-of-state students. The top panel of Figure 15.6 shows the two lines (in-state solid, out-of-state dashed) for the predicted logits (substituting 0 for *Gen*, which was coded +1/2 and –1/2). Nonparallel lines are characteristic of an interaction. Note that the slope relating *PGPADev* to the logit of graduation rate is steeper for in-state students. The middle panel transforms the logits to odds of graduating. Consistent with predicted GPA being more potent for in-state students, the odds curve accelerates more dramatically for the in-state students. Finally, the lower panel transforms to probability of graduating. In this graph the graduation rates for in-state students are lower than those for out-of-state students when predicted GPA is low but at about the mean of predicted PGA (i.e., *PGPADev* = 0), the probability of graduating for in-state students overtakes that of out-of-state students. Again, the enhanced potency for predicted GPA for in-state students is reflected in the S-shaped curve that changes more rapidly than for out-of-state students.

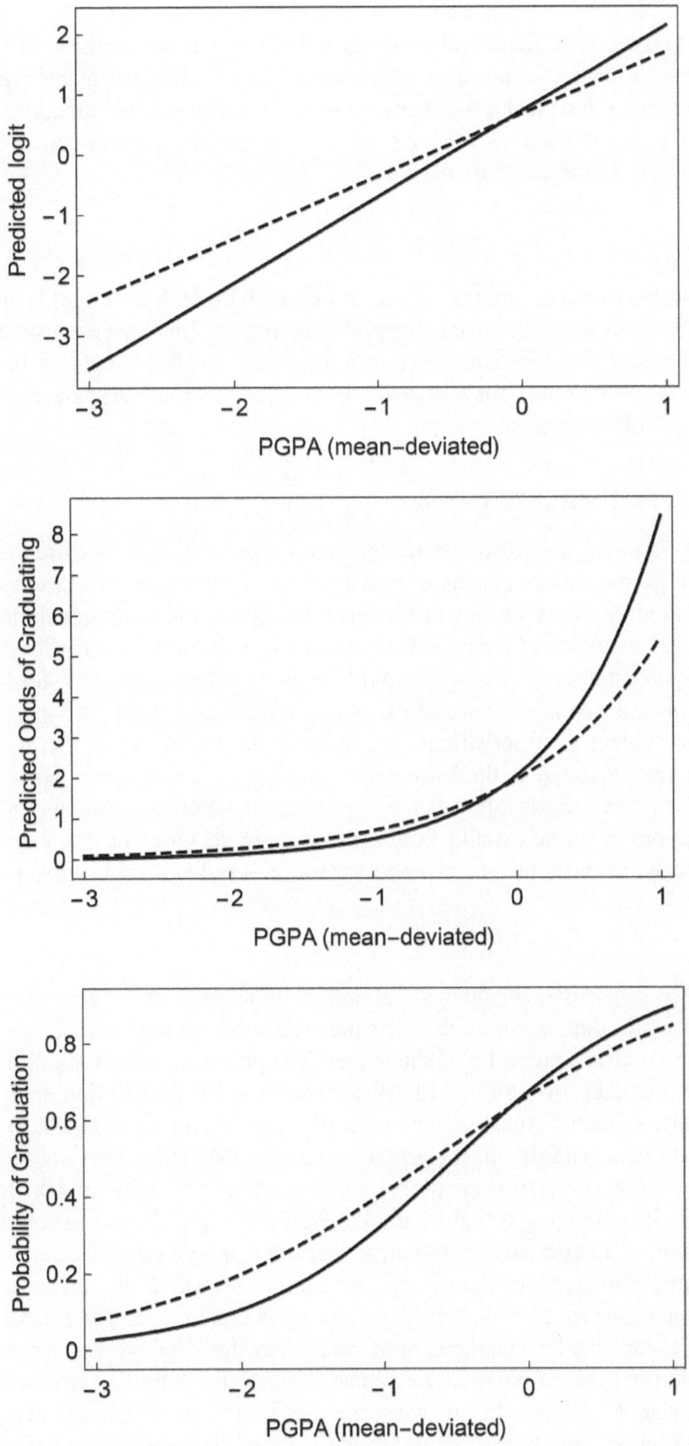

Figure 15.6 Plots of the interaction model predictions in terms of logits, odds, and probability: Solid lines are in-state students; dashed lines are out-of-state students

Other Model Comparisons

The examples in Figure 15.5 are intended merely to indicate the possibilities. The data analyst using logistic regression may use the same variety of models as in least-squares regression. Models involving combinations and products of continuous and contrast-coded categorical predictors may be used. Again, model comparisons when $PA - PC = 1$ are preferred because there is less ambiguity when interpreting the model comparison.

Outliers

Checking for unusual predictor values (i.e., observations with high leverage) is the same as it was for least-squares models and described in the previous chapter. There are corresponding techniques for identifying unusual Y values and observations having unusual impact on the overall model. However, they are too complex for discussion here. Most computer programs provide tools for identifying unusual observations.

Statistical Power for Logistic Regression

Statistical power is important for the same reasons it was in least-squares regression. With low power, there will be increased chances of making Type II errors. Just as the computation of the maximum likelihood solutions for logistic models is best left to the computer algorithms, so too the computation of statistical power for logistic regression model comparisons is best left to the computer. Most computer statistical packages provide tools for estimating power for logistic analyses. The same guidelines for enhancing statistical power apply. Collect data of better quality (less error), increase the number of observations, and increase the variability of the predictor variables. Additionally, all else being equal, the least-squares analysis of a continuous outcome variable will be more powerful than a logistic regression analysis of a dichotomous outcome. So whenever possible use a continuous outcome variable instead of a categorical outcome. For example, it would be more powerful to use actual scores from a performance test rather than a simple pass/fail measure.

Summary

In this chapter, we considered the analysis of dichotomous categorical dependent variables, that is, dependent variables that rely on either/or categories, such as pass versus fail or voted versus did not vote. Such analyses present two challenges: The predicted values are difficult to interpret, and the errors are unlikely to be normally distributed. Logistic regression analyses solve these problems by modeling distributions that theoretically generate the data, not the dichotomous data themselves. What this essentially means is that we convert the dichotomous observations to probabilities given the model, convert these probabilities to odds, and finally take the log of those odds to derive the logit. It is these logits that we model. Predictor variables can be continuous, categorical, or both. Maximum likelihood estimation, as described at the end of Section A, is then used to estimate parameters and error (or deviance). The amount by which the deviance for Model A is smaller than the deviance for Model C represents the reduction in error. The proportional reduction in deviance (PRD) can also be calculated in the same way that PRE would be calculated. We also described the interpretation of parameter estimates, which uses familiar procedures, but is more complicated because the relationship between the predictor variable and the dependent categorical variable is nonlinear. Finally, we concluded with a brief discussion of statistical power, noting that ordinary least squares analyses of a continuous dependent variable are generally more desirable than logistic regression analyses of a dichotomous dependent variable as the former is more statistically powerful.

Section C Recap

This marks the end of Section C and the end of the book. The final section dealt with a variety of situations that violate the standard assumptions for ordinary least squares regression: independence, normality, and homoscedasticity. We introduced a number of diagnostic techniques to help researchers identify potential violations, including the intraclass correlation, normal quantile-quantile plots, spread-location plots, and Cook's D. We also explored tools to deal with these issues, including repeated measures analysis of variance, linear mixed models, power transformations, and logistic regression. With the exception of linear mixed models and generalized linear models, these tools involve relatively simple transformations of the data, which yield an outcome variable that satisfies our original assumptions. Once the data are appropriately transformed, the researcher can thus employ all of the tools developed in Sections A and B.

DOI: 10.4324/9781003438335-24

References

Bakker, M., van Dijk, A., & Wicherts, J. M. (2012). The rules of the game called psychological science. *Perspectives on Psychological Science, 7*, 543–554.

Baron, R. M., & Kenny, D. A. (1986). The moderator–mediator variable distinction in social psychological research: Conceptual, strategic, and statistical considerations. *Journal of Personality and Social Psychology, 51*, 1173–1182.

Belsley, D. A., Kuh, E., & Welsch, R. E. (1980). *Regression diagnostics: Identifying influential data and sources of collinearity*. Wiley.

Brauer, M., & Curtin, J. J. (2018). Linear mixed-effects models and the analysis of nonindependent data: A unified framework to analyze categorical and continuous independent variables that vary within-subjects and/or within-items. *Psychological Methods, 23*(3), 389–411. https://doi.org/10.1037/met0000159

Campbell, D. T., & Stanley, J. C. (1963). Experimental and quasi-experimental designs for research on teaching. In N. L. Gage (Ed.), *Handbook of research on teaching* (pp. 171–246). Rand McNally. (Also, published as Experimental and quasi-experimental designs for research. Rand McNally, 1966).

Clark, H. H. (1973). The language-as-fixed effect fallacy: A critique of language statistics in psychological research. *Journal of Verbal Learning and Verbal Behavior, 12*, 335–359.

Cohen, J. (1977). *Statistical power analysis for the behavioral sciences* (rev. ed.). Academic Press.

Cohen, J. (1978). Partialed products are interactions: Partialed powers and curve components. *Psychological Bulletin, 85*, 858–866.

Cook, R. D. (1977). Detection of influential observations in linear regression. *Technometrics, 19*, 15–18.

Cook, R. D. (1979). Influential observations in linear regression. *Journal of the American Statistical Association, 74*, 169–174.

Correll, J., Mellinger, C., McClelland, G. H., & Judd, C. M. (2020). Avoid Cohen's "small", "medium", and "large" for power analysis. *Trends in Cognitive Sciences, 24*(3), 200–207. https://doi.org/10.1016/j.tics.2019.12.009

Henderson, H., & Velleman, P. F. (1981). Building multiple regression models interactively. *Biometrics, 37*, 391–411.

Hoffman, L. (2015). *Longitudinal analysis: Modeling within-person fluctuation and change*. Routledge.

Hox, J. J. (2010). *Multilevel analysis: Techniques and applications* (2nd ed.). Wiley.

Judd, C. M., & Kenny, D. A. (1981a). *Estimating the effects of social interventions*. Cambridge University Press.

Judd, C. M., & Kenny, D. A. (1981b). Process analysis: Estimating mediation in treatment evaluations. *Evaluation Review, 5*, 602–619.

Judd, C. M., Westfall, J., & Kenny, D. A. (2012). Treating stimuli as a random factor in social psychology: A new and comprehensive solution to a pervasive but largely ignored problem. *Journal of Personality and Social Psychology, 103*, 54–69.

Judd, C. M., Yzerbyt, V., & Muller, D. (2014). Mediation and moderation. In H. Reis & C. M. Judd (Eds.), *Handbook of research methods in social and personality psychology* (2nd ed., pp. 653–676). Cambridge University Press.

Kenny, D. A., & Judd, C. M. (1986). Consequences of violating the independence assumption in analysis of variance. *Psychological Bulletin, 99*, 422–431.

Kenny, D. A., & Judd, C. M. (2014). Power anomalies in testing mediation. *Psychological Science, 25*(2), 334–339. https://doi.org/10.1177/0956797613502676

Lakens, D. (2014). Performing high-powered studies efficiently with sequential analyses. *European Journal of Social Psychology, 44*(7), 701–710. https://doi.org/10.1002/ejsp.2023
MacKinnon, D. P. (2008). *Introduction to statistical mediation analysis*. Taylor & Francis.
MacKinnon, D. P., Lockwood, C. M., Hoffman, J. M., West, S. G., & Sheets, V. (2002). A comparison of methods to test mediation and other intervening variable effects. *Psychological Methods*, *7*, 83–104.
McNeish, D. (2023). A practical guide to selecting and blending approaches for clustered data: Clustered errors, multilevel models, and fixed-effect models. *Psychological Methods*. Advance online publication. https://doi.org/10.1037/met0000620
McNeish, D., & Kelley, K. (2019). Fixed effects models versus mixed effects models for clustered data: Reviewing the approaches, disentangling the differences, and making recommendations. *Psychological Methods*, *24*(1), 20–35. https://doi.org/10.1037/met0000182
McNeish, D., Stapleton, L. M., & Silverman, R. D. (2017). On the unnecessary ubiquity of hierarchical linear modeling. *Psychological Methods*, *22*(1), 114–140. https://doi.org/10.1037/met0000078
McNiesh, D. M., & Stapleton, L. M. (2016). The effect of small sample size on two-level model estimates: A review and illustration. *Educational Psychology, 28*(2), 295–314. https://doi.org/10.1007/s10648-014-9287-x
O'Rourke, H. P., & MacKinnon, D. P. (2018). Reasons for testing mediation in the absence of an intervention effect: A research imperative in Prevention and Intervention Research. *Journal of Studies on Alcohol and Drugs*, *79*(2), 171–181.
Rosenthal, R., & Rosnow, R. L. (1985). *Contrast analysis: Focused comparisons in the analysis of variance*. Cambridge University Press.
SAS Institute. (2015). *SAS/STAT software*. SAS Institute, Inc. www.sas.com/stat
Shadish, W. R., Cook, T. D., & Campbell, D. T. (2002). *Experimental and quasi-experimental designs for generalized causal inference*. Houghton Mifflin.
Snijders, T. A. B., & Bosker, R. J. (2012). *Multilevel analysis: An introduction to basic and advanced multilevel modeling* (2nd ed.). Sage.

Appendix

	α							
n – PA	*.5*	*.2*	*.1*	*.05*	*.025*	*.01*	*.005*	*.001*
1	.500	.905	.976	.994	.998	1.000	1.000	1.000
2	.250	.640	.810	.902	.951	.980	.990	.998
3	.163	.472	.649	.771	.853	.919	.949	.982
4	.121	.370	.532	.658	.753	.841	.887	.949
5	.096	.303	.448	.569	.667	.765	.820	.904
6	.079	.257	.386	.499	.595	.696	.756	.855
7	.067	.222	.339	.444	.536	.636	.699	.807
8	.059	.196	.302	.399	.486	.585	.647	.761
9	.052	.175	.272	.362	.445	.540	.602	.717
10	.047	.158	.247	.332	.410	.501	.562	.678
11	.042	.145	.227	.306	.379	.467	.526	.642
12	.039	.133	.209	.283	.353	.437	.495	.608
13	.036	.123	.194	.264	.330	.411	.467	.578
14	.033	.114	.181	.247	.310	.388	.441	.550
15	.031	.107	.170	.232	.292	.367	.419	.525
16	.029	.100	.160	.219	.277	.348	.398	.502
17	.027	.095	.151	.208	.262	.331	.379	.480
18	.026	.090	.143	.197	.249	.315	.362	.461
19	.024	.085	.136	.187	.238	.301	.346	.443
20	.023	.081	.129	.179	.227	.288	.332	.426
22	.021	.074	.118	.164	.208	.265	.307	.395
24	.019	.067	.109	.151	.192	.246	.285	.369
26	.018	.062	.101	.140	.179	.229	.266	.346
28	.016	.058	.094	.130	.167	.214	.249	.325
30	.015	.054	.088	.122	.157	.201	.234	.307
35	.013	.046	.075	.105	.135	.175	.204	.269
40	.011	.041	.066	.093	.119	.155	.181	.240
45	.010	.036	.059	.083	.107	.138	.162	.216
50	.009	.033	.053	.075	.096	.125	.147	.196
55	.008	.030	.048	.068	.088	.115	.135	.180
60	.008	.027	.044	.063	.081	.106	.124	.166
80	.006	.020	.033	.047	.061	.080	.094	.127
100	.005	.016	.027	.038	.049	.065	.076	.103
150	.003	.011	.018	.025	.033	.043	.051	.070
200	.002	.008	.013	.019	.025	.033	.039	.053
500	.001	.003	.005	.008	.010	.013	.016	.021

Figure A.1 Critical values of PRE for $PA - PC = 1$

	PA – PC								
n – PA	*1*	*2*	*3*	*4*	*5*	*6*	*7*	*8*	*9*
1	.994	.997	.998	.999	.999	.999	.999	.999	1.000
2	.902	.950	.966	.975	.980	.983	.985	.987	.989
3	.771	.864	.903	.924	.938	.947	.954	.959	.964
4	.658	.776	.832	.865	.887	.902	.914	.924	.931
5	.569	.698	.764	.806	.835	.856	.872	.885	.896
6	.499	.632	.704	.751	.785	.811	.831	.847	.860
7	.444	.575	.651	.702	.739	.768	.791	.810	.825
8	.399	.527	.604	.657	.697	.729	.754	.775	.792
9	.362	.486	.563	.618	.659	.692	.719	.742	.761
10	.332	.451	.527	.582	.624	.659	.687	.711	.731
11	.306	.420	.495	.550	.593	.628	.657	.682	.703
12	.283	.393	.466	.521	.564	.600	.630	.655	.677
13	.264	.369	.440	.494	.538	.574	.604	.630	.653
14	.247	.348	.417	.471	.514	.550	.580	.607	.630
15	.232	.329	.397	.449	.492	.527	.558	.585	.608
16	.219	.312	.378	.429	.471	.507	.538	.564	.588
17	.208	.297	.361	.411	.452	.488	.518	.545	.569
18	.197	.283	.345	.394	.435	.470	.501	.527	.551
19	.187	.270	.331	.379	.419	.454	.484	.510	.534
20	.179	.259	.317	.364	.404	.438	.468	.495	.518
22	.164	.238	.294	.339	.377	.410	.439	.466	.489
24	.151	.221	.273	.316	.353	.385	.414	.440	.463
26	.140	.206	.256	.297	.332	.363	.391	.417	.440
28	.130	.193	.240	.279	.314	.344	.371	.396	.418
30	.122	.181	.226	.264	.297	.326	.353	.377	.399
35	.105	.157	.198	.232	.262	.289	.314	.336	.357
40	.093	.139	.176	.207	.234	.259	.282	.304	.323
45	.083	.125	.158	.186	.212	.235	.257	.277	.295
50	.075	.113	.143	.170	.194	.215	.235	.254	.272
55	.068	.103	.131	.156	.178	.198	.217	.235	.252
60	.063	.095	.121	.144	.165	.184	.202	.219	.234
80	.047	.072	.093	.111	.127	.142	.157	.171	.184
100	.038	.058	.075	.090	.103	.116	.128	.140	.151
150	.025	.039	.051	.061	.070	.080	.088	.096	.104
200	.019	.030	.038	.046	.053	.060	.067	.074	.080
500	.008	.012	.015	.019	.022	.025	.028	.030	.033

Figure A.2 Critical values of PRE for $\alpha = .05$

	α							
n – PA	*.5*	*.2*	*.1*	*.05*	*.025*	*.01*	*.005*	*.001*
1	1.00	9.47	39.9	161.00	648.00	4052.00	16,210.00	405,284.00
2	0.67	3.56	8.53	18.51	38.51	98.50	198.00	998.00
3	0.59	2.68	5.54	10.13	17.44	34.12	55.55	167.00
4	0.55	2.35	4.54	7.71	12.22	21.20	31.33	74.14
5	0.53	2.18	4.06	6.61	10.01	16.26	22.78	47.18
6	0.51	2.07	3.78	5.99	8.81	13.75	18.63	35.51
7	0.51	2.00	3.59	5.59	8.07	12.25	16.24	29.25
8	0.50	1.95	3.46	5.32	7.57	11.26	14.69	25.41
9	0.49	1.91	3.36	5.12	7.21	10.56	13.61	22.86
10	0.49	1.88	3.29	4.96	6.94	10.04	12.83	21.04
11	0.49	1.86	3.23	4.84	6.72	9.65	12.23	19.69
12	0.48	1.84	3.18	4.75	6.55	9.33	11.75	18.64
13	0.48	1.82	3.14	4.67	6.41	9.07	11.37	17.82
14	0.48	1.81	3.10	4.60	6.30	8.86	11.06	17.14
15	0.48	1.80	3.07	4.54	6.20	8.68	10.80	16.59
16	0.48	1.79	3.05	4.49	6.12	8.53	10.58	16.12
17	0.47	1.78	3.03	4.45	6.04	8.40	10.38	15.72
18	0.47	1.77	3.01	4.41	5.98	8.29	10.22	15.38
19	0.47	1.76	2.99	4.38	5.92	8.18	10.07	15.08
20	0.47	1.76	2.97	4.35	5.87	8.10	9.94	14.82
22	0.47	1.75	2.95	4.30	5.79	7.95	9.73	14.38
24	0.47	1.74	2.93	4.26	5.72	7.82	9.55	14.03
26	0.47	1.73	2.91	4.23	5.66	7.72	9.41	13.74
28	0.47	1.72	2.89	4.20	5.61	7.64	9.28	13.50
30	0.47	1.72	2.88	4.17	5.57	7.56	9.18	13.29
35	0.46	1.71	2.85	4.12	5.48	7.42	8.98	12.90
40	0.46	1.70	2.84	4.08	5.42	7.31	8.83	12.61
45	0.46	1.69	2.82	4.06	5.38	7.23	8.71	12.39
50	0.46	1.69	2.81	4.03	5.34	7.17	8.63	12.22
55	0.46	1.68	2.80	4.02	5.31	7.12	8.55	12.09
60	0.46	1.68	2.79	4.00	5.29	7.08	8.49	11.97
80	0.46	1.67	2.77	3.96	5.22	6.96	8.33	11.67
100	0.46	1.66	2.76	3.94	5.18	6.90	8.24	11.50
150	0.46	1.66	2.74	3.90	5.13	6.81	8.12	11.27
200	0.46	1.65	2.73	3.89	5.10	6.76	8.06	11.15
500	0.46	1.65	2.72	3.86	5.05	6.69	7.95	10.96

Figure A.3 Critical values of F for $PA - PC = 1$

	PA – PC								
n – PA	*1*	*2*	*3*	*4*	*5*	*6*	*7*	*8*	*9*
1	161.45	199.50	215.71	224.58	230.16	233.99	236.77	238.88	240.54
2	18.51	19.00	19.16	19.25	19.30	19.33	19.35	19.37	19.38
3	10.13	9.55	9.28	9.12	9.01	8.94	8.89	8.85	8.81
4	7.71	6.94	6.59	6.39	6.26	6.16	6.09	6.04	6.00
5	6.61	5.79	5.41	5.19	5.05	4.95	4.88	4.82	4.77
6	5.99	5.14	4.76	4.53	4.39	4.28	4.21	4.15	4.10
7	5.59	4.74	4.35	4.12	3.97	3.87	3.79	3.73	3.68
8	5.32	4.46	4.07	3.84	3.69	3.58	3.50	3.44	3.39
9	5.12	4.26	3.86	3.63	3.48	3.37	3.29	3.23	3.18
10	4.96	4.10	3.71	3.48	3.33	3.22	3.14	3.07	3.02
11	4.84	3.98	3.59	3.36	3.20	3.09	3.01	2.95	2.90
12	4.75	3.89	3.49	3.26	3.11	3.00	2.91	2.85	2.80
13	4.67	3.81	3.41	3.18	3.03	2.92	2.83	2.77	2.71
14	4.60	3.74	3.34	3.11	2.96	2.85	2.76	2.70	2.65
15	4.54	3.68	3.29	3.06	2.90	2.79	2.71	2.64	2.59
16	4.49	3.63	3.24	3.01	2.85	2.74	2.66	2.59	2.54
17	4.45	3.59	3.20	2.96	2.81	2.70	2.61	2.55	2.49
18	4.41	3.55	3.16	2.93	2.77	2.66	2.58	2.51	2.46
19	4.38	3.52	3.13	2.90	2.74	2.63	2.54	2.48	2.42
20	4.35	3.49	3.10	2.87	2.71	2.60	2.51	2.45	2.39
22	4.30	3.44	3.05	2.82	2.66	2.55	2.46	2.40	2.34
24	4.26	3.40	3.01	2.78	2.62	2.51	2.42	2.36	2.30
26	4.23	3.37	2.98	2.74	2.59	2.47	2.39	2.32	2.27
28	4.20	3.34	2.95	2.71	2.56	2.45	2.36	2.29	2.24
30	4.17	3.32	2.92	2.69	2.53	2.42	2.33	2.27	2.21
35	4.12	3.27	2.87	2.64	2.49	2.37	2.29	2.22	2.16
40	4.08	3.23	2.84	2.61	2.45	2.34	2.25	2.18	2.12
45	4.06	3.20	2.81	2.58	2.42	2.31	2.22	2.15	2.10
50	4.03	3.18	2.79	2.56	2.40	2.29	2.20	2.13	2.07
55	4.02	3.16	2.77	2.54	2.38	2.27	2.18	2.11	2.06
60	4.00	3.15	2.76	2.53	2.37	2.25	2.17	2.10	2.04
80	3.96	3.11	2.72	2.49	2.33	2.21	2.13	2.06	2.00
100	3.94	3.09	2.70	2.46	2.31	2.19	2.10	2.03	1.97
150	3.90	3.06	2.66	2.43	2.27	2.16	2.07	2.00	1.94
200	3.89	3.04	2.65	2.42	2.26	2.14	2.06	1.98	1.93
500	3.86	3.01	2.62	2.39	2.23	2.12	2.03	1.96	1.90

Figure A.4 Critical values of F for $\alpha = .05$

	α						
df	*0.5*	*0.2*	*0.1*	*0.05*	*0.01*	*0.005*	*0.001*
1	0.45	1.64	2.71	3.84	6.63	7.88	10.83
2	1.39	3.22	4.61	5.99	9.21	10.60	13.82
3	2.37	4.64	6.25	7.81	11.34	12.84	16.27
4	3.36	5.99	7.78	9.49	13.28	14.86	18.47
5	4.35	7.29	9.24	11.07	15.09	16.75	20.52
6	5.35	8.56	10.64	12.59	16.81	18.55	22.46
7	6.35	9.80	12.02	14.07	18.48	20.28	24.32
8	7.34	11.03	13.36	15.51	20.09	21.95	26.12
9	8.34	12.24	14.68	16.92	21.67	23.59	27.88
10	9.34	13.44	15.99	18.31	23.21	25.19	29.59
11	10.34	14.63	17.28	19.68	24.72	26.76	31.26
12	11.34	15.81	18.55	21.03	26.22	28.3	32.91
13	12.34	16.98	19.81	22.36	27.69	29.82	34.53
14	13.34	18.15	21.06	23.68	29.14	31.32	36.12
15	14.34	19.31	22.31	25.00	30.58	32.80	37.70
16	15.34	20.47	23.54	26.30	32.00	34.27	39.25
17	16.34	21.61	24.77	27.59	33.41	35.72	40.79
18	17.34	22.76	25.99	28.87	34.81	37.16	42.31
19	18.34	23.9	27.2	30.14	36.19	38.58	43.82
20	19.34	25.04	28.41	31.41	37.57	40.00	45.31

Figure A.5 Critical values of χ^2_{df}

Author Index

Subject Index